Lecture Notes of the Institute
for Computer Sciences, Social Informatics
and Telecommunications Engineering   76

Editorial Board

Ozgur Akan
  Middle East Technical University, Ankara, Turkey
Paolo Bellavista
  University of Bologna, Italy
Jiannong Cao
  Hong Kong Polytechnic University, Hong Kong
Falko Dressler
  University of Erlangen, Germany
Domenico Ferrari
  Università Cattolica Piacenza, Italy
Mario Gerla
  UCLA, USA
Hisashi Kobayashi
  Princeton University, USA
Sergio Palazzo
  University of Catania, Italy
Sartaj Sahni
  University of Florida, USA
Xuemin (Sherman) Shen
  University of Waterloo, Canada
Mircea Stan
  University of Virginia, USA
Jia Xiaohua
  City University of Hong Kong, Hong Kong
Albert Zomaya
  University of Sydney, Australia
Geoffrey Coulson
  Lancaster University, UK

Martin Griss Guang Yang (Eds.)

Mobile Computing, Applications, and Services

Second International ICST Conference, MobiCASE 2010
Santa Clara, CA, USA, October 25–28, 2010
Revised Selected Papers

Volume Editors

Martin Griss
Carnegie Mellon University
Silicon Valley, Bldg. 23
Moffett Field, CA 94035, USA
E-mail: martin.gris@sv.cmu.edu

Guang Yang
Nokia Research Center
955 Page Mill Rd
Palo Alto, CA 94304, USA
E-mail: guang.g.yang@nokia.com

ISSN 1867-8211 e-ISSN 1867-822X
ISBN 978-3-642-29335-1 e-ISBN 978-3-642-29336-8
DOI 10.1007/978-3-642-29336-8

Springer Heidelberg Dordrecht London New York

Library of Congress Control Number: 2012934858

CR Subject Classification (1998): C.2, H.4, I.2, D.2, H.3, H.5

© ICST Institute for Computer Science, Social Informatics and Telecommunications Engineering 2012

This work is subject to copyright. All rights are reserved, whether the whole or part of the material is concerned, specifically the rights of translation, reprinting, re-use of illustrations, recitation, broadcasting, reproduction on microfilms or in any other way, and storage in data banks. Duplication of this publication or parts thereof is permitted only under the provisions of the German Copyright Law of September 9, 1965, in its current version, and permission for use must always be obtained from Springer. Violations are liable to prosecution under the German Copyright Law.
The use of general descriptive names, registered names, trademarks, etc. in this publication does not imply, even in the absence of a specific statement, that such names are exempt from the relevant protective laws and regulations and therefore free for general use.

Typesetting: Camera-ready by author, data conversion by Scientific Publishing Services, Chennai, India

Printed on acid-free paper

Springer is part of Springer Science+Business Media (www.springer.com)

Preface

Welcome to the proceedings of MobiCASE 2010, the Second International Conference on Mobile Computing, Applications, and Services, held in the heart of the Silicon Valley. Built on last year's inaugural success, MobiCASE provided great opportunities to researchers and practitioners, both from academia and industry, to present and discuss their work on mobile applications and service provisioning.

We were thrilled to have a very strong program this year. Philippe Lucas, Senior Vice President of France Telecom Orange, Marc Davis, Partner Architect of Microsoft Online Services Division, and Arpit Joshipura, Vice President of Ericsson Silicon Valley, delivered morning keynote speeches. Tony Wasserman, Professor at Carnegie Mellon University, and Adam Blum, CEO of Rhomobile, were invited to talk in the afternoons. Panelists explored a variety of topics at three discussion sessions, including one sponsored by Wireless Communications Alliance, and we teamed up with Ericsson Silicon Valley and Sprint Developer Conference for joint exhibit sessions in the evenings.

The program was single-track with six regular sessions, covering a wide range of topics including Web and mash-ups, software engineering and development tools, cross-layer approaches, location-based services, healthcare, and social networking. Each submission went through a rigorous evaluation process with at least two independent reviews from the TPC, after which a total of 15 papers were selected for publication. All submissions and reviews were read by the TPC Chair to insure quality. The program also had a demonstration/poster session to allow better interactivity and hands-on experiences among attendees.

The conference could not be held without the tremendous work done by the Technical Program Committee (TPC). Each TPC member reviewed an average of five papers and provided in-depth critique and analysis. Special thanks go out to the General Chair, Martin Griss, for his unyielding efforts in putting together the conference. Additional thanks go to Christian Martin for working on industry submissions, Wendy Fong and KRS Murthy for handling publicity and communications, Yingen Xiong and Djamal Zeghlache for chairing workshops, Angela Nicoara for managing demonstrations and posters, and Stacy Marshall for taking care of local arrangements. Yet more thanks go to the IEEE Computer Society and CREATE-NET for their technical co-sponsorships.

Finally, my particular thanks and appreciation must be put forth for the MobiCASE Steering Committee Chair, Imrich Chlamtac, and committee members Thomas Phan and Petros Zerfos, whose vision has steered and shaped this effort.

Guang Yang

Organization

Steering Committee

Imrich Chlamtac	CREATE-NET, Italy
Thomas Phan	Microsoft
Petros Zerfos	IBM Research

Organizing Committee

Martin Griss	Carnegie Mellon Silicon Valley, USA
Imrich Chlamtac	CREATE-NET, Italy
Guang Yang	Nokia Research Center, USA
Christian Martin	Institut Telecom, France
Angela Nicoara	Deutsche Telekom Inc. R&D Lab, USA
David Kagan	Kagan Media Net, USA
Yingen Xiong	Nokia Research Center, USA
Djamal Zeghlache	Institut Telecom, France
Wendy Fong	Carnegie Mellon Silicon Valley, USA
KRS Murthy	I Cubed, USA
Stacy Marshall	Carnegie Mellon Silicon Valley, USA

Technical Program Committee

Paolo Bellavista	University of Bologna, Italy
Nina Bhatti	HP Labs, USA
Jeff Burke	UCLA, USA
Ling-Jyh Cheng	Academica Sinica, Taiwan
Jerry Cheng	Yahoo!, USA
Angela Dalton	Johns Hopkins Univerity, USA
Anind Dey	Carnegie Mellon University, USA
Hanni Jamjoom	IBM Research, USA
Lukas Kencl	Czech Technical University, Czech Republic
Satya Mallya	Orange Labs, USA
Gerard Memmi	Institut Telecom, France
Tommi Mikkonen	Tampere University, Finland
April Mitchell	HP Labs, USA
Rebecca Montanari	University of Bologna, Italy
Thomas Phan	Microsoft, USA
Calicrates Policroniades	Telenor, Norway

Michel Simatic   Telecom & Management SudParis, France
Martin Svensson   Ericsson Research, USA
Pablo Vidales   Deutsche Telekom Labs, Germany
Yingen Xiong   Nokia Research Center, USA
Djamal Zeghlache   Institut Telecom, France
Petros Zerfos   IBM Research, USA
Yu Zheng   Microsoft Research, China

Workshop Committees

International Workshop on Mobile Security 2010

Philippe Proust   Security labs, Gemalto
Patrick Tague   Carnegie Mellon University, USA
Yafei Yang   Qualcomm
Lujo Bauer   Carnegie Mellon University, USA
Hao Chen   University of California - Davis, USA
Songqing Chen   George Mason University, USA
Yuecel Karabulut   SAP
Angelos Keromytis   Columbia University, USA
Sridhar Machiraju   Google
Patrick Traynor   Georgia Institute of Technology, USA
Zheng Yan   Nokia Research Center

International Workshop on Mobile Computing and Clouds 2010

Calton Pu   Georgia Institute of Technology, USA
Weidong (Larry) Shi   Nokia Research Center, Palo Alto, USA
Sujoy Basu   HP Lab, Palo Alto, USA
Ivona Brandic   Vienna University of Technology, Austria
Rajkumar Buyya   University of Melbourne, Australia
Keke Chen   Wright State University, USA
Wenguang Chen   Tsinghua University, China
Byung-Gon Chun   Intel Labs, Berkeley, USA
Frdric Desprez   INRIA, France
Guofei Gu   Texas A&M University, USA
Zhu Li   Hong Kong Polytechnic University, Hong Kong
Shivajit Mohapatra   Motorola Applied Research Center, USA
Nitya Narasimhan   Motorola Applied Research Center, USA
Guillaume Pierre   VU University, The Netherlands
Kyung Dong Ryu   IBM T.J. Watson Research Center, USA
Weisong Shi   Wayne State University, USA
Dilma Da Silva   IBM T.J. Watson Research Center, USA
Sivasubramanian Swami   Amazon Inc, USA
Zhenlin Wang   Michigan Technological University, USA

Shouhuai Xu	University of Texas at San Antonio, USA
Jun Yang	Nokia Research Center, Palo Alto, USA
Sheng Zhong	State University of New York (SUNY) at Buffalo, USA
Dong Zhou	NTT DoCoMo Research Lab, Palo Alto, USA
Suiping Zhou	Nanyang Technological University, Singapore

Workshop on Mobile Software Engineering 2010

Sarah Allen	Blazing Cloud
Ray Bareiss	Carnegie Mellon Silicon Valley, USA
Adam Blum	Rhomobile
David Brittain	Motorola
Martin Griss	Carnegie Mellon Silicon Valley, USA
David Witkowski	Wireless Communications Alliance

Table of Contents

Mobile Web and Mashups

3D Audio Interface for Rich Mobile Web Experiences 1
  Victor K.Y. Wu and Roy H. Campbell

Secure, Consumer-Friendly Web Authentication and Payments with a
Phone 17
  Ben Dodson, Debangsu Sengupta, Dan Boneh, and Monica S. Lam

Tool Support for Constructing Mobile Mashups 39
  Lasse Holmstedt, Tommi Mikkonen, and Mikko Terho

Software Engineering and Development Tools

Cuckoo: A Computation Offloading Framework for Smartphones 59
  Roelof Kemp, Nicholas Palmer, Thilo Kielmann, and Henri Bal

Debugging Tools for MIDP Java Devices 80
  Olli Kallioinen and Tommi Mikkonen

Dynamic Reduction of Rollbacks in Wireless Multi-user Virtual
Environments 100
  Abdul Malik Khan, Sophie Chabridon, and Antoine Beugnard

Cross-Layer Approaches

Handling the M in MANet: An Algorithm to Identify Stable Groups of
Peers Using Cross-Layering Information 117
  Hoa Dung Ha Duong and Isabelle Demeure

Small World VoIP 137
  *Xiaohui Yang, Angelos Stavrou, Ram Dantu, and
  Duminda Wijesekera*

Location-Based Services

When Will You Be at the Office? Predicting Future Locations
and Times 156
  Ingrid Burbey and Thomas L. Martin

PosQ: Unsupervised Fingerprinting and Visualization of GPS
Positioning Quality 176
  Mikkel Baun Kjærgaard and Kay Weckemann

SensOrchestra: Collaborative Sensing for Symbolic Location
Recognition 195
    *Heng-Tze Cheng, Feng-Tso Sun, Senaka Buthpitiya, and
Martin Griss*

Mobile Healthcare

Activity-Aware Mental Stress Detection Using Physiological Sensors 211
    *Feng-Tso Sun, Cynthia Kuo, Heng-Tze Cheng, Senaka Buthpitiya,
Patricia Collins, and Martin Griss*

Dr. Droid: Assisting Stroke Rehabilitation Using Mobile Phones 231
    Andrew Goodney, Jinho Jung, Scott Needham, and Sameera Poduri

Mobile Social Networking

Open Transaction Network: Connecting Communities of Experience
through Mobile Transactions 243
    Kwan Hong Lee, Dawei Shen, Andrew Lippman, and Erik Ross

Mobile Lifelogger – Recording, Indexing, and Understanding a Mobile
User's Life 263
    Snehal Chennuru, Peng-Wen Chen, Jiang Zhu, and Joy Ying Zhang

Demos and Posters

Activity-Aware Mental Stress Detection Using Physiological Sensors 282
    *Feng-Tso Sun, Cynthia Kuo, Heng-Tze Cheng, Senaka Buthpitiya,
Patricia Collins, and Martin Griss*

A Decentralized Decision Support System for Mobile Devices 302
    *Gert Scholten, Nicholas Palmer, Roelof Kemp, Thilo Kielmann, and
Henri Bal*

Inferring Complex Human Behavior Using a Non-obtrusive Mobile
Sensing Platform... 306
    *Bruce DeBruhl, Michele Cossalter, Roy Want, Ole Mengshoel, and
Pei Zhang*

Magic Wand: A Framework for Developing Remote Controlled Web
Applications.. 311
    Vibhor Nanavati

International Workshop on Mobile Computing and Clouds

Bringing the Cloud Down to Earth: Transient PCs Everywhere 315
  Mahadev Satyanarayanan, Stephen Smaldone, Benjamin Gilbert, Jan Harkes, and Liviu Iftode

VStore++: Virtual Storage Services for Mobile Devices 323
  Sudarsun Kannan, Karishma Babu, Ada Gavrilovska, and Karsten Schwan

On Economic Mobile Cloud Computing Model 329
  Hongbin Liang, Dijiang Huang, and Daiyuan Peng

The Smartphone and the Cloud: Power to the User 342
  Roelof Kemp, Nicholas Palmer, Thilo Kielmann, and Henri Bal

Towards Cloud Mobile Hybrid Application Generation Using Semantically Enriched Domain Specific Languages 349
  Ajith Ranabahu, Amit Sheth, Ashwin Manjunatha, and Krishnaprasad Thirunarayan

Augmenting Pervasive Environments with an XMPP-Based Mobile Cloud Middleware 361
  Dejan Kovachev, Yiwei Cao, and Ralf Klamma

Elastic HTML5: Workload Offloading Using Cloud-Based Web Workers and Storages for Mobile Devices 373
  Xinwen Zhang, Won Jeon, Simon Gibbs, and Anugeetha Kunjithapatham

International Workshop on Mobile Security 2010

Cost-Sensitive Detection of Malicious Applications in Mobile Devices ... 382
  Yael Weiss, Yuval Fledel, Yuval Elovici, and Lior Rokach

On-line Signature Verification on a Mobile Platform 396
  Nesma Houmani, Sonia Garcia-Salicetti, Bernadette Dorizzi, and Mounim El-Yacoubi

Google Android: An Updated Security Review 401
  Yuval Fledel, Asaf Shabtai, Dennis Potashnik, and Yuval Elovici

SAVED: Secure Android Value addED services 415
  Antonio Grillo, Alessandro Lentini, Vittorio Ottaviani, Giuseppe F. Italiano, and Fabrizio Battisti

Author Index 429

3D Audio Interface for Rich Mobile Web Experiences

Victor K.Y. Wu[1] and Roy H. Campbell[2]

[1] Department of Electrical Engineering
[2] Department of Computer Science
University of Illinois at Urbana-Champaign, USA
{vwu3,rhc}@illinois.edu

Abstract. We propose a novel paradigm of consuming rich web content in a mobile setting (which are often eyes-free), through a predominantly 3D audio interface. Web content, formatted in audio, is streamed via the mobile device's network connection, and placed virtually in a 3D audio space. The user moves in the virtual space, using a variety of human computer interaction (HCI) means, such as voice input, touching, rotating, and shaking the device, as well as hand and head gesturing. We provide applications benefitting from this paradigm of rich mobile 3D audio web consumption. We provide system designs, and a system architecture for mobile 3D audio. Finally, we implement our ideas in a system prototype using the Apple iOS mobile platform.

Keywords: 3D audio, mobile HCI, mobile web.

1 Introduction

We propose a novel paradigm of consuming rich digital content in a mobile setting (which are often eyes-free), through a predominantly 3D audio interface. In this paper, we focus on web content, since it is the most relevant and interesting in a variety of applications. Other types of digital content, such as GPS signals or personal music collections are still important, and we highlight them as well.

In a typical scenario, web content, formatted in audio, is streamed via the mobile device's network connection, and placed virtually in a 3D audio space. The 3D audio space is readily achieved through stereo headphones and head related transfer functions (HRTFs). That is, the user perceives sound streams coming from different locations in her surrounding 3D volume. The user moves in the virtual space (thus changing her perceived soundscape), using a variety of human computer interaction (HCI) means, such as voice input, touching, rotating, and shaking the device, as well as hand and head gesturing. This creates a rich audio mobile web experience where users can consume multiple streams of content *simultaneously*.

For example, consider a person standing in a crowded subway train carrying an Internet-connected smart phone. She cannot look at her device, but multiple web services are streaming her email messages, financial news, stock updates, and social network updates in audio format. She is even talking over a VOIP connection. These and other sound items are virtually located at various positions in the 3D space, relative to the user, as shown in Fig. 1. Initially, she is close to the email stream (Gmail), and listens to important messages. The finance stream (Yahoo! Finance) is further away, and is relatively quieter. Later, this stream announces the user's heavily-invested stock price, grabbing her attention. She rotates her smart phone (without having to look at it), moving closer virtually to the finance stream, to listen to the stock's news more carefully. The Skype sound item automatically stays close to the user, moving along with her in the 3D space, allowing the VOIP conversation to continue unhindered.

Our simple example illustrates that 3D mobile audio allows for rich mobile web experiences, which are otherwise not possible in visually based mobile systems characterized by small screen sizes and eyes-free settings. Our paradigm is also readily applied for the visually impaired.

The rest of the paper is organized as follows. In Section 2, we motivate our paradigm and discuss related work. In Section 3, we provide applications that benefit greatly from mobile 3D audio. We provide system designs in Section 4, and a system architecture in Section 5. In Section 6, we provide a system prototype implementing our ideas. Finally, we conclude and discuss future work in Section 7.

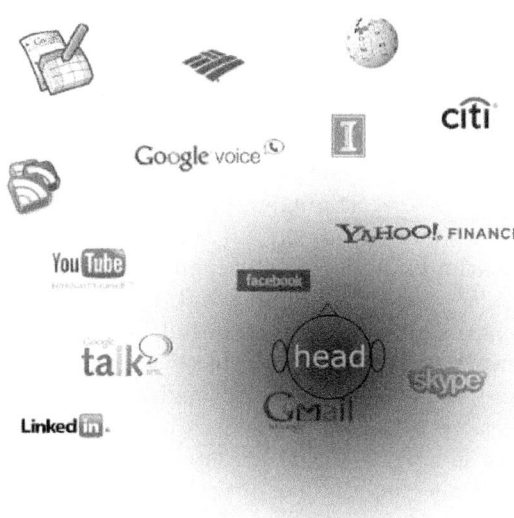

Fig. 1. Consuming multiple audio web streams in 3D audio space. The "head" indicates the user in the virtual audio space. The gradient shading shows that closer web streams are relatively louder.

2 Motivations and Related Work

Existing mobile web experiences are still far inferior than those enjoyed in traditional desktop (non-mobile) settings. This is the primary motivation for mobile 3D audio. We explain how our ideas address the problems associated with the current mobile web paradigm and provide a rich mobile web experience. Furthermore, we argue that research in mobile 3D audio advances digital systems aimed at visually impaired users.

2.1 Rich Mobile Web Experiences from a Truly Pervasive Web

Existing mobile devices have an inherent design flaw. They are modeled after desktop computers. In fact, other than the few and relatively new interfaces such as touch and voice input [1], mobile devices are just small versions of desktop computers, each equipped with a smaller screen and often, a smaller keyboard. This results in a very poor web experience. Small screen sizes prevent users from consuming content efficiently. (Large screen sizes on the other hand, reduce the mobility of the device. We are inherently plagued with this tradeoff.) Furthermore, mobile scenarios are often eyes-free. That is, a user who is driving, jogging, or standing in a crowded subway train cannot access the web because she cannot view her device. (In contrast, she has no problem listening to music or any other audio content in these situations.)

A *truly pervasive* web should not be constrained by access technologies, such as visually based mobile web interfaces. Rather, it should be constrained by how many processes a user can cognitively handle concurrently. Our mobile 3D audio paradigm fulfills this requirement by providing a rich web experience wherever network connectivity is available. For example, drivers rightfully should not be allowed to send text messages, because the distractions are proven to be dangerous [2], [3]. However, a long distance driver should be able to safely listen to her emails through an audio interface, using her Internet-connected smart phone. This is the true pervasive web. Furthermore, when the user does have the choice to view her mobile device, we argue that innovative 3D audio interfaces are still better, since we can achieve much richer web experiences than that of a small screen.

2.2 Enabling Technologies

Existing mobile hardware is already very powerful. (For example, the Qualcomm Snapdragon chipset used in smart phones has a 1 GHz processor [4].) Furthermore, mobile broadband Internet access is quickly being deployed all over the world [5]. Many cities are already equipped with 3G connectivity, with 4G already in a few others. The economics of the mobile industry are also allowing increasingly more users to have Internet-connected devices. In other words, supporting technologies (and the economics therein) already allow for mobile 3D audio.

2.3 Enabling Human Factors

The *cocktail party effect* allows a person to focus on an individual sound source, even in the presence of other sounds [6] [7]. In cognitive psychoacoustic terms, this is known as *selective attention*. *Alternating attention* refers to the ability to switch focus between different sound sources. *Divided attention* refers to the ability to simultaneously focus on multiple sources. Research in this area of cognitive listening shows that listeners achieve better divided attention if the sound sources are spatially separated, which is rather intuitive [8], [9], [10].

Recent research in HCI has demonstrated these human characteristics in modern mobile scenarios. In [11], [12], [13], sound items are placed in a virtual audio circle around the user's head, and the user selects the items using hand or head gestures. Once the user selects an item (by a head nod in the corresponding direction for example), that content is streamed to her. This idea has even been customized to a mobile application that allows a user to interface to her music in an entirely eyes-free manner, by placing selection menus in the 3D space, and using device rotations for input [14]. We propose extending this idea to a larger space, which we call the *movable 3D audio space*. Many sound items are placed in a large virtual 3D audio space. The items are constantly streaming audio content. Instead of selecting a particular item, a user chooses to move (using hand or head gestures, or device movements) to different locations in the virtual space, allowing her to consume multiple sound streams at once. This is similar to the "audio minimization" techniques in [15].

2.4 Digital Systems for the Visually Impaired

Computer web browsers targeted for visually impaired users are readily available. However, they do not innovate very much beyond screen reading [16], [17], [18]. In this paper, we focus on sighted users and the mobile web. We argue that technologies initially developed for sighted users in mobile environments (which are often eyes-free) will eventually be applied to visually impaired users. Developing visually impaired technology is difficult because the small market economics do not justify the engineering design costs. By developing rich 3D audio mobile web interfaces for the sighted in general, visually impaired users gain the added benefit of easily adaptable (if not just transferrable) technologies to them.

3 Applications

We provide several categories of applications that our paradigm supports. Obviously, many of these overlap. And in a typical situation, a user may be engaged in multiple applications from multiple web services at the same time. (We discuss this further in Sections 4 and 5.)

3.1 Media Converters and Players

Media converters and players take web content, convert them to audio format and play them for the user. For example, users can listen to music streamed from various content websites, in various formats, including video (by extracting only the audio portion of course). Other types of content include lectures or presentations and text-based content (e.g. news websites and blogs), which can all be easily converted to audio format. We envision that a media converter and player installed locally on a user device has a generic way to parse web content and convert it into audio (similar to screen readers for the visually impaired). However, a web service or web content provider might provide more friendly interfaces to audio web users, allowing them richer experiences. For example, after a user queries a music streaming website for a song, the website could play short audio preview clips of search results, which is arguably better than just a spoken text description.

3.2 Productivity

Productivity programs are often the killer applications that drive a technology to maturity. We envision that mobile 3D audio is no different. Obviously, many office applications, such as word processing or spreadsheets do not work well in an audio setting for document creation or editing. But they can be easily accessed through text-to-speech means. Organizational productivity applications are very important, since users benefit greatly when they are accessed in eyes-free settings. These include email or messaging, voice calling, calendar or organizers, and mapping or navigation. For example, a driver can talk on the phone (when it is safe to do so) with a coworker, and at the same time, check her calendar to set up a meeting by accessing her meeting management tool (by accessing the associated stream in 3D audio space). She may even check her email conversations (to listen to previous meeting minutes) to set up an agenda for the meeting. We classify another type of productivity application as informational. These include web streams such as news, stock prices, and weather reports. Since they require minimal interactivity, these informational web streams are easily implemented. In fact, web feed or subscription technologies such as RSS (Really Simple Syndication) and podcasting can be readily incorporated.

3.3 Location-Based

With the advent of modern web services, the mobile web and positioning technologies such as GPS are becoming extremely important. One notable example is overlaying the virtual world onto the physical world. Many services provide some type of geotagging feature. Users associate (that is, "tag") physical locations with virtual locations on the web. As a result, a plethora of information can possibly be written to and read from the virtual locations by users. This allows for a variety of existing location-based services to be readily incorporated in our designs.

For example, consider a mapping application such as Google Maps used on a regular computer, where users pan around a map view to look at user-generated reviews (possibly in different media formats) on nearby restaurants and entertainment. In a navigation setting inside a vehicle, the experience is significantly compromised. The screen on the device is typically smaller, making panning difficult. Many systems provide some point of interest (POI) search functionality. However, the experience is still worsened, since spatial awareness (e.g. relative distances between locations) is lacking. As well, the entire system is dangerous to use when driving (an eyes-bound activity). Conversely, consider a mapping and navigation application using our paradigm. In addition to POI search, a user can move around virtually in 3D audio space, listening for POIs streaming audio information (and maybe even user-generated content). This allows for some spatial awareness (e.g. relative distances between POIs) to be recovered, even in an eyes-free setting. For example, using simple hand or head gestures, a driver can move in 3D audio space corresponding to the vicinity near her real physical location, listening to user-generated reviews on POIs. This is shown in Fig. 2.

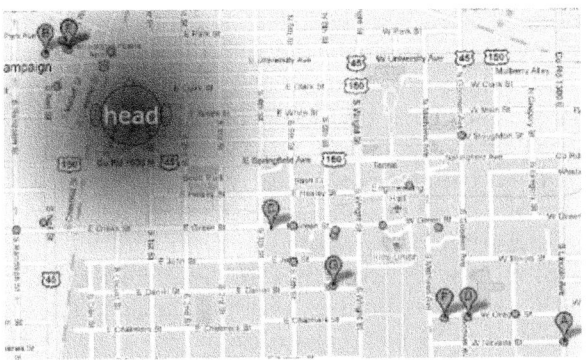

Fig. 2. The Google Map is traversed in movable 3D audio space. The red dots are locations determined by Google after a query for "coffee".

4 System Designs

We discuss several system designs in general. In Section 5, we provide a system architecture that organizes many of these ideas in a coherent manner.

4.1 Movable 3D Audio Spaces

The mobile 3D audio web consumption paradigm is based largely on a virtual 3D audio space in which multiple sound sources are streaming content. In the most general setting, both the user and sound items move in multiple spaces. For example, a user might switch between a work space and a personal space, and thus come into contact with different services (and possibly even other users who share the same spaces). An item in a 3D space can be generalized to any entity that has interaction. An item might be as simple as a YouTube video stream, or as complex as a group of users playing a virtual board game. Items are movable and their trajectories can depend on each other.

4.2 Sound Stream Prioritization

One design that affects sound item trajectories in 3D space is prioritization of the items themselves. For example, a user may wish to have access to her email at all times. Then her email item may be preset to stay with her, even if she moves in the space. As well, she may not want too many items in her immediate area, distracting her focus. Other items may thus be dynamically pushed aside as she moves. For instance, in Fig. 1, as the user moves up, her Gmail item moves along with her, and the Facebook item is automatically pushed to the left.

Notifications and alerts are a form of prioritization, and are very natural in an audio setting. For example, in Fig. 1, the Google Talk and Citibank items are far away from the user. If a message is received through Google Talk, a short notification sound plays close to the user. She may then choose to ignore it, or move towards the Google Talk item, and start a chat conversation. Similarly, if Citibank suspects somebody has stolen the user's credit card credentials, it issues an alert sound close to the user. Since this is critical, all other items may be automatically disabled until the user responds to the alert.

Instead of using sounds, we can use keywords or phrases that a user can hear and process, even if the audio streams producing these are far away in the 3D space. In particular, studies have shown that a user is able to recognize somebody speaking her name, even if that audio source is out of focus and far away [19]. (For example, in Fig. 1, the user likely hears her name, even if it is streamed by the Google Reader item far away.) In our design, we could allow a user to first tag several interest words or phrases. For example, these may be, "New York Yankees", "Yo-Yo Ma", "Simpsons", and "Vinton Cerf", representing the user's interests in sports, music, television, and technology. Whenever one of these words is about to be streamed, the system automatically replaces it with the user's name, capturing her attention, even if the audio stream is far away. The user then locates the audio stream in virtual space and moves to it. The audio stream may replay the user's name for several moments, allowing her time to locate it. (In the meantime, the actual stream may be buffered if it is a real-time stream.)

In an alternative design, if an audio stream comes across a tagged interest word or phrase, it automatically moves near the user. The user then focuses in on that stream or pushes it away in virtual space (using some input mechanism). This alternative design removes the need for a user to locate a stream of potential interest to her. However, it may create an unpleasant experience if many streams are frequently appearing in her vicinity.

4.3 Sound Encoding

Thus far, we have assumed that sounds themselves are relatively agnostic to the system. However, in general, we can encode sound according to where the user is in 3D audio space. We have already assumed that further away sounds should be relatively quieter (by definition of the virtual space mimicking a real-world scenario where sound volume decays over distance). Therefore, a simple design is to just stop streaming a sound source (saving computing and networking resources) if the user is far away from it. (In this case, we consider "muting" as a degenerate form of sound encoding.)

Another observation is that users tend to be able to recover less information from a far away sound item, since the sound is relatively softer, and the user is likely focused on another sound nearby. If we model this situation as a channel (in the communications theory sense) between the sound source and the user, we see that the channel capacity is reduced as the sound is positioned further away in 3D audio space. Thus, if we try to push too many "bits" through this narrow pipe, some of them are inevitably dropped. Therefore, we should encode the sound with a lower "information rate", in a psychoacoustic sense. (We do not define this rigorously. However, it is the subject of future work with regard to mobile 3D audio.) Practically, low-rate encoding can be achieved by reducing the speed of sounds. For example, if a news website sound item is far away, it can stream the text of articles slower than usual by allowing more delay between audio samples. Alternatively, the sound source may choose instead to stream a summary or a simpler version of its content. In this case, the news item streams only headlines when it is far away. As the user moves closer to the news item, the news item begins to stream detailed article text. As well, the encoding can also be in space, in addition to time. That is, when the user is far away from the news item, it streams only headlines from various news sections. As the user moves closer, the item actually splits into multiple sound items, which locate themselves in the vicinity of the original item in virtual space. Each of the new sound items represents a different news section, and streams its own headlines. Finally, the user can move close to the one of these new items, causing it to stream detailed news article text.

5 System Architecture

We provide a system architecture in Fig. 3. Components to the right of the vertical dashed line are all contained within a mobile device. Components to the left are remote entities from the web.

5.1 Overview

The audio spaces engine (ASE) maintains all the state information of the various audio streams in all the audio spaces. This includes the locations of the audio streams in 3D audio spaces, and the user's location. The input/output module

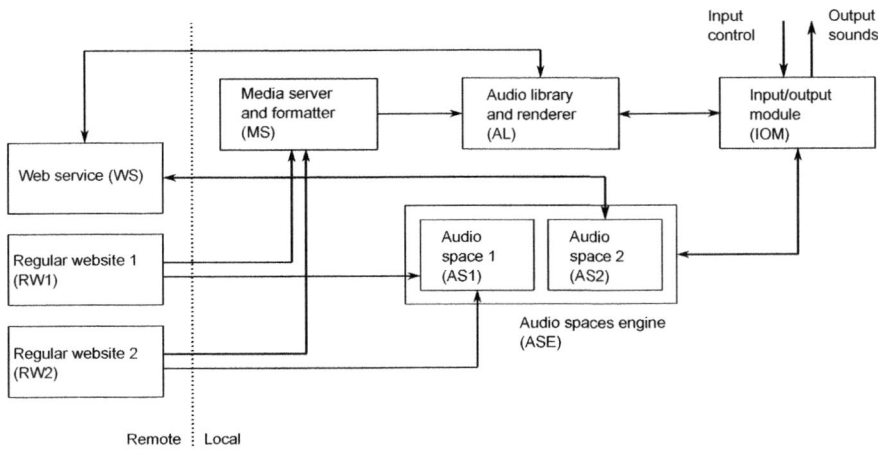

Fig. 3. Mobile 3D audio system architecture

(IOM) is an interface between the user and the underlying system components. That is, the user gives input control signals such as touch or shake gestures, and voice inputs. IOM then translates these signals into commands such as moving within and switching between audio spaces. The commands may be even more interactive, such as leaving an audio comment on a blog post. The underlying system components update themselves based on these commands and report back to IOM. IOM then collects the resulting output audio and sends it back to the user.

5.2 Audio Spaces Engine (ASE)

ASE contains all the relevant audio information to model how a user perceives a soundscape, given her location. Note that this component does not contain any audio content. (Actual audio content is stored in the audio library.) ASE is composed of distinct audio spaces. Each audio space encapsulates information such as sound item locations, number of sound items, and user location, if the user is in that particular audio space. Furthermore, the audio spaces are also interactive. As we discuss later, a web service may directly control an audio space, moving sound items around, perhaps reacting to user movement or other actions. Note that each audio space is sandboxed apart from each other in ASE, for a stable and secure design.

5.3 Non-optimized Websites

Consider two cases when a user visits a website. In the first, the website is a traditional visually based site, unaware of the audio capabilities of the user's device. In the second, the website is optimized to provide the user a rich audio experience.

In the first case, the mobile device first establishes a connection with the remote server. This is RW1 in Fig. 3. RW1 sends visually formatted content (such as text, audio, or video), which is relayed to the media server (MS) on the device. MS converts the content to audio and stores it in the audio library (AL). ASE then organizes the sound locations in an audio space (AS1 in Fig. 3). This is done according to local device preferences set by the user, since RW1 has no idea that its content is being formatted into audio. ASE determines the perceived soundscape of the user. This information is given to IOM. IOM uses the soundscape information to ask AL to render and aggregate the sounds accordingly. The result is returned to IOM, and passed to the user as a single stereo signal. The perceived soundscape is constantly updated as the user moves in sound space. If the user visits another website, the parsed content may also be assigned to the same audio space, as shown by RW2 in Fig. 3.

For example, suppose a user with a mobile device first visits a local news website, represented by RW1 in Fig. 3. The website returns a webpage containing three news articles and the local weather, in text format. The text content is sent to MS where it is converted to audio, and then stored in AL. In particular, the news articles are stored as three audio files, and all of them are tagged with the same library identifier, ID_{news}. The weather report audio file is tagged with $ID_{weather}$. This is shown in Table 1. ASE creates two sound items corresponding to the news and weather, and places them in AS1. Each sound item is characterized by its associated library ID, and its virtual location in 3D audio space. Thus, AS1 only stores a table of sound items, shown in Table 2. It does not contain any actual audio information. (Audio files are stored in AL.) In this case, the two sound items are located at $(100, 0, 0), (-100, 0, 0)$, two preset locations customized by the user. The movable user location (x, y, z) is also stored in the table. (Note the user might even be outside AS1, and in AS2 for example.) Suppose the user is located closer to the news item. With a simple audio propagation model, ASE determines that 80% and 30% maximum volume of the news and weather audio files, respectively, should be used as part of the user's perceived soundscape (ultimately her modeled head related transfer function). This information (along with the respective library IDs, ID_{news} and $ID_{weather}$) are passed to IOM. IOM requests the actual audio from AL (using the library IDs as keys). AL thus produces an aggregate stereo audio signal that combines 80% maximum volume of the three news audio files (looped in series) and 30% maximum volume of the weather audio file. As the user moves in AS1, the volume proportions change, so that AL is constantly re-rendering the aggregate output stereo audio signal. Suppose later the user visits another website, RW2 in Fig. 3 (without "leaving" RW1). RW2 is a finance website with stock quotes and an investment advice program video stream. Two more sound items are added to AS1, with the corresponding audio files added to AL. Now, a total of four sound items are in AS1. So as before, depending on the location of the user in AS1, ASE sends the appropriate information to IOM. IOM then requests the aggregate stereo audio signal from AL as before, but now with the output signal composed of up to four components. Note that since stock quotes are updated regularly, and the investment advice program is a video stream, the device has to periodically query RW2 for new content, which eventually updates the audio files in AL. This situation mimics existing web browsers having multiple tabs open, connecting the user to multiple websites. The audio content stored in AL and the sound items in AS1 for this case of two simultaneous connections to RW1 and RW2 are shown in Tables 3 and 4.

Table 1. Audio content stored in AL for connection to RW1

Audio Content	Library ID
News article 1	ID_{news}
News article 2	ID_{news}
News article 3	ID_{news}
Weather report	$ID_{weather}$

Table 2. Sound items in AS1 for connection to RW1

Sound Item (or User)	Library ID	Virtual Location in Audio Space
News	ID_{news}	$(100, 0, 0)$
Weather	$ID_{weather}$	$(-100, 0, 0)$
User	-	(x, y, z)

Table 3. Audio content stored in AL for simultaneous connections to RW1 and RW2

Audio Content	Library ID
News article 1	ID_{news}
News article 2	ID_{news}
News article 3	ID_{news}
Weather report	$ID_{weather}$
Stock quotes	ID_{stocks}
Investment video	ID_{invest}

Table 4. Sound items in AS1 for simultaneous connections to RW1 and RW2

Sound Item (or User)	Library ID	Virtual Location in Audio Space
News	ID_{news}	$(100, 0, 0)$
Weather	$ID_{weather}$	$(-100, 0, 0)$
Stock quotes	ID_{stocks}	$(0, 100, 0)$
Investment video	ID_{invest}	$(0, -100, 0)$
User	-	(x, y, z)

5.4 Audio-Optimized Websites

In the second case, when the device connects to an audio-optimized website, a web service replies, asking for additional, device-specific information and resource permissions in order to create an optimized mobile 3D audio web experience. This may include the web service requesting multiple shared or not shared audio spaces, a limit on the allowed total number of sound items, and the allocated bandwidth for audio streams. As well, the web service may ask for available and allowable user interfaces. The device may reply with, for example, a touch screen with a certain resolution, and an accelerometer with certain sensitivity settings. After providing this initial information, the device assigns the web service direct access to possibly multiple audio spaces. In Fig. 3, we show one web service (WS) with direct access to one audio space (AS2). WS directly sends formatted audio content to AL, bypassing MS, easing the computational load of MS and AL. With this level of control, web services can provide customizable and rich experiences for users

For example, suppose a user visits a social networking website, represented by WS in Fig. 3. After requesting for device information and permissions, the social network has total control of AS2 and is allowed to place up to 10 sound items in it. It is allowed to store up to 30 audio files in AL, with each file not exceeding 3 MB in size, and 3 minutes in duration. With these constraints, the social network creates a customized 3D audio space in AS2 for the user. That is, on the server side, the audio space is represented by a large virtual space consisting of all the user's friends as sound items. Since WS can only place up to 10 sound items, only a portion of that large virtual space is replicated in AS2, corresponding to the 10 closest friends with respect to the user's current location, in AS2. This is shown in Fig. 4. As the user moves in AS2, its location is fed back to WS directly. This is indicated by the arrow pointing from AS2 to WS in Fig. 3. Thus, WS can update the sound items in AS2 accordingly in real-time. (That is, the 10 closest friends change as the user moves.) WS sends the audio files (of social networking status updates and other social information) corresponding to the 30 closest friends of the user's current location, to AL, and updates them as the user moves. The larger subset of friends in AL (audio files) than in AS (sound items with no audio information inherently) enables the system to react fast enough, since there is a delay for WS to send audio files to AL, as the user moves quickly in AS2. WS also modifies the entire structure of friend placement in the audio space, over multiple sessions, depending on which friends the user is near more often. These friends are placed closer together, and nearer to the user when she visits the social networking website in subsequent visits to the social network.

6 Implementation

We integrate many of these ideas in a system prototype. We design and program a mobile application on the Apple iOS mobile operating system platform. It is currently available on the Apple App Store [20]. The application is an audio

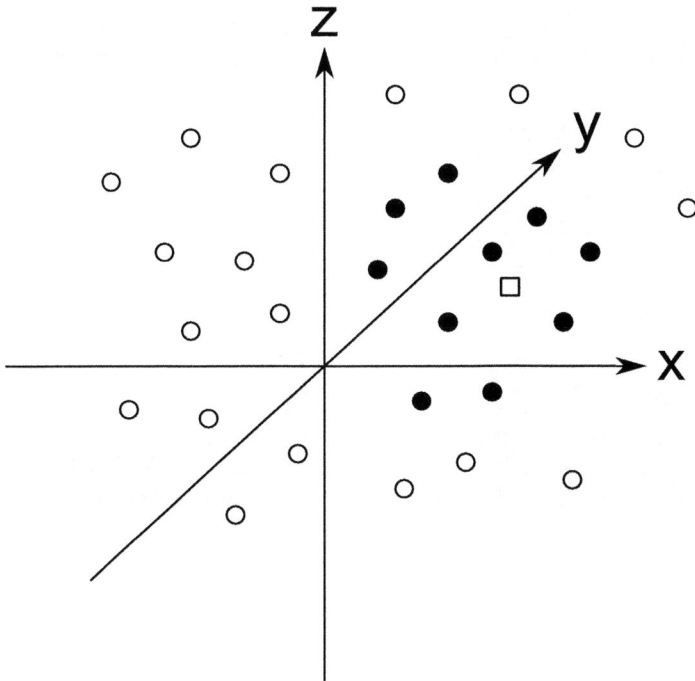

Fig. 4. 3D audio space representation of a social network. The user is represented by the square. The circles are the user's friends. They are all placed in 3D audio space. The filled circles indicate the 10 closest friends. The locations of the user and of these 10 friends are stored in AS2.

RSS reader. It retrieves RSS feeds, converts them to audio, and plays them in 3D audio space. The user moves in the space with touch gestures and shaking the device.

6.1 Prototype Design and Functionality

The application is composed of two screens, as shown in Fig. 5. The first is the navigation screen, where the user moves in 3D audio space. The second is the content selection screen, where the user can select up to six RSS feeds (and locally stored music files). If the user is in the first screen, a simple two-finger tap anywhere brings up the second screen. A back button labelled "3D" on the second screen returns the user to the first screen.

The application downloads up to six RSS feeds and converts them to audio files. (A feed is re-downloaded and re-converted after a user selects it on the second screen if it is more than 15 minutes old.) The audio files are placed in fixed positions in 3D audio space and played (in a looping manner). The six positions are {north, south, east, west, up, down}, equidistant from the origin. In

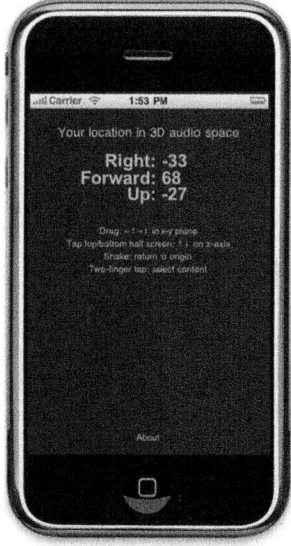 

(a) 3D audio navigation screen (b) Content selection screen

Fig. 5. System prototype on the Apple iOS mobile platform

the first screen, the user controls her movement in audio space. Touch dragging forward or backward and left or right on the screen moves her x-y coordinates. Tapping the top half of the screen moves her up on the z-axis, and tapping the bottom half moves her down. Shaking the device returns her to the origin.

The navigation interface is entirely eyes-free. The user senses her location by listening to how relatively loud the audio feeds are, and how the soundscape changes as she moves (a feedback mechanism). (The screen does display the user's current coordinates nonetheless.) This prototype demonstrates a rich mobile RSS experience. The user can pre-select her desired content, and listen to them in an eyes-free situation. Switching between feeds is extremely simple. The user can even listen to feeds simultaneously, as well as listen to "background" music.

6.2 Audio Implementation Details

In our prototype, the downloaded RSS feed content is in text format. Therefore, we convert the text to audio using a text-to-speech (TTS) engine. This is part of MS in Fig. 3. We have one audio space in ASE. The dragging, tapping, and shaking gestures are passed from IOM into the audio space, which calculates the position of the user. Based on the user location, AL renders the audio files, and aggregates them into a single stereo sound. This is accomplished with OpenAL, a cross-platform API for simulating 3D audio [21]. Basically, we pass in the relative audio locations and user location (and orientation) in 3D space, and the API returns the perceived stereo sound. This is constantly updated as the user moves in the 3D audio space.

7 Conclusion and Future Work

In this paper, we propose a new paradigm of 3D audio mobile web consumption. This allows the user a rich digital experience, even on a small device with a small screen. We motivate our paradigm and provide applications. We provide system designs and a system architecture. Finally, we implement many of our ideas in a system prototype on the Apple iOS mobile platform.

Future work includes further developing our system designs. There are still many interactive audio web interfaces that we have yet to explore. These include designing standardized audio sounds to help the user navigate in the 3D audio spaces. As well, we predominantly discuss the spatial aspect of sound in this paper. This is inherently related to a sound's volume. We wish to explore other aspects such as pitch (frequency), quality (the harmonics of a tone), and reverberation (echoes). We also plan to exploit the temporal aspect of sound, in addition to spatial.

We plan to incorporate more components in our system architecture (Fig. 3) in a more complete implementation. This includes both integrating a more general audio spaces system on the local device side, as well as developing audio-aware web services on the remote side. In particular, there needs to be many additional web protocols that cater to 3D audio information flow. This will allow web services to provide personalized audio web experiences in a standardized manner.

We also plan to do extensive user studies to test our designs. In particular, we want to know how the cocktail party effect depends on the vastly different types of content on the web. For example, background music likely requires very little attention. But we want to characterize the differences in attention required for listening to emails versus listening to social network streams, for example.

Acknowledgments. We acknowledge many people who contributed at various stages to this work. Dr. Virginia Way Tong Chu was heavily involved early on in conceptualizing movable 3D audio spaces, as being analogous to the real physical world. Professor Alex Kirlik, an expert in audio perception, provided a lot of context to the human side of the problem. Professor Yih-Chun Hu helped frame the problems and issues from an engineering design standpoint. And Mirko Montanari provided valuable feedback on the structure and flow of the manuscript.

References

1. Announcing Eyes-Free Shell For Android (April 2009),
   http://google-opensource.blogspot.com/2009/04/announcing-eyes-free-shell-for-android.html
2. Parker-Pope, T.: What Clown on a Unicycle? Studying Cellphone Distraction, New York Times (October 2009),
   http://well.blogs.nytimes.com/2009/10/22/what-clown-on-a-unicycle-studying-cell-phone-distraction/

3. Richtel, M.: Forget Gum. Walking and Using Phone Is Risky, New York Times (January 2010), http://www.nytimes.com/2010/01/17/technology/17distracted.html
4. The Snapdragon Platform. Qualcomm, http://www.qualcomm.com/products_services/chipsets/snapdragon.html
5. Evolution to LTE report. Global Mobile Suppliers Association (April 2010), http://www.gsacom.com/news/gsa_298.php4
6. Cherry, E.C.: Some Experiments on the Recognition of Speech, with One and with Two Ears. Journal of the Acoustical Society of America 25(5), 975–979 (1953)
7. Bronkhorst, A.W.: The Cocktail Party Phenomenon: A Review of Research on Speech Intelligibility in Multiple-Talker Conditions. Acta Acustica united with Acustica 86, 117–128 (2000)
8. Shinn-Cunningham, B.G., Ihlefeld, A.: Selective and Divided Attention: Extracting Information from Simultaneous Sound Sources. In: International Conference on Auditory Display (ICAD), Sydney, Australia (July 2004)
9. Best, V., Gallun, F.J., Ihlefeld, A., Shinn-Cunningham, B.G.: The Influence of Spatial Separation on Divided Listening. Journal of the Acoustical Society of America 120(3), 1506–1516 (2006)
10. Ihlefeld, A., Shinn-Cunningham, B.G.: Spatial Release from Energetic and Informational Masking in a Divided Speech Identification Task. Journal of the Acoustical Society of America 123(6), 4380–4392 (2008)
11. Lumsden, J., Brewster, S.: A Paradigm Shift: Alternative Interaction Techniques for Use with Mobile and Wearable Devices. In: Conference of the Centre for Advanced Studies on Collaborative Research (CASCON), Toronto, Canada, pp. 197–210 (October 2003)
12. Marentakis, G., Brewster, S.: A Study on Gestural Interaction with a 3D Audio Display. In: Brewster, S., Dunlop, M.D. (eds.) Mobile HCI 2004. LNCS, vol. 3160, pp. 180–191. Springer, Heidelberg (2004)
13. Brewster, S., Murray-Smith, R., Crossan, A., Vasquez-Alvarez, Y., Rico, J.: The GAIME project: Gestural and Auditory Interactions for Mobile Environments. In: Whole Body Interaction Workshop, ACM Conference on Human Factors in Computing Systems (CHI), Boston, MA (April 2009)
14. Hipui Funkyplayer, http://www.hipui.com/funkyplayer/
15. Vasquez-Alvarez, Y., Brewster, S.: Audio Minimization: Applying 3D Audio Techniques to Multi-Stream Audio Interfaces. In: Poster Session International Workshop on Haptic and Audio Design (HAID), Dresden, Germany (September 2009)
16. Parente, P.: Audio Enriched Links: Web Page Previews for Blind Users. In: International ACM SIGACCESS Conference on Computers and Accessibility (ASSETS), Atlanta, GA, pp. 2–8 (October 2004)
17. Yu, W., Kuber, R., Murphy, E., Strain, P., McAllister, G.: A Novel Multimodal Interface for Improving Visually Impaired People's Web Accessibility. Virtual Reality 9(2), 133–148 (2006)
18. Bigham, J., Prince, C., Ladner, R.: WebAnywhere: A Screen Reader On-the-Go. In: ACM International Cross-disciplinary Conference on Web Accessibility (W4A), Beijing, China, pp. 73–82 (April 2008)
19. Moray, N.: Attention in Dichotic Listening: Affective Cues and the Influence of Instructions. The Quarterly Journal of Experimental Psychology 11, 56–60 (1959)
20. Wu, V.K.Y.: 3D Audio RSS/Music Player. Apple App. Store, http://itunes.apple.com/us/app/3d-audio-rss-music-player/id363753578?mt=8
21. OpenAL, http://connect.creativelabs.com/openal/

Secure, Consumer-Friendly Web Authentication and Payments with a Phone

Ben Dodson, Debangsu Sengupta, Dan Boneh, and Monica S. Lam[*]

Computer Science Department
Stanford University, Stanford, CA 94305
{bjdodson,debangsu,dabo,lam}@cs.stanford.edu

Abstract. This paper proposes a challenge-response authentication system for web applications called Snap2Pass that is easy to use, provides strong security guarantees, and requires no browser extensions. The system uses QR codes which are small two-dimensional pictures that encode digital data. When logging in to a site, the web server sends the PC browser a QR code that encodes a cryptographic challenge; the user takes a picture of the QR code with his cell phone camera which results in a cryptographic response sent to the server; the web server then logs the PC browser in. Our user study shows that authentication using Snap2Pass is easy to learn and considerably faster than existing one-time password and challenge-response systems. By implementing our solution as an OpenID provider, we have made this scheme available to over 30,000 websites that use OpenID today. This paper also proposes Snap2Pay, an extension of Snap2Pass, to improve the usability and security of online payments. Snap2Pay allows a consumer to use one-time credit cards as well as the Verified by Visa or Mastercard SecureCode services securely and easily with just a snap of a QR code.

1 Introduction

This paper introduces two consumer-friendly techniques, Snap2Pass and Snap2Pay, that leverage a mobile phone to improve the security of web logins and web payments on a PC. They require no special hardware beyond a cell phone with a camera. With Snap2Pass, there is no more memorization of passwords and the login process is faster and less error-prone than with existing systems such as one-time passwords. With Snap2Pay, consumers no longer need to let retailers keep their passwords because it eliminates the credit card entry by hand. In addition, it allows users to conveniently take advantage of one-time use credit cards for additional security.

---

[*] This research is supported in part by the NSF POMI (Programmable Open Mobile Internet) 2020 Expedition Grant 0832820, Stanford Clean Slate Program, Deutsche Telekom, and Google.

1.1 Authentication on the Web

Passwords are the predominant form of authentication system used by today's websites. It is not because the password system is secure; quite the contrary, they are known to have many problems. Passwords are vulnerable to dictionary attacks and can be easily phished using a spoofed web site. For example, a recent breach at a large web site showed that close to 1% of users choose "123456" as their password [1]. Moreover, since users tend to use the same password at many sites, a single server compromise can result in account takeover at many other sites. Florêncio and Herley found that a single password is typically used to access over five sites [10]. As more sites support email addresses as a username, this poses a significant risk–if an account is breached at one site, others are at risk as well. The study also indicates that on the order of 0.4% of users fall victim to a phishing attack each year. Despite these limitations passwords are widely used.

Over the years, many enhancements have been proposed, including smart cards, one-time password tokens (such as RSA SecurID) and challenge-response authentication. To date, none of these have been widely adopted on the Web.

Challenge-response is a good case study. While it prevents some attacks that defeat basic passwords, it is rarely used on the Web due to the cumbersome user experience. For example, a system called CRYPTOCard uses a smartcard with a screen and a keyboard where users key in the challenge and then copy the response to the desktop. Authentication using CRYPTOCard takes far longer than authentication using a simple password. As a result, CRYPTOCard is primarily used in corporate settings where the additional hardware cost and the extra inconvenience is acceptable.

1.2 Fast Secure Login on the Web Browser

The web has become the dominant platform for modern applications. Perhaps the largest contribution to the web's success as a platform is the ability for users to visit any web page or application from a standard web browser, found on *any* modern computer today, with no configuration. Simply entering a unique name for that web application is enough to download the necessary code and launch the application. A web application's authentication system must support this interaction — a user should be able to authenticate against a web application from any available browser, with no additional configuration. In particular, the authentication mechanism is restricted to using generic browser components combined with information supplied by the user.

By leveraging the mobile device, our proposed Snap2Pass technique improves security for web applications on a standard web browser, without preconfiguration. The phone is always with us and switched on. It is a personal device–we do not use others' phones, and nobody uses our phone, except in rare circumstances. In fact, the phone is an ideal device for keeping personal, private, information. In other words, it serves to identify the owner, and can be used as a second-factor authentication. Browsers on the PC, on the other hand, are not personal. We often drift between browsers, on different PCs, at home, at work and on the road. Since the PC has a larger screen and a big keyboard, we can make the best of both worlds by pairing the phone as a personal device with a generic PC browser.

At a high level, the user experience using Snap2Pass is as follows. The user navigates his PC browser to the login page of a web site. The login page displays a QR code containing a cryptographic challenge, among other things. The user takes a picture of the QR code using his cell phone camera. No other user interaction is needed to log in. Under the hood, a pre-shared secret key stored on the phone is used to compute a response to the cryptographic challenge which is then transmitted to the site via the cellular (or wi-fi) network. The site checks the response and if it verifies, triggers the PC browser to successfully complete the login process, and load secured pages. The use of both the phone and PC provides an added security benefit, as checking the co-location of these devices can mitigate man-in-the-middle attacks.

1.3 Secure Payment on the Web Browser

When shopping at an online retailer for the first time, the checkout page asks users to enter all of their information (e.g. credit card number, billing address, shipping address, etc.) before the transaction can complete. This step is generally cumbersome and can cause shopping cart abandonment. In addition, there is some risk in sending sensitive information to a relatively unknown retailer. Moreover, consumers are often faced with the dilemma of either letting the retailer keep their credentials hence increase the risk or suffering the inconvenience of having to enter their credit card information repeatedly.

We can apply the same basic technique used in Snap2Pass to submit the credit card number with just a snap of a QR code. There is no need for manual entry of credit card information, and consequently no need to let the web site retain credit card information. In addition, we can provide added security by combining this approach with one-time use credit card numbers.

1.4 Contributions

This paper makes the following contributions:

Snap2Pass: A Consumer-Friendly Challenge-Response Authentication System. This technique is easy to use, requiring users to only take a picture of the QR code with a camera on their cell phones. The website displays a QR code that embeds a challenge. The cell phone sends the response to the challenge directly to the web server.

OpenID Implementation. We have implemented a custom OpenID provider that uses Snap2Pass, and a mobile client for the Android environment. This provider can be used immediately to log onto over 30,000 existing websites that use OpenID today [21]. We demonstrate that our techniques can be implemented today with minimal changes to legacy services. No changes are required on browsers.

Snap2Pay: Secure Online Payments. We combine Snap2Pass with one-time credit card numbers to obtain a payment system providing some user privacy from online merchants. The Snap2Pass concept can also be used to improve the security and usability of the Verified by Visa or Mastercard SecureCode services.

User Study. Our user studies suggest that Snap2Pass is easy to use and is preferred to existing mechanisms like RSA SecureID and CRYPTOCard.

The Snap2 Technology. Snap2 is a general technique based on the ability to create quickly a secure three-way connection between a server, a PC browser and a phone. The browser connects to the server with a web page visit, which is then connected to the phone via a QR code that embeds a session key. This enables a server to engage in secure sessions with the browser and the phone simultaneously. The server acts as a secure message router between the phone and the browser.

1.5 Paper Organization

Section 2 presents the threats addressed by this paper. Section 3 describes the core Snap2Pass algorithm based on both symmetric keys and public/private keys. Next, Section 4 describes how accounts for multiple web sites are managed. We describe how we simplify OpenID sign-on by extending it with Snap2Pass. Section 5 shows how the Snap2Pass concept can be extended to improve the experience and security of online payments. Sections 6 and 7 present some extensions and a security analysis, respectively. We describe our implementation as an OpenID provider in Section 8 and the results of our user study in Section 9. Section 10 presents related work and Section 11 concludes.

2 Threat Model

Snap2Pass is an authentication system designed for ease of use while providing stronger security than traditional passwords or one-time passwords. Our design is intended to protect against the following types of adversaries:

- *Phishing.* Phishing targets users who ignore the information presented in the browser address bar. A phishing attacker sets up a spoof of a banking site and tries to fool the user into authenticating at the spoofed site. Furthermore, we allow for online phishing where the phishing site plays a man-in-the-middle between the real banking site and the user. The phisher can wait for authentication to complete and then hijack the session. One-time-password systems, such as SecurID over SSL, cannot defend against online phishing. With Snap2Pass this attack is considerably harder, as discussed in Section 7.2.
- *Network attacker.* We allow the attacker to passively eavesdrop on any network traffic. Moreover, we allow for a wide class of *active* network attackers discussed in Section 7.
- *Phone theft.* Snap2Pass enables quick revocation in case of phone theft.

Snap2Pass does not provide security against malware on the user's machine. Indeed, a sophisticated transaction generator could, in principle, execute transactions on the user's behalf once authentication completes. A good example is the stealthy transaction generator described by Jackson et. al [11]. Similarly, Snap2Pass does not protect against malware on the phone itself.

3 The Snap2Pass System

We now present the core algorithms in the Snap2Pass system. Recall that challenge-response authentication comes in two flavors. The first is a system based on symmetric cryptography. It uses little CPU power and generates very short messages, however it requires that the server possess the user's secret authentication key. As a result, the user must maintain a different secret key for each server where she has an account. The second is a system based on public-key cryptography. It requires considerably more CPU power to generate responses to challenges, but the server only keeps the user's public-key. Consequently, the same user secret key can be used to authenticate with many servers.

We describe both challenge-response systems as implemented in Snap2Pass. We present the basic work flow, including account creation, login, and revocation.

3.1 Symmetric key Challenge Response Authentication

In a symmetric key based challenge-response system, the client(s) and web server communicate using a pre-shared secret key. Our implementation uses a key length of 128 bits. This key is used in the HMAC-SHA1 algorithm to compute responses to server-issued challenges. The challenges are 128-bit length nonces embedded in a QR code, while the responses are 160 bits long and are sent over the wireless network.

3.1.1 Account Creation. The account creation web page invites a new user to submit a username. Upon receiving an acceptable username, the server generates a shared secret for the account and then sends a QR code to the web page encoding the account information. The user launches the Snap2Pass application on the phone and selects the "Set Account" button to activate the camera, consume and decode the QR code. The web site then confirms that the account was created successfully. To avoid adding spurious entries in the provider's database, it should require a user login to complete the creation process.

Figure 2 shows the Snap2Pass application running on the Android phone. We create a QR code representing a created account by encoding the following contents:

```
{   protocol:   "V3",
    provider:   "goodbank.com",
    respondTo:  "https://login.goodbank.com/response",
    username:   "mr_rich",
    secret: "2934bab43cd29f23a9ea"
}
```

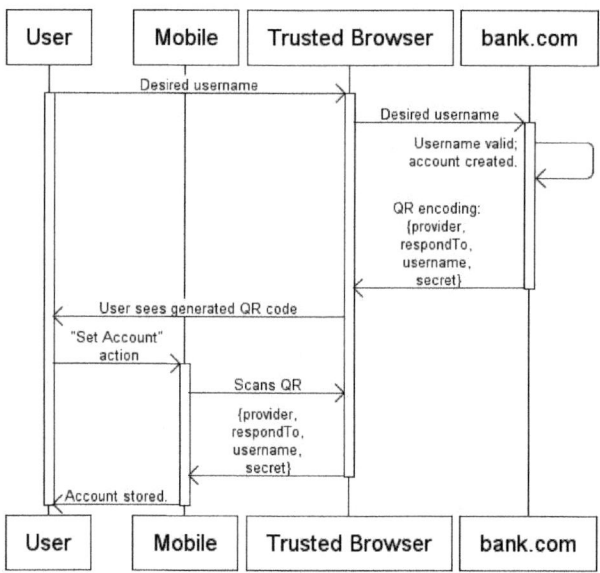

Fig. 1. A sequence diagram for creating an account in Snap2Pass

Fig. 2. The mobile client running on Android. (a) The home screen, with a single button to log in and with "Set Account" accessible as a menu entry. (b), scanning into a browser session. (c), confirmation of a login.

Note that this process eliminates the need for a user to create and remember a password, which is not just cumbersome but extremely insecure as discussed in Section 1. As a matter of fact, it is not even necessary for the user to have a friendly user name; however it is important for the sake of addressing and reassuring the user that the website recognizes him.

Instead of using a user-supplied password, our scheme allows the web server to generate a random key as a shared secret between the web site and the user.

The shared secret is presented in a QR code and saved on the phone's password manager once scanned. The user can present it for subsequent logins without needing to know its value.

The QR code also specifies the endpoint where the phone will send responses to challenges as part of the login procedure. A sequence diagram showing the account creation protocol is shown in Figure 1.

In principle, account creation can be done entirely on the phone, without the need for an interaction between the PC and the phone. We chose not to use this approach in Snap2Pass since during account creation the user is often required to supply account details such as a physical address, email address, etc. Typing all this information on the phone can be cumbersome. Instead, with Snap2Pass the user enters all account details on the PC and uses a QR code to move the corresponding credentials to the phone.

3.1.2 Account Login

On the login page, a website displays a QR code and asks the user to snap the picture with his phone's camera to log in. The page should also provide an alternative login method in case the user's phone is not available.

The QR code on the login page, unique per session, encodes a random challenge nonce to be used in the symmetric challenge-response authentication. This is generally presented within the context of an SSL session between the browser and the web server. An example of the contents contained in a challenge QR code shown at the time of login:

```
{   protocol:    "V3",
    provider:    "goodbank.com",
    challenge:   "59b239ab129ec93f1a"
}
```

By binding the challenge nonce to the browser session, the server ensures that only one browser session can make use of its authorization.

To log in, the user launches the mobile Snap2Pass application and selects "Log In". By using the phone's camera, the application consumes the challenge QR code and extracts the challenge within. The application finds a shared secret key and response endpoint that match the provider name and desired user account. It computes a response comprising of the HMAC-SHA1 hash of the entire challenge message using the pre-shared secret as key and sends it to the response endpoint, as well as the original challenge and account identifier. The provider verifies this response and, if successful, the browser session is authenticated with the appropriate account. An example of a response message is shown below:

```
{   protocol:    "V3",
    challenge:   "59b239ab129ec93f1a",
    response:    "14432nafdrwe2443af",
    username:    "mr_rich"
}
```

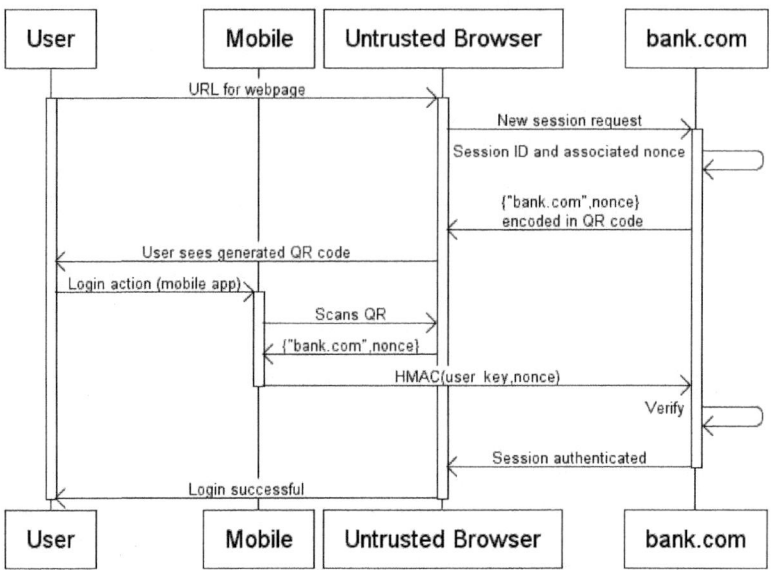

Fig. 3. A sequence diagram for logging in to a web application using Snap2Pass

The challenge and response flows occur within SSL sessions. A sequence diagram showing the login protocol is shown in Figure 3.

3.2 Public-key Based Challenge Response Authentication

Key proliferation is a prevalent problem with a challenge response system that utilizes symmetric keys. The user needs to negotiate and manage a shared secret with each web site he visits. We describe a public-key based challenge response systems to combat this problem.

Instead of using symmetric keys, the Snap2Pass mobile application can generate private/public key pair for the user upon installation (and on-demand). The account creation step is modified such that the user presents his public key to the web site instead of having the site generate a secret. The challenge process proceeds as before. The Snap2Pass application generates a response by signing the challenge with the private key. The web server verifies the response by matching it against the user's public key. The user's public key can be used across all the sites that he wishes to sign in at.

There is an alternative solution to the key proliferation problem in symmetric challenge systems. The user can take advantage of a Snap2Pass-enabled OpenID provider and benefit from OpenID's single sign-on properties across multiple web sites. Thus, the user's Snap2Pass application needs to maintain a single shared secret between the user and his OpenID provider. The number of keys is limited to the number of OpenID providers he uses. He may even use the same private/public key pair across his OpenID providers enabling Snap2Pass to maintain fewer keys.

Note that this protocol requires no certificate authority (CA) infrastructure. Client certificates are entirely avoided in either solution, while the first solution also avoids a CA. The OpenID-based solution is a centralized component necessary to enable Snap2Pass use with unmodified websites while mitigating the key proliferation problem. The OpenID provider may use a CA; our scheme interoperates with this design but does not require it. Phone loss and recovery scenarios are address in Section 7.

4 Supporting Multiple Websites

Section 3 describes the core Snap2Pass algorithm for logging onto one website. We now describe how we use a single mobile Snap2Pass client to log onto multiple websites, potentially with multiple personas. We also describe how by leveraging OpenID, we can enable the adoption of this technology immediately across a large number of existing websites.

4.1 Independent Accounts

In practice, we wish to carry only one Snap2Pass client on our phone to log onto multiple websites. The Snap2Pass client maintains a mapping from providers to accounts. It may also maintain multiple accounts per provider, allowing the user to select their desired identity during a login attempt. The response message from the phone to the web site contains the user's identity that is logging in along with the corresponding cryptographic response.

To minimize the risk of phishing when extending Snap2Pass to multiple sites, we associate a recognizable image with each web site, obtained at account creation time. The image is displayed on the phone during login, as illustrated in Figure 2(c).

4.2 OpenID

We have implemented Snap2Pass as a custom OpenID provider. Many web sites today have adopted the use of OpenID, enabling single sign-on using their OpenID credentials. The key advantage is that all of the websites that support OpenID, known as relying parties, can enable Snap2Pass based login without requiring any code changes on their end. The user's credentials reside with an OpenID provider that uses Snap2Pass. We used Snap2Pass to log into several websites supporting the standard, including: Slashdot, ProductWiki, ccMixter, and LiveJournal. Upon typing in an OpenID account name to the web site of a relying party, the web page automatically redirects the user to the login page of the OpenID provider. In this case, our OpenID provider presents the user with the Snap2Pass QR code, and the login process proceeds as described in Section 3. Once the login process completes, the OpenID provider signals the result to the relying party web site.

4.3 OpenPass Integrates Snap2Pass into OpenID

A benefit of Snap2Pass is its simple user interaction — a user no longer needs to type in any credentials at a participating website. Unfortunately, the first step of an OpenID login is to type in the user's OpenID address, so they may log in using their chosen identity provider. This defeats our goal of logging in with a single snap.

To address this issue we use a modified version of challenge-response. Now, the relying party is charged with creating the challenge. The phone sends its response to this challenge to a pre-configured identity provider, which then notifies the relying party of the transaction.

OpenPass works as follows:

- When a user visits an OpenPass-enabled website (relying party), it generates and displays a QR code containing a challenge created by this relying party, as well as an endpoint for handling responses.
- The phone computes the response to the embedded challenge and sends it to the user's pre-configured identity provider. The message also contains the reliant party's response endpoint.
- The provider, possessing the user's shared secret, verifies the response and notifies the relying party using the given endpoint. It also forwards the username and challenge associated with the authentication attempt.
- Finally, as in OpenID, the relying party verifies a token with the identity provider, using a shared secret. If the authentication is successful and this token is valid, the relying party notifies the user's browser of a successful login. The user's browser asynchronously waits for this response, and a page refresh completes the authentication.

5 Payments

With Snap2Pass, we use the private storage of a mobile phone to create a secure web authentication experience. We can generalize the technique to improve our browsing experience in other ways.

5.1 Snap2Pay

A generalization of Snap2Pass can help with both usability and security of online payments. The system, called Snap2Pay, functions as a digital wallet on the phone and interacts with the web site using QR codes. We first describe the user experience assuming the Snap2Pay application already has the user's payment information. We later explain how to automatically populate the phone with this data.

When making payments with Snap2Pay, the phone automatically contacts the user's bank and requests a one-time credit card number specific to the current retailer. This greatly reduces the risk of giving out the credit card number to an unknown retailer. Moreover, it enhances user privacy since it is more difficult for the retailer to track the user via credit card numbers. Combining this with other private browsing mechanisms, such as TorButton, gives the user a convenient way to shop online in private.

One-time credit card numbers greatly reduce the risks involved in giving a credit card number to an unknown retailer. While one-time credit card numbers were introduced some time ago, they have had limited use primarily due to the manual labor required to generate them. With Snap2Pay, one-time credit card numbers are built in and generated automatically by the system. As a result, the system is highly effective for interacting with small retailers or other questionable sites on the Internet.

Using Snap2Pay the checkout process works as follows:

- When the user's PC browser arrives at the retailer's checkout page, the page displays a QR code encoding transaction details, in addition to normal shopping cart information. The QR code encodes a response channel URL.
- Instead of manually entering personal information at the standard checkout page, the user can simply snap a picture of the QR code via her Snap2Pay phone application.
- Once the QR code is snapped the user is asked to confirm the transaction on the phone. Next, the phone securely obtains a one-time credit card number from the user's bank specific to that retailer.
- Next, the phone contacts the response channel URL on the retailer's site, and provides one-time payment information.
- The retailer completes the transaction, and redirects the user's PC browser to the transaction completed page.

The Snap2Pay checkout process requires the user to a) snap a picture of the checkout QR code and b) confirm the transaction on the phone. No other action is required. The main reason for doing this on the phone (as opposed to in the browser) is mobility: the user's payment data is available to use on any computer and any browser. No special hardware or software is required on the PC.

To complete the discussion of Snap2Pay, we explain how to populate the phone with the user's payment data. Past experience with digital wallets (e.g. Microsoft's Digital Wallet) suggests users do not take the time to enter their payment information into the wallet. Instead, with Snap2Pay, every time the user manually enters credit card information at an online retailer, the retailer displays a QR code containing that data. The user can simply snap the QR code to bootstrap the Snap2Pay database. Future transactions can use this data as explained above. We believe this process will make adoption much easier, but this can only be verified through a massive user study with the support of many online retailers.

5.2 Verified by Visa

Verified by Visa [24] and Mastercard SecureCode [13] are, in effect, single sign-on services run by Visa and Mastercard that let merchants obtain user confirmation

Fig. 4. A mockup of Snap2Pay in an ecommerce checkout page

on requested transactions. When the user visits a merchant's checkout page, the browser is redirected to the user's bank where the user is asked to confirm the transaction with a password. The browser is then redirected back to the merchant where the transaction completes, provided a valid confirmation token is supplied by the bank. The resulting transaction is considered a "card present" transaction which is a strong incentive for merchants to adopt this system. This architecture is highly vulnerable to phishing and received much criticism [17].

Combining Snap2Pass with Snap2Pay can help improve the usability and security of Verified by Visa and Mastercard SecureCode. The mechanism is similar to how we integrate Snap2Pass with OpenID, as discussed in Section 4.2 and works as follows:

- In addition to standard transaction details, the merchant's checkout page includes a QR code that encodes the transaction amount plus a random challenge for a challenge-response protocol. The challenge also uniquely identifies the merchant.
- The user snaps the QR code with the Snap2Pass application and approves the transaction on the phone. The phone then sends a message to the user's bank containing the transaction amount, the random challenge from the merchant, and the response to that challenge (computed using the user's secret key stored on the phone). The message also includes account information such as the user's credit card number. Note that Snap2Pass is pre-configured at account setup to only send this message to the user's bank and nowhere else.
- The bank checks that the challenge from the merchant and the response from the phone, both contained in the message from the phone, are valid; namely, that the response from the phone is a valid response for the challenge. If so, it uses a merchant response channel URL (a well-known endpoint) to send to the merchant the Verified by Visa confirmation token, which includes the random challenge contained in the message from the phone in addition to the standard fields.

– The merchant verifies the token from the bank and also verifies that the challenge in the token is the challenge that the merchant supplied in the QR code — this verification is needed to ensure that the phone answered the correct challenge. If all is valid, the merchant completes the transaction and transitions the browser from the checkout page to the transaction completed page.

Using this approach the random challenge is provided by the merchant (in the QR code), but is verified by the bank. The improved user experience is very simple: snap a picture of the QR code on the checkout page, confirm the transaction on the phone, and wait for the transaction to complete. Nothing needs to be typed in and no confusing redirections take place.

Since the user never supplies a credential to the merchant, this approach prevents offline phishing by a malicious merchant. Online phishing, discussed in Section 6.1, is still possible, but our geolocation-based defense described in that section applies here too.

6 Extensions

We now discuss several extensions to Snap2Pass to improve its security and to cope with the scenarios when the user forgets his phone, or forgets to log out.

6.1 Active Man in the Middle

The basic Snap2Pass system does not prevent an active man in the middle attack such as online phishing. In an online phishing attack, the attacker creates a spoofed web site that constantly scrapes the target web site. The phisher lures users to the spoofed site and uses their responses to immediately login to their account at the target site. Once in, the phisher can take any action on the user's account. This attack easily defeats one-time password mechanisms and many phishing toolkits now work this way [7].

We minimize this attack vector by using geolocation information of the phone relative to the user's PC. Recall that in Snap2Pass the target web site communicates with the user's PC and with the user's phone. In normal use, the two are in close proximity. In an online phishing attack, the site communicates with the phishing server and the user's phone. The two are very likely to be far apart. Thus, the web site can use geolocation information to test if the two IP addresses it is seeing are in close proximity. If so, it allows the connection and if not it rejects it. Thus, for the phisher to succeed he must identify a victim user's location, find a compromised host close to the victim and place the phishing server there. While not impossible, in most phishing settings, this will be quite challenging for the phisher. Importantly, the phone's location measurement is not known to the web browser.

The above example works well when both the cell phone and user's PC operate are addressable, such as on wifi or wired networks. Commercial systems such as MaxMind [14] offer geolocation databases claiming over 90 percent accuracy for resolving IP addresses to city locations. However, the cell phone is often not addressable, operating from the cellular provider's data network with an external gateway IP address. Cell phone IP addresses change frequently, and geographically diverse locations may operate under the same IP ranges. For example, a test user's Palo Alto, CA location resolved to one of T-Mobile's gateway IP addresses in Seattle, WA. The user's phone aids the geolocation system by providing GPS or cell tower ID data at transaction time. Furthermore, a complimentary approach involves exploiting application latency measurements to disambiguate cities operating under the same IP address range within a cellular data network [3].

We also may not have to rely on the IP network to determine the phone's location — most modern platforms can provide applications with relatively accurate location information. We expect the phone to cooperate with the authentication provider.

Another safeguard against the man in the middle is to require that sensitive transactions be verified on the mobile device. Here, the attacker gains access to the user's account and attempts to make a malicious transaction. The web site only allows this transaction to complete with confirmation from the phone, which the man in the middle cannot access. Using phones for transaction confirmation was previously studied in other projects [23,11] and nicely complements Snap2Pass.

6.2 Alternative Logins

Although users will typically have their phones with them, an additional login method allows users with missing phones to gain entry to a webpage. This backup login method is treated as a password reset request. That is, to login without a phone requires solving a Captcha [6], responding to a selection of security questions, and retrieving a link sent to a primary email address.

In some cases, a user's phone may not have network connectivity, but is still available. Here, the phone displays a truncated version of the HMAC response, which the user enters directly into the webpage to complete the authentication.

6.3 Signing Out

It is difficult for a web site to know if a user has walked away from an authenticated session [22]. With Snap2Pass we can use the phone as a proximity sensor, powered by the device's location sensors or accelerometers. For example, when the phone detects motion above a threshold after authentication on the PC completes, it notifies the site. The site can then require re-authentication for subsequent requests. Thus, upon leaving an internet café, the user's session is immediately terminated. For web users on a moving train, the site may request one re-authentication and subsequently ignore motion notifications from the phone for the duration of the session.

More generally, with Snap2Pass a user can manually log out of all of her active sessions from her mobile phone, without returning to the abandoned terminal.

7 Security Analysis

We describe a number of attacks on the system and how they are addressed. Throughout the section, we assume that the login process and the subsequent session on the PC are served over SSL so that basic session hijacking (i.e. the attacker waits for authentication to complete and then hijacks the session) is not possible.

We first observe that with Snap2Pass, unlike passwords, a compromise at one web site does not affect the user's account at other sites. To see why, recall that in the symmetric scheme, Snap2Pass maintains a different shared secret with each site. In the public-key scheme, the site never stores the phone's secret key. Thus, in neither case does a compromise of one site affect another.

It is also worth noting that since the user never types in their password, Snap2Pass protects users against present day keylogging malware installed on the user's PC. Nevertheless, more sophisticated malware on the user's PC (e.g. [11]) can defeat Snap2Pass.

7.1 Offline Phishing

An offline phishing attack refers to a phisher who sets up a static spoofed web site and then waits for users to authenticate at the site. The term "offline" refers to the fact that the phisher scrapes the target web site's login page offline. For sites using password authentication, an offline phisher obtains a list of username/password that can be sold to others. We note that users who fall victim to this attack typically ignore information displayed in the address bar [8]. Consequently, the SSL lock icon or the extended validation colors in the address bar do not prevent this attack.

Snap2Pass clearly prevents offline phishing since the phisher does not obtain a credential that can be used or sold. In fact, the offline phisher gets nothing since the phone sends its response directly to the target web site. Recall that during account creation Snap2Pass records the target web site's address on the phone. During login, it sends the response to that address. Consequently, the offline phisher will never see the response.

7.2 Online Phishing

Online phishing is an example of an active man in the middle discussed in Section 6.1. The end result of the attack is that the phisher's browser is logged into the user's account at the target site. As in the offline phishing case, we cannot rely on security indicators in the browser chrome to alert the user to this attack. In Section 6.1, we discussed how Snap2Pass uses geolocation to defend against this attack.

It is also worth noting that this attack is easily defeated using a PC browser extension. The extension would retrieve the SSL session key used in the connection to the web site (i.e. the phishing site) and embed a hash of this key in the QR code (if the connection is in the clear the data field would be empty) along with the extension's digital signature on the hash. The phone would verify the signature and then send the hash to the real site along with its response to the challenge. The web site would now see that the browser's SSL session key (used to communicate with the frontend of the phishing server) is different from its own SSL key (used to communicate with the backend of the phishing server) and would conclude that a man in the middle is interfering with the connection. The reason for the extension's signature on the hashed key is to ensure that the phisher cannot inject its own QR code onto the page with the "correct" key in it. An alternative to a digital signature is to place the QR code containing the hashed key in the browser chrome (e.g. in the address bar) where the phisher cannot overwrite it with its own data.

7.3 Phone Theft and Key Revocation

If a phone is lost or stolen, that phone can potentially be used to impersonate the user at all websites where the user has an account. Snap2Pass mitigates this issue in two ways.

Firstly, the Snap2Pass application can require the user to authenticate to the phone before the application can be used. Rather than implement an unlock feature in Snap2Pass we rely on the phone's locking mechanism for this purpose. Users who worry about device theft can configure their phones to require a pass code before applications like Snap2Pass can be launched. This forces a thief to first override the phone's locking mechanism. Moreover, several phone vendors provide a remote kill feature that destroys data on the phone in case it is lost or stolen.

Secondly, when a phone is lost, users can easily revoke the Snap2Pass credentials on the phone by visiting web sites where they have an account and resetting their Snap2Pass credentials at those sites. This results in a new keying material generated for the user thus invalidating the secrets on the lost phone.

8 Implementation

Our implementation of Snap2Pass includes server-side code, called a provider, and a mobile client. The provider and the client provide a reference implementation for the server and client ends of the Snap2Pass protocol, respectively.

8.1 Provider

The provider is implemented as a custom OpenID provider and offers server-side challenge/response functionality as described above. OpenID is a popular protocol for federated identity management and single sign-on. With the addition of this layer of indirection, we enable tens of thousands of existing OpenID consumer web sites to use Snap2Pass without requiring modification of their server-side login protocols.

The provider implementation makes use of the Joid open source project, and is written in Java. It is loosely coupled to Joid; thus it can be plugged into other standard OpenID providers. The custom provider consists of a symmetric-key based challenge response system, account management and a web portal. The challenge response modules are written in Java using built-in cryptography libraries. It includes modules for symmetric key generation, and HMAC-based challenge/response creation and verification. The account management modules manage user accounts, provide persistence and include a cache for fast lookup of incoming responses.

The web portal adds QR code features to the OpenID provider. It includes custom registration and login pages, implemented as Java Server Pages (JSP) to support the Snap2Pass account creation and login protocol. On completion of the login protocol, the web portal integrates with the provider backend to signal the result using the OpenID protocol. This enables existing OpenID consumer sites to support Snap2Pass with no code changes. The provider module has approximately 1,600 source lines of code (SLOC).

8.2 Mobile Client

The mobile client is written in Java for the Android environment. It implements the client-side Snap2Pass protocol, and offers functionality for credential management and symmetric key challenge/response computation. We use Android's SharedPreferences API to store and manage credentials retrieved from the provider. In a production implementation, the credentials are managed using a secure credential manager. The login module uses built-in APIs to compute responses to challenges. We use Android's intent system and the ZXing project to scan and consume QR codes. For improved security, the scanning functionality will be embedded directly in the application. The mobile client has approximately 400 SLOC.

8.3 Multi-party Connectivity

Our Snap2 applications involve multiple devices interacting to perform a common task. We make use of the Junction [9] platform for device pairing and messaging in a multi-party session.

In both Snap2Pass and Snap2Pay, the session contains agents representing a server, a mobile device, and a web page. The server instantiates a Junction session, generating a unique session identifier. The server then passes the identifier to the web page, which then executes Javascript code to join the session. The web page also encodes this session identifier in a QR code. After scanning the QR code, the mobile client joins the session and begins the transaction.

With Junction, messaging occurs asynchronously. Junction uses XMPP for its messaging infrastructure, with the BOSH extension [26] supporting messaging to standard web clients.

9 Experimental Results

We compare our implementation against existing secure authentication mechanisms. Our goal is to provide a system that is more consumer-friendly than what is currently available without negatively affecting security.

9.1 Using QR Codes

We presented our approach to new users to gauge their reaction. Our subjects ranged from savvy smartphone users to those who are considerably less technical.

We found that the UI design for pinpointing a QR code is important. The visual feedback on the phone's screen was helpful in guiding the user to a successful login. Sometimes, a user would believe a code was recognized by the software before it was actually able to locate it, causing some confusion during early attempts. Subsequent uses were much faster and free of this confusion. Usually two or three sample trials were enough for the user to become comfortable with the system. Most users, including the non-technical, agree that the approach is "very simple" and usable. Several even found it fun to use.

The amount of time required for the software to locate the barcode is important for the system's usability. We compared the user experience for the same application across three hardware platforms: the HTC G1, the Motorola Droid, and the HTC Nexus One. As the platform's power increased, the scanning process became noticeably quicker, improving overall user experience. We did not experiment with other barcode formats or software solutions — scanning QR codes with the ZXing application proved sufficiently usable.

9.2 Comparing Authentication Techniques

We wanted to gauge both the usability and perceived security of Snap2Pass as compared to other authentication methods. We ran a user study across 30 users of different backgrounds. 17 of our users were students, and the remaining worked at a technology company. 15 of our users were female; 16 of our users had a technical background.

Our study compared two web authentication methods enhanced with a cell phone — Snap2Pass, and a system much like RSA SecurID. With the SecurID system, participants used a mobile application to retrieve six randomly generated digits. They were told that, in deployment, this number would be synchronized between the phone and the authentication server. To log in to a web page, then, the participant entered his or her username, password, and this six-digit string.

We demonstrated the two systems to each user once before having them try for themselves, twice each. We then had them fill out a short survey about their experience.

To determine the perceived security of authentication systems, we asked how safe they would feel using a given system with their primary bank, from 1 (not safe at all) to 10 (completely safe). As a baseline, we asked about using a standard username/password scheme. The average score was 5.9. We then asked about

SecurID and Snap2Pass; SecurID came out slightly better than Snap2Pass—7.7 as compared to 7.3. Both were perceived to be noticeably more secure than username/passwords.

Of Snap2Pass, one user writes, "It was really cool. Felt very secure, like something out of a James Bond movie." She also agreed that SecurID felt secure, "but entering another field was very tedious, and I imagine would be pretty rough to have to do frequently."

To evaluate the usability of each system, we first asked how easily users found learning each system, on a scale from 1 (very difficult) to 10 (trivial). We found that users had no trouble learning either system; Snap2Pass was given an average of 9.2, and SecurID a 9.1. We then asked how they found repeated logins with each system, from 1 (very tedious) to 10 (very easy). Here, Snap2Pass fared better than SecurID — 8.5 as compared to 6.9. A user who was skeptical of the system's security commented that "Snap2Pass does not seem to be safe but given the fact that it's SO easy to use, I will be happy to use it as main login method."

Next, we asked how a participant would feel about using each system in practice. For the two systems, participants were asked, assuming they had a capable phone, if they would prefer to use each system with their primary bank as compared to username/passwords. They were given three choices: "I would not want to use this system," "I would use this system only if my bank made me," and "I would prefer to use this system." The resulting numbers are:

	Snap2Pass	SecurID
prefer the system over pwds:	15	13
would only use if required:	11	12
would prefer passwords:	4	5

Finally, we asked participants how they would feel about using a Snap2Pass-like system for making purchases on the web; in such a system, a user would point their phone at their shopping cart's checkout page to complete the transaction, without typing in billing information. We posed this question to 22 participants, and 13 indicated they would prefer such a system to entering billing details; 2 would do so only if required by their bank, and the remaining 7 would not want to use such a system.

By far, the most common concern with both demonstrated systems was an unavailable or stolen phone. Indeed, a system in deployment must be robust to such incidents, as we discuss in Sections 6.2 and 7.3 respectively. Many users also stressed that they would only trust the security of Snap2Pass with suitable on-phone security. A participant writes, "(I am) concerned about phone security. I love the idea of using my phone to store/share personal info, but I do not trust the current security on my phone." Another says, "I feel that if someone stole your phone, they could easily gain access with this technology to whatever password protected account you have." For a system in deployment, this concern must be suitably addressed.

10 Related Work

Using cell phone for authenticating people is an old idea. Wu et al. [25] and Oprea et al. [18] use phones to help establish a secure session on an insecure device. Snap2Pass addresses a different threat scenario where the user is contacting a remote server on his own PC. Our threat model is very common and is simpler than in [25,18] which enables us to design a more efficient solution that is easier for the user.

Phoolproof [19] is designed as an anti-phishing authentication mechanism. With Phoolproof, Parno et al. require custom software on the PC as well as a bluetooth connection. Snap2Pass in contrast, requires no modification to the PC and provides security using information available on smartphones that was not available at the time Phoolproof was designed.

Aloul and Zahidi discuss the use of one time passwords generated on a mobile phone for use with ATMs [2]. By using modern smart phones, we are able to provide a system that is easier to use without compromising security.

Bellovin et al. [4] propose EKE, a protocol for shared key exchange, which has been refined further. In contrast, we rely on QR codes over SSL to transfer the shared secret between the PC browser and the phone to manage account credentials.

Mccune et al. [16] have previously combined phone cameras and two dimensional barcodes for transmitting public keys from one device to another. Other clever approaches to device pairing use the accelerometer for generating a shared key [15,12] or for proving proximity [5].

In Snap2Pass and Snap2Pay, we present a novel system and applications that are based on using a visual communication channel, to present and transfer a challenge from one device to another. More importantly, the visual channel contains a reference to a communication endpoint that the mobile device uses to respond to the challenge. The combined effect is that the act of scanning a barcode becomes a user interaction with the site hosting that barcode.

In [20], Pierce et al. explore the possible uses of pairing a mobile with a PC, presenting several useful utilities. Their model uses a centrally managed account to discover services, as opposed to our ad-hoc pairing technique.

11 Conclusions

We described Snap2Pass, an easy-to-use authentication system that defeats many of the attacks on traditional password schemes on the Web. Snap2Pass is implemented as a custom OpenID provider, thereby immediately enabling usage on the tens of thousands of websites that accept OpenID-based authentication, without any server-side code changes. We have extended the OpenID protocol so that the user can simply snap a QR code presented by a relying party without having to enter user credentials on the login page.

We also presented Snap2Pay, which allows consumers to use one-time credit card numbers with just a snap of a QR code. One-time credit card numbers are useful for reducing the risk of interacting with small retailers or questionable sites on the Internet. Snap2Pay eliminates the manual labor involved, which has so far limited the adoption of the technique. Similarly, Verified by Visa and Mastercard SecureCode are single sign-on services which have not been adopted because they are highly vulnerable to phishing. We showed in this paper how the Snap2Pass technology improves both their usability and security.

Our user studies show that the system is fast to use and easy to learn. Even without much experience with smartphones, users can easily use Snap2Pass after seeing it done only once. The comparison with RSA tokens indicates that users tend to perceive more cumbersome techniques to be more secure; however, users care more about ease of use than security. Snap2Pass has the advantage that it is both secure and easy to use.

Snap2Pass can be used in an off-the-shelf PC browser with no modifications, and works well with all popular browsers today. We have an open-source implementation of Snap2Pass at http://mobisocial.stanford.edu/snap2pass.

References

1. T.I.A.D.C. (ADC). Consumer password worst practices (2009), http://www.imperva.com/download.asp?id=239
2. Aloul, F., Zahidi, S.: Two factor authentication using mobile phones. In: Proceedings of the IEEE International Conference on Computer Systems and Applications, pp. 641–644 (2009)
3. Balakrishnan, M., Mohomed, I., Ramasubramanian, V.: Where's that phone?: geolocating IP addresses on 3G networks. In: Proceedings of the Internet Measurement Conference, pp. 294–300 (2009)
4. Bellovin, S.M., Merritt, M.: Encrypted key exchange: Password-based protocols secure against dictionary attacks. In: Proceedings of the IEEE Symposium on Research in Security and Privacy, pp. 72–84 (1992)
5. Bump technologies, http://bumptechnologies.com
6. Captcha, http://www.captcha.net
7. Cova, M., Kruegel, C., Vigna, G.: There is no free phish: An analysis of "free" and live phishing kits. In: Proceedings of the 2nd Usenix Workshop on Offensive Technologies 2008, pp. 1–8 (2008)
8. Dhamija, R., Tygar, D., Hearst, M.: Why phishing works. In: Proceedings of ACM CHI 2006 Conference on Human Factors in Computing Systems, pp. 581–590 (2006)
9. Dodson, B., Nguyen, C., Huang, T.-Y., Lam, M.S.: Junction: a decentralized platform for ad hoc mobile social applications (2010), http://mobisocial.stanford.edu
10. Florencio, D., Herley, C.: A large-scale study of web password habits. In: Proceedings of the 16th International Conference on World Wide Web, pp. 657–666 (2007)
11. Jackson, C., Boneh, D., Mitchell, J.: Transaction generators: Root kits for the web. In: Proceedings of the 2nd USENIX Workshop on Hot Topics in Security (2007)
12. Kirovski, D., Sinclair, M., Wilson, D.: The martini synch. Technical report, Microsoft Research Technical Report, MSR-TR-2007-123 (2007)
13. Mastercard securecode, http://www.mastercard.com/securecode
14. Maxmind, http://maxmind.com

15. Mayrhofer, R., Gellersen, H.: Shake Well Before Use: Authentication Based on Accelerometer Data. In: LaMarca, A., Langheinrich, M., Truong, K.N. (eds.) Pervasive 2007. LNCS, vol. 4480, pp. 144–161. Springer, Heidelberg (2007)
16. Mccune, J.M., Perrig, A., Reiter, M.K.: Seeing-is-believing: Using camera phones for human-verifiable authentication. In: Proceedings of the IEEE Symposium on Security and Privacy, pp. 110–124 (2005)
17. Murdoch, S.J., Anderson, R.: Verified by Visa and MasterCard SecureCode: Or, How Not to Design Authentication. In: Sion, R. (ed.) FC 2010. LNCS, vol. 6052, pp. 336–342. Springer, Heidelberg (2010)
18. Oprea, A., Balfanz, D., Durfee, G., Smetters, D.: Securing a remote terminal application with a mobile trusted device. In: Proceedings of the 2004 Annual Computer Security Applications Conference, pp. 438–447 (2004)
19. Parno, B., Kuo, C., Perrig, A.: Phoolproof Phishing Prevention. In: Di Crescenzo, G., Rubin, A. (eds.) FC 2006. LNCS, vol. 4107, pp. 1–19. Springer, Heidelberg (2006)
20. Pierce, J.S., Nichols, J.: An infrastructure for extending applications' user experiences across multiple personal devices. In: Proceedings of the 21st Annual ACM Symposium on User Interface Software and Technology, pp. 101–110 (2008)
21. Rafter, M.V.: A breakout year for openid (2009), http://technology.inc.com/security/articles/200902/openID.html
22. Schneier, B.: Unauthentication (2009), http://www.schneier.com/blog/archives/2009/09/unauthenticatio.html
23. Steeves, D.: Securing online transactions with a trusted digital identity. In: First TIPPI (Trustworthy Interfaces for Passwords and Personal Information) Workshop (2005), http://crypto.stanford.edu/TIPPI/first/program.html
24. Verified by visa, http://www.visa.com/verified
25. Wu, M., Garfinkel, S., Miller, R.: Secure web authentication with mobile phones. In: DIMACS Workshop on Usable Privacy and Security Software (2004)
26. XEP-0206: XMPP over BOSH, http://xmpp.org/extensions/xep-0206.html

Tool Support for Constructing Mobile Mashups

Lasse Holmstedt[1], Tommi Mikkonen[2], and Mikko Terho[3]

[1] Nokia Qt Development Frameworks
Invalidenstrasse 117, Berlin, Germany
lasse.holmstedt@nokia.com
[2] Tampere University of Technology
Korkeakoulunkatu 1, Tampere, Finland
tommi.mikkonen@tut.fi
[3] Nokia Devices, Visiokatu 5, Tampere, Finland
mikko.j.terho@nokia.com

Abstract. The ability to instantly publish software worldwide, and the ability to dynamically combine data, code and other content from numerous web sites all over the world has opened up entirely new possibilities for software development. In web terminology, a web site that combines ("mashes up") content from more than one source into an integrated experience is referred to as a *mashup*. At present, the development of mashups usually relies on the tools for composing server-side software, and off-the-shelf browser is commonly assumed as the runtime environment. However, when considering client-side mashups that are well-suited for mobile devices due to local processing and associated interactivity, numerous complications exist. One of these problems is available tool support, which is commonly targeted to desktops and browsers. In this paper, we introduce a tool for developing client-side mashup applications. In spirit, the tool is similar to tools available for mainstream mashup development, but all the actual processing is done on the client side using a special purpose runtime environment.

Keywords: Web applications, mashup development.

1 Introduction

In the past few years, the Web has become a popular deployment environment for new software systems and applications such as word processors, spreadsheets, calendars and games. In the new era of web-based software, applications live on the Web as services. They consist of data, code and other resources that can be located anywhere in the world.

The ability to instantly publish software worldwide, and the ability to dynamically combine data, code and other content from numerous web sites all over the world will open up entirely new possibilities for software development. In web terminology, a web site that combines ("mashes up") content from more than one source is commonly referred to as a *mashup*. Mashups are content aggregates that leverage the power of the Web to support instant, worldwide sharing of content.

Today, mashups are run inside a web browser. However, because the web browser was originally designed to be a document viewing tool – not an environment for highly interactive applications – there are challenges when running web applications and mashups that behave in a highly interactive fashion, especially in mobile devices. Support for user interface widgets can also be limited. Furthermore, poor performance of the web browser can be a major issue especially when running mashups in mobile devices. On the other hand, success stories of application stores by Apple and others have shown that users wish to use content and applications that have tight integration with the mobile platform. Consequently, it seems more fruitful to run mobile mashups predominantly on the client side, as already proposed in [1, 2], to combine the best possible performance and benefits of the services residing in the web.

In this paper, we introduce a tool created for easing the development of mobile mashups that are predominantly run on the client-side. The tool has been constructed using Qt, an industry-scale cross-platform environment, which has been rapidly extending into mobile devices. In terms of background, the work is based on our previous research results [1, 2], and on extending tools that the Qt environment provides for composing compelling applications.

The paper is structured as follows. Section 2 discusses mashup development in general. Section 3 introduces Qt and associated tools that were used in our implementation. Section 4 provides the tools for composing client-side mashups, and Section 5 introduces some sample mashups composed using these tools. Section 6 provides a discussion on our experiences, and Section 7 finally concludes the paper with some final remarks.

2   Mashup Development

Mashups – systems that amalgamate existing content from the web – can be composed manually using the classic DHTML technologies. However, since the actual representation of data, behavior and content can vary dramatically between different web sites, manual mashup construction can be extremely tedious, fragile and error-prone.

A number of tools are available for mashup development. To begin with, mashups can be developed using general-purpose web application development platforms. Unfortunately, the capabilities of such general-purpose web programming environments are still somewhat limited in features, especially when considering flexible extraction and combination of data from different web sites, which is important for mashup development. Most general-purpose web content development tools bear the same (or similar) shortcomings. Finally, there are also tools that have been intended for mashup development, many of which are more or less experiments. In such tools, there are some common emerging themes and trends [2]:

Using the web browser not only to run applications/mashups but also to develop them. For instance, and Yahoo Pipes (http://pipes.yahoo.com/pipes/) and Google Mashup Editor (http://code.google.com/gme/) use the web browser to host the development environment and to provide seamless transition between the development and use of the mashups.

Using visual programming techniques to facilitate end-user development. Visual "tile scripting" and "program by wire" environments are provided by Yahoo Pipes, for example.

Using the web server to host and share the created mashups. Many mashup development tools store the created mashups on a web server that is hosted by the service provider.

Direct hook-ups to various existing web services. Since the Web itself does not provide enough semantic information or well-defined interfaces to access information in web sites in a generalized fashion, most of the mashup development tools include custom-built hook-ups to existing web services such as Twitter (http://twitter.com/), Digg (http://digg.com/), Facebook (http://www.facebook.com/), Flickr (http://www.flickr.com/), Yahoo Traffic (http://developer.yahoo.com/traffic/), Google Maps (http://maps.google.com), and various RSS news feeds.

Despite the ever-increasing role of the web browser as target platform for mashups, the availability of a web browser is not an essential requirement for mashup development. On the contrary, there are technologies that utilize custom-built, special-purpose web runtimes that can bypass security limitations associated with the browser and can offer better performance, e.g., by performing the client-side mashup generation using native processing capabilities. For instance, mashups intended for mobile devices often utilize a custom-built client environment. Similarly, tools used for composing mashups need not be run inside the browser, but they can also be implemented as extensions of generic software development environments.

Our approach to the development of client-side mashups is based on a runtime environment for client-side mashups. The fashion we have implemented the system builds on top of Qt and our earlier activity, Lively for Qt (http://lively.cs.tut.fi/qt/) [1], as well as experiences on mobile mashups discussed in [2].

3 Qt as a Mashup Environment

Qt (http://www.qtsoftware.com/) is an industry-scale cross-platform environment that supports a rich set of APIs, widgets and tools that run on most commercial software platforms. In addition, Qt has been used in various embedded devices and applications, including in particular mobile phones, but also PDAs, GPS receivers and handheld media players.

From the technical viewpoint, Qt is primarily a GUI framework. It comprises a rich set of widgets, graphics rendering APIs, layout and stylesheet mechanisms. In addition, tools are provided that can be used for creating compelling user interfaces that run in a wide array of target platforms.

Qt has been gradually expanded to offer a range of connectivity facilities as well. Networking, filesystem access and web browsing capabilities are all available, together with scripting, multimedia and XML processing frameworks. While Qt is primarily targeted for C++ development, bindings to its libraries are also available for other languages, officially for Java and unofficially for several others, perhaps most notably Python.

3.1 Qt Creator

Intended to make Qt development easier, Qt Creator is a multi-platform IDE for several languages. While it primarily supports C++, other languages such as ECMAScript are also supported. The architecture is entirely plug-in based, including the components shipped with the editor itself. Some of the most important features are integrated help for the Qt Framework, a code editor and Designer, a tool for visual GUI design. Designer is intended for building traditional desktop interfaces and provides limited support for creating complex, non-standard interfaces. It is best used for its original purpose and requires a fair amount of understanding how interfaces created with Qt work, as its widget and tool names are derived from their Qt class counterparts. Qt also has a concept of styles akin to CSS, but only a text editor is provided to the user in this respect. However, the most important part of the tool, laying out widgets, is made easily approachable with a standard drag-and-drop interface, shown in Figure 1.

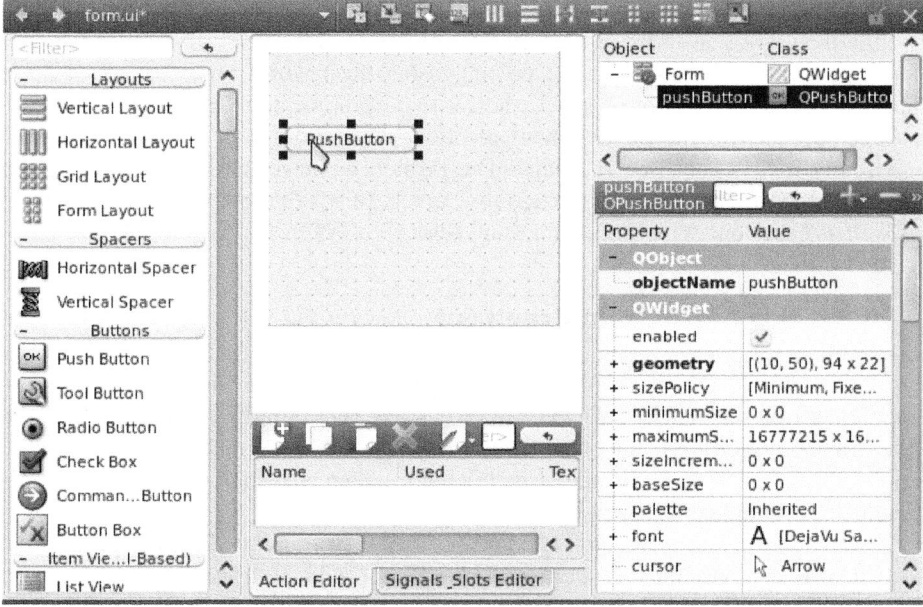

Fig. 1. Qt Designer default interface

From the technical point of view, Qt Creator is very extensible with its plug-in interface, but the interfaces are not documented well. In effect, a plug-in author has to study the interfaces through the numerous existing plug-ins and examples provided by the framework. The Designer plug-in, on the other hand, has only a handful of plug-in interfaces. While there is decent documentation for all of them, there is not enough built-in extensibility. Due to such technical restraints, implementation of extra functionality had to be made by directly modifying the Qt Designer code, in effect branching it. Additionally, a separate Qt Creator plug-in was created to facilitate creation of mashups through a combination of visual programming and traditional scripting. Combined, these plug-ins enable loading, modification and display of data with minimal effort.

3.2 Qt for Client-Side Mashups

There is a strong connection between Qt and web applications. Qt libraries include a complete web browser based on the *WebKit* (http://webkit.org/) browser engine. Moreover, the necessary DOM and XML APIs are included to parse, manipulate and generate new web content easily. In addition, Qt includes a fully functional ECMAScript [3] (JavaScript) engine called *QtScript*. The presence of a JavaScript engine is important, since JavaScript – along with XML – is the *lingua franca* of the Web that is used by popular web service APIs such as the Google Maps API (http://code.google.com/apis/maps). Powerful debugging tools are also available for QtScript, making it more attractive for developers. However, these debugging tools have not yet been fully integrated to the Qt Creator IDE, making their usage more difficult than the ones intended for C++ debugging.

Qt also offers excellent connectivity facilities to network resources, as well as an SQL database interface and filesystem access classes. In particular, network access is made as simple as making requests, without having to care for the protocol, provided that it is supported. HTTP, FTP, TCP and UDP are supported out of the box, as well as SSL-secured connections.

Based on the above, Qt lends itself to a client-side mashup environment. The mashups can leverage the rich Qt APIs for information visualization and processing. In addition to binary image and video formats such as GIF, JPEG, PNG and MPEG-4, textual representations such as XML, CSV (Comma-Separated Value format), JSON (JavaScript Object Notation) and plain JavaScript source code play a central role in enabling the reuse of web content and scripts in new contexts. Qt provides excellent capabilities for processing such information.

In terms of pure functionality, Qt has a definite upper hand over pure web-based mashup development. Its WebKit component enables the usage of any web-based mashups, and allows for their further manipulation through the WebKit's DOM interface and JavaScript engine. Existing mashups can also be reinforced with local filesystem content and results can be locally cached and saved, allowing for faster access on subsequent runs. Entirely new possibilities also realize – building mashups using data sources readily available on the operating system, such as address book data, GPS location and user-created content such as images, video, documents and other such content, becomes possible. Examples of mashups utilizing this information could be fetching user-relevant content from the web, like music, while the user's personal preferences could be analyzed locally from their music library. Other examples include using user's location data as an input for a web service offering travel information of the surroundings, such as famous sights.

Qt's main advantages are its high performance and clear API, which encapsulates most of the difficult and tedious tasks involved with network connectivity or filesystem access. Additionally, the cross-platform nature of Qt enables deployment to a wide range of platforms. However, to make development faster, a higher concentration on QtScript is needed instead of vanilla C++, while still maintaining the native performance of UI rendering and data access through script bindings.

4   Client-Side Mashup Tools

The mashup creation process for Qt Creator could be described as follows – take data, manipulate it and assign it to a display widget. Because Qt Creator already offers a powerful GUI design tool, Qt Designer, mashup integration with it was decided upon. A data model editor, similar to that of Yahoo! Pipes with a visual programming interface, was also created. In the following, this visual editor is called Mashup Editor. Although some extensions had to be built into Qt Designer to enable building mashups, the modifications are not called by any name in particular. Figure 2 illustrates the intended mashup creation process.

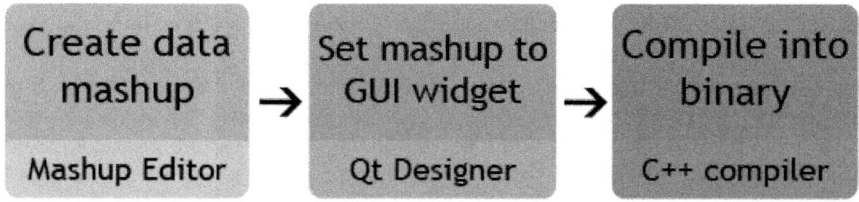

Fig. 2. Creating a mashupping application

The mashup creation part is not always necessary, as previously created mashups are reusable. Also, Qt Creator is able to use the compilers installed into the system and offers and simple button interface for compilation, further easing the process. Therefore, creating simple data-displaying programs can be as simple as dragging and dropping content to widgets and pressing a button to compile the generated code. Illustrations of both the Mashup Editor and the Qt Designer extensions can be found in the following sections.

4.1   Mashup Editor

Mashup Editor is a visual programming tool that enables loading, modification and display of data in a graph-like interface. The editor was created from the ground up as a plug-in for Qt Creator, so that existing functionality of the IDE and interaction between form design and mashup creation could be harnessed. The tool creates two types of XML files. *Mashup files* describe how various data and script elements interact with each other, and *Mashup views*, which describe how the visual mashup elements, denoting either a source of data or a script, are laid out on the screen. This separation was made because the content of the view files is not needed in order to process the mashup data. A screenshot of the editor used to create this markup is shown in Figure 3.

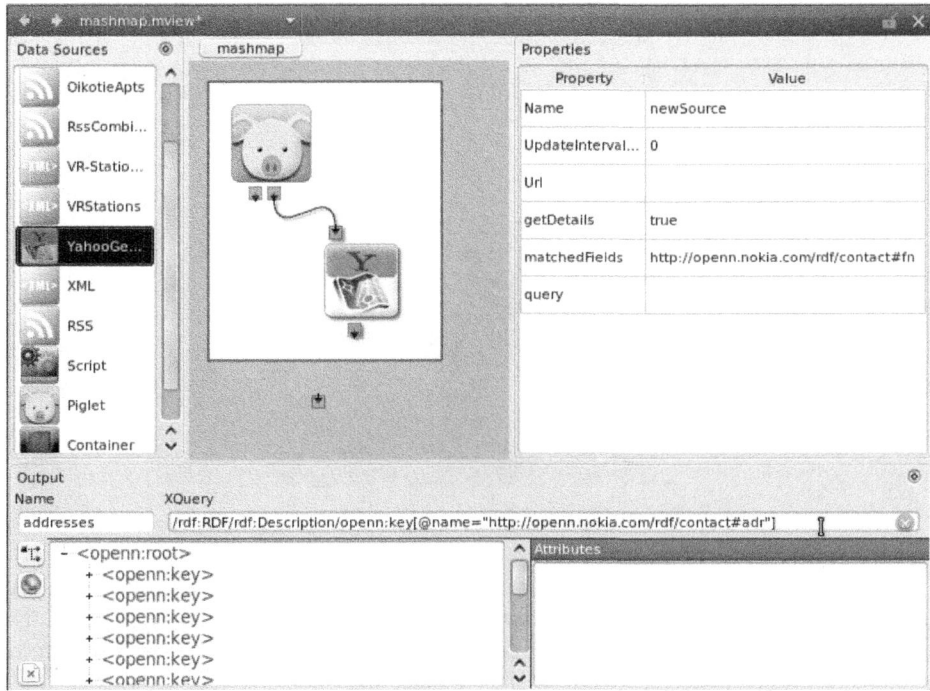

Fig. 3. Screenshot of the Mashup Editor for the Qt Creator

As the tool was originally intended to be used to aid in software prototyping, the data type was limited to text. XML based documents are preferred, because Qt contains good XML support, and because a lot of the Web content is XML based. Furthermore, converting other types of data to valid XML is usually a rather straightforward process. However, no limitations to the document contents are made, and the data processor of the mashups could just as well process binary files. Regarding the scripting, QtScript – essentially JavaScript with some minor deviations regarding libraries – is used, as it is an integral part of Qt itself. It is enhanced with a collection of libraries created with the experimental QtScriptGenerator, a tool which creates bindings from most of the available Qt objects into QtScript, exposing more Qt functionality to the script engine.

In Figure 4, a simple mashup, which combines the contents of two RSS feeds into one XML listing, is shown. In the row above, two RSS icons have their output nodes *connected* to the input nodes of a script processor. A single output can have multiple outgoing connections, but an input can have only a single incoming connection. Additionally, an output node of an XML element, such as the RSS feed, can be modified with XQuery expressions, which are natively supported by Qt. The script element contents can be either written by the user or a previously saved script can be assigned. When creating a new script, all boilerplate code is automatically generated to assist the user. New inputs and outputs can be created to script elements, and to other appropriate elements as well.

Fig. 4. A mashup system concatenating two RSS feeds

A powerful feature of structuring mashups and building them from pre-made parts is that mashup elements can be placed inside containers, and containers can be also placed inside each other. The containers act like other mashup elements – they have input and output nodes, and allow such connections to be made. Otherwise, they behave like black boxes – only the information returned as output from them is seen by the rest of the elements. This considerably facilitates building complex mashups or reusing existing components.

After mashups are saved, they can be set as templates, after which they are available as an otherwise limited list of building blocks. The mashup elements available by default are XML, RSS, Container, Script and YahooGeoCoder. Additionally, support for a triple-store database called Piglet exists. The capabilities of the most primitive items, XML and RSS, are simple. They are able to load data from an URL understood by QNetworkAccessManager, or local files, specified by their full path. YahooGeoCoder is an example implementation of a REST-based element, which is able to receive place names or addresses as inputs, sends data to Yahoo's Geocoding service [4], and returns GPS coordinates.

A developer can create new native items, and new script templates can be introduced as well by installing them in the appropriate directory.

To facilitate manipulation of XML based data, a drag-and drop-based XQuery editor exists. While not able to handle complicated logic, several common actions such as element and attribute selection are in place. Drag and drop is extensively used in other parts of the Mashup Editor interface, too. Examples of such behavior include dragging and dropping web browser bookmarks and text containing URLs into the editor, which automatically creates XML elements with URLs pointing to the given sources.

With the features listed above, the mashup editor is able to produce XML-compliant files that describe how data is to be processed. Next, we take a look at the extensions built into Qt Designer, which enable attaching these mashup instructions to an UI created with Qt Designer.

4.2 Qt Designer Extension

Extensions built on top of Qt Designer enable the user to set data on a number of item view widgets [5]. Due to the lack of an extension interface built for such a purpose, the whole Qt Designer code was branched and necessary modifications were built on top of that. While this posed some new challenges, such as maintainability and a large overhead due to lack of interface documentation, a relatively easy-to-use drag-and-drop interface could be built.

The extension implements a similar interface for data models as seen earlier in Figure 1. This data model list can be accessed through a toolbar button or a keyboard shortcut, in similar fashion with the existing tools. The mashup files that are enumerated in the data model list come from two different locations: user's home directory and the current Qt project's directory. The user's home directory contains mashup templates, which can also be utilized directly from within Designer, while the current project directory typically contains all project-specific mashups. While the user cannot directly modify mashup templates, an editor for project-specific ones can be opened by double-clicking the desired item.

To assign a mashup into a widget, the user needs to first create an item view widget that inherits the QAbstractItemView class on the form canvas by dragging and dropping it there. Qt ships with a number of such widgets, and the user is free to subclass their own, provided that they follow the Model-View-Delegate pattern used in Qt [6]. Next, the user is able to drag and drop a mashup on top of the view item. Provided that the mashup engine can load the data as instructed in the XML files, the Designer interface will display the XML based per-item output in the view.

As an example of subclassing QAbstractItemView, a Nokia Maps Item View widget was created, which is able to parse location data in several XML based formats and project these points as points of interest (POI) on the map. Such a widget cannot display most text-based data in a sensible manner, but for location-based content, it is very effective.

The Model-View-Delegate pattern also allows for the user to set delegates into views, which is also implemented in drag-and-drop style in the extension. Delegates are simply components that instruct how the data is displayed for per widget, or by its rows and columns. The Designer extension implements the ability to set a per-widget delegate, which essentially means that the whole widget will be rendered with the instructions of a single delegate. More detailed delegate processing can be done on the code level.

When a user double-clicks on an item view widget or presses a button in the toolbar of the editor, the data and list toolbox is replaced with a list of available delegates. The delegates are stored as binary files in the QPicture format provided by Qt [7], which lists 2D painting instructions in a sequence.

Creating these delegates is difficult, however, since no editor is available for creating QPicture content. While delegates are fairly straightforward to create in C++ with an average lines of code less than 50, it still requires a knowledge that is out of place for a GUI editor. The reasons for not building a delegate editor are addressed later in the paper.

5 Sample Applications

To demonstrate two different types of applications that can be created with the aid of the mashup editor, two samples are provided. One uses a map widget to display data, while the other relies on standard item view widgets and delegates to display latest news items from an RSS feed.

5.1 RSS Newsfeed Combiner Mashup

The first sample application is a simple list of news items from two RSS services – BBC World and AP News. While RSS is a very popular and fairly simple XML based syndication format, it offers a comparison with already existing tools that allow users to take existing content, modify it, and display it on a widget. Because wizards to create new projects are built into the mashup editor, we concentrate on how the mashup is built with the editor. To add both of the BBC World and AP News feeds to the mashup canvas, the user drags and drops two RSS elements to the canvas (Figure 5, left). To assign URLs to them, the user can copy the URL into each element's respective panel (Figure 5, right).

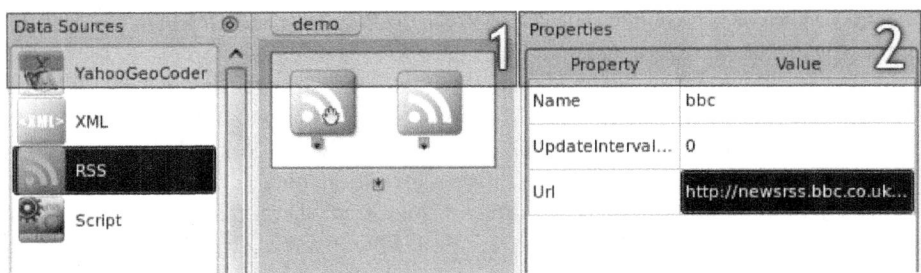

Fig. 5. Adding RSS feeds to the canvas and modifying their properties

Next, the user needs to combine the data of the two feeds by concatenating them. For this duty, the user can add a new script element (Figure 6) and select which template to use by double-clicking on it (Figure 7). A template for concatenating XML content exists, and is logically called Concatenator. The user can also write their own script and freely modify the script after selecting the template, as a copy is made out of them.

Fig. 6. Adding a script processor to the canvas

Fig. 7. Creating a new script out of a template

Next, two input and a single output are created for the script element to provide it with data. The output is connected with the output of the container mashup itself (Figure 8, left side) and the inputs are similarly connected with the outputs of the RSS items (Figure 8, right side). The connection happens in a drag-and-drop fashion, starting from either end of the node pair and finishing at the other end.

Finally, the user simply drags and drops the data model on top of the list view, double clicks it and assigns a delegate which is able to render the RSS data in a sensible manner. As creating delegates themselves is somewhat difficult, the editor currently ships with a number of delegate templates, including renderers for the popular RSS and Atom feed types. This process is shown in Figure 9.

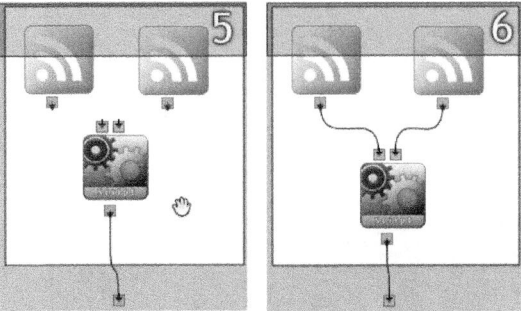

Fig. 8. Connecting mashup subsystems together

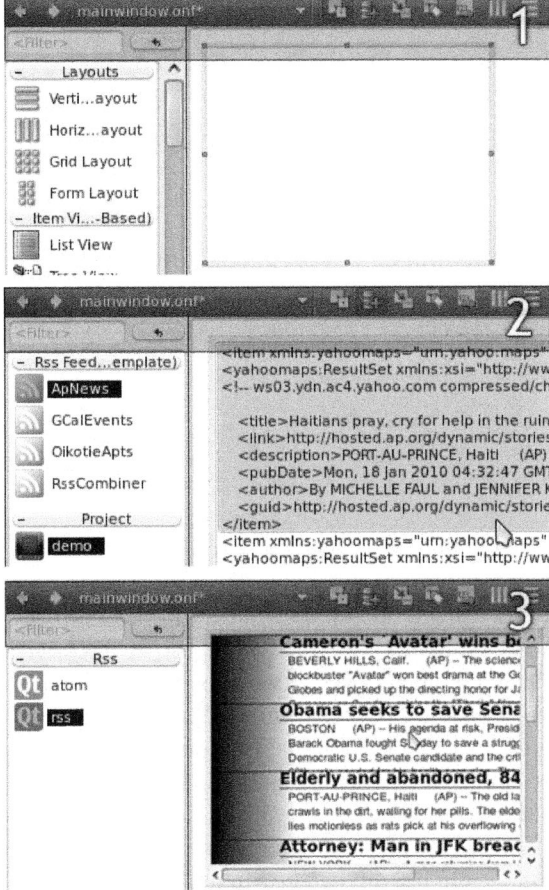

Fig. 9. Dragging and dropping a data model on a widget

To briefly analyze this process, it can be still enhanced by removing the need to add inputs and outputs manually – it is only required because the current implementation lacks functionality in this respect. However, even at the prototype stage of the tool, this sample outlines that only a few minutes of work is required to build a simple but fully working program that is able to display data. Experience has shown that most cases are not so simple, but even in non-trivial cases, the majority of time was spent in fetching the desired data from a broken XML-like file and reformatting it to make it compliant. Such examples included generating route suggestions from a Finnish national railway web service (http://www.vr.fi), as well as using a Finnish combined public transit website (http://www.matka.fi) for a similar task. Writing scripts to parse the data took several hours in both cases, which illustrates one of the standing problems with the whole concept of mashups.

5.2 GPS Coordinates from Addresses of Contacts in Local Database

The second sample application uses a database called Piglet, which is a triple-store database originating from Wilbur [8], a toolkit for programming web applications with XML and Resource Description Framework (RDF) [9] support. Piglet is also based on RDF and as such, its user can add any kind of metadata into any kind of object. Out Piglet database included a contact book, in which each contact had an address in textual format. This text could then be sent to a Yahoo's Geocoding service in order to fetch the location in GPS coordinates. Each user is also associated with a distinctive URI within Piglet, which allows users to point and click on the points of interest on the map, opening up more information about each user. However, a similar scenario is loading up any data with a postal address on a map, so this example is not restricted to a certain data engine. The illustrations below display creating this mashup.

The user first adds a Piglet element to the canvas and adds an XML namespace filter to select only contacts (Figure 10). Then, the user can add a new output and create an XQuery for it to filter out everything but the addresses. Next, a Geocoder is added, and connected to the Piglet's address output (Figure 11). Finally, a script element is added, using a template XMLMerge, which takes two XML inputs, called *inner* and *outer*, and inserts the *inner* input's element into the *outer* data source. The element names can be specified in the properties (Figure 12).

After connecting the Geocoder result node with the *inner* input and the unmodified data from Piglet in the *outer* input (Figure 13), the mashup is finished. The differences lie just in usage of data sources and scripts. Figure 14 shows what a compiled program using the created mashup looks like.

The mashup engine library heavily relies on XML parsing and QtScript, both of which are likely to cause a performance penalty. The first sample application, combining two RSS feeds into one and displaying it, was implemented with standard Qt for comparison. Because the performance hit happens almost exclusively on initialization when plugins have to be loaded and mashup files have to be parsed, the initialization time for a mashup-engine powered feed displayer was measured – 2227ms on average. A pure C++ implementation took only 1149ms, so the execution time is nearly doubled. Most of the extra time accounted for is spent loading external libraries that handle additional QtScript functionality provided by QtScriptGenerator and parsing XML and script files. Were these additional QtScript libraries part of Qt itself, this loading time would not be experienced. In the light of these numbers, a more effective instruction storage format is also desirable over XML, together with optimizations to the mashup engine itself, in particular regarding library loading.

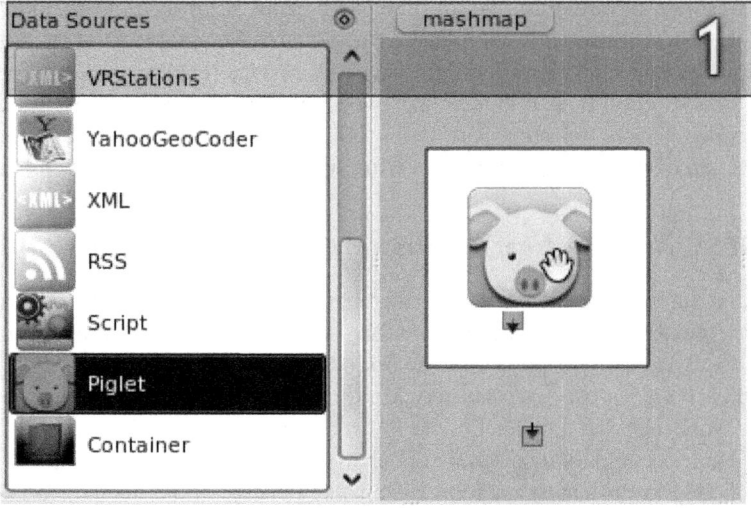

Fig. 10. Creating Piglet data source

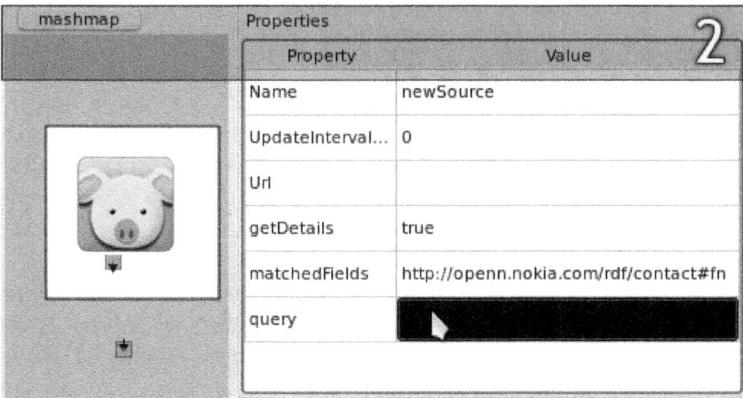

Fig. 11. Modifying Piglet data source properties

The vanilla implementation of the RSS feed reader was approximately 500 lines of code. Most of the code deals with XML processing and delegate rendering that are non-reusable by themselves. With the mashup engine and the Qt Creator extensions, the required amount of C++ programming is reduced to *zero* lines, provided that the wizards in Qt Creator are used. Even without the wizards, the user only has to write about 4 code lines per program, consisting of boilerplate initialization code.

Tool Support for Constructing Mobile Mashups   53

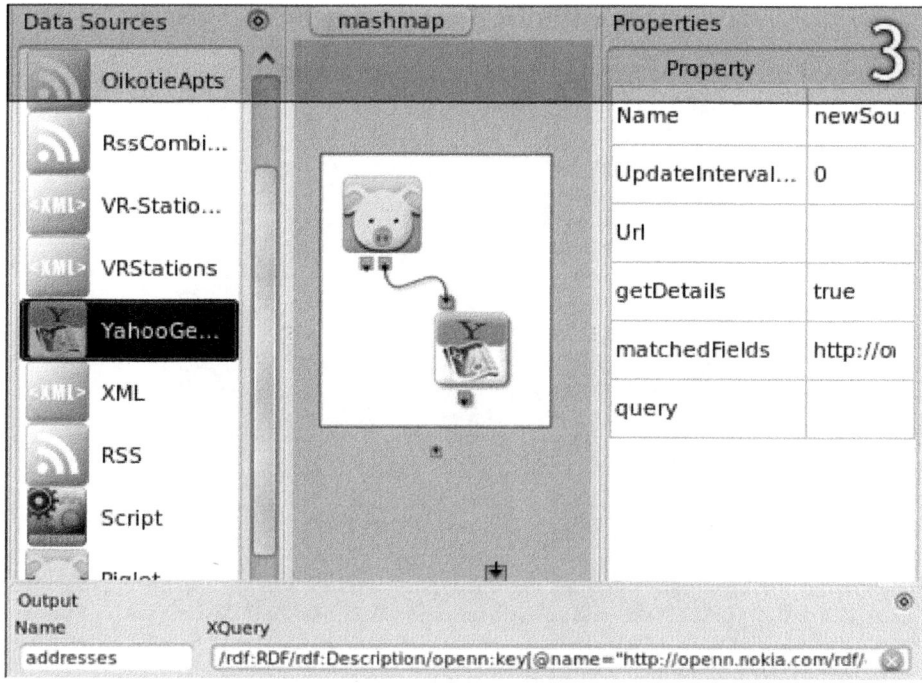

Fig. 12. Adding a Geocoder and fetching address data from the Piglet

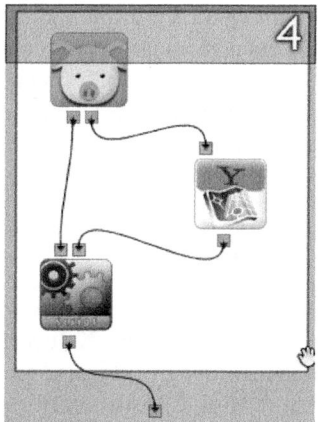

Fig. 13. Mashup that takes address data of contants, gets GPS coordinates, and appends them

Fig. 14. Finished application using contacts address mashup

6 Discussion

At present the tool we have implemented is experimental software, and it is not ready for prime time use. However, even in its present condition it demonstrates that one can rapidly develop client-side mashups that combine data from the Internet. Moreover, since we are using a special-purpose runtime environment, the tool is not bound to the restrictions of the browser but can freely access different web sites and internal data that is only available in the device.

6.1 Implementation Restrictions

There are certain use cases to which even the current prototype of the mashup editor is more than suitable for. Modifying data with XQuery and ECMAScript is easy for anyone with a rudimentary understanding of the said technologies, and for those who are not, templates are available for use. Consequently, extending functionality is easy, which is something that most of the existing mashup systems have also done well. Map mashups could be argued to be easier than with competing technologies, as coordinate-containing data can be simply dragged and dropped on the widget. For interaction, normal programming is still required, but Qt Designer provides some ailments to this respect as well, namely visual connection of signals and slots. Displaying valid XML-compliant data is also very easy, provided that there is a

delegate available. However, therein lays one of the biggest drawbacks with the current approach - the difficulty around creating delegates for item view widgets.
Some item views, such as a map widget, do not even require additional delegates to display data, but for the majority of cases, this poses a greater problem as displaying even text is difficult. Although initially planned, no delegate editor was built as Nokia changed the development direction of Qt, including next-generation item views, which did not follow the Model-View-Delegate pattern.
In addition, a declarative UI toolkit especially targeting mobile devices, called Qt Quick, is being built as well, which allows for creating interfaces with a simple markup, in addition to a GUI editor, reuse of created interfaces, as well as programming of interaction within the UI. One of the core differences between traditional and declarative development is that the latter is based on 2D graphics rendering, while the former is based on standard widgets provided by the underlying operating system. As such, the declarative approach allows for far more flexibility and possibilities, while the widgets make it possible for the UI to adapt to the operating system being deployed to. As a language, Qt Quick resembles JSON notation of JavaScript and it has been perceived as easy to learn also to those with no prior JavaScript experience, due to its very hierarchical structure.
Currently, Qt Quick seems to be the most interesting future development direction. The mashup editor and Qt Quick seem to complement each other as technologies, as Qt Quick currently supports displaying data from models, and building UIs around already available data requires no C++ knowledge, further easing the development of mashup applications. In addition, making mashups even with plain Qt Quick is possible, thanks to its XML and XQuery-processing capabilities. In real-world stress tests, Qt Quick performs 60 FPS on devices such as N900 without additional optimization work, so the CPU power of the device is freed for data acquisition and processing for the need of mashups and application business logic.
For what comes to the implemented Qt Creator extensions, they are on the prototype stage. While they work reasonably well on both OS X 10.6 and Ubuntu 9.10 with distribution packages also created, a fair amount of work is to be done for productization. The groundwork is laid out, however, and the mashup engine itself is only in need of performance optimizations.

6.2 Mashup-Related Restrictions

Another, much more difficult problem is related to the nature of mashups. That is, while manipulating data from various sources, there is little to no guarantee that the structure of the data would not suddenly change due to an upgrade in the corresponding web site. As most of the cases where mashups are desired, are fetching data from a Web source, a widget could stop working if the web author even slightly modifies the structure of the site markup. The mashup templates were created to reduce the influence of this problem and the idea was that such templates could be easily shared through a web service between users of the mashup tool. Upon a content change, any community member could then proceed to fix the affected mashup files and other users could then update their repositories respectively. While this approach does not even attempt to defeat the original problem, there is no definite solution – for instance, a web page author might move the content to an entirely different location, rendering even advanced heuristic-based content retrieval attempts futile.

A further typical mashup related problem associated with the nature of the web sites is that they are usually not valid HTML or XHTML. Research shows that only a fraction of all pages on any given site are completely valid [10], which results in errors in parsers that do not offer error processing facilities. While the native XQuery support for Qt is otherwise adequate, it offers minimal error handling facilities, making it difficult to use with a great many test cases. Experience has shown that regular expressions and even concatenated substring matches are much more effective at finding the desired data. While a visual editor for generating these does not exist yet, it is one of the future development targets.

6.3 Tools and Technologies

Comparable tools exist in terms of visual data creation, like Yahoo Pipes. The difference lies in web-based interface of Yahoo Pipes, and the fact that Pipes is concentrated solely on data mashups, as opposed to the mashup tool built on top of Qt Creator, which also enables usage of created data in user interfaces. A limited comparison can be also made to tools that help creating desktop widgets, such as Dashcode for OS X. Dashcode offers powerful tools modifying the visual appearance of widgets and programming them with JavaScript. While web-based data can be used in the interface, such as RSS feeds, the data itself cannot be modified in a visual way as in the Qt Creator extensions. However, it can be argued that Dashcode's UI tools are easier to use for novices than those in Qt Designer – a test with Dashcode to create an RSS feed with a personalized interface took less than half an hour, with no prior experience with the tool. On the other hand, creating a similar application with vanilla Qt Designer amounts to hours even from an experienced developer, due to the fact that data has be set into the widgets programatically. The mashup extensions set Qt Creator on par with tools like Dashcode, as it is only a matter of minutes to build a data model from web sources and assign it to a view.

Compared to other technologies, client-side mashups created with Qt are associated with higher performance and a more complete access to the operating system if needed through the powerful Qt API. As any existing web applications can be readily utilized through Qt's WebKit API with similar performance to web browsers utilizing the WebKit, such as Apple Safari, Qt seems to be an ideal choice, if additional client-side functionality is desired. Obviously, when considering a client-side mashup runtime environment, it is clear that security features need attention before large-scale use. At present, considerations on how a convenient model could be created have been based on J2ME [11], but we do of course acknowledge that in that context the distribution model of applications is completely different.

Another approach for a mixture of web and native content would be implementation of a plug-in for web browsers that is able to render Qt content. Building a basic version of such a plugin is not difficult, but problems arise in particular with security. Furthermore, using web browser as a platform again somewhat defeats the purpose of client-side mashups, although gained performance that arises from the use of Qt can always be seen as a positive side. Additionally, constructing websites with an unlimited set of tools, as opposed to the current approach of modifying document-based content with scripts to make websites look more functional, is attractive by itself.

Competing technologies such as Flash, often used for mashup as well as web application development, are thought of as an easier alternative than C++. Research supports that C++ is difficult to learn [12] and because of its complexity, it is also less productive. With the Mashup Editor, new UI technologies, scripting and other advances, these problems may be eliminated or at least become less pronounced. Even with the current, experimental QtScript bindings for the GUI widgets and such, it is possible to build complete applications with pure QtScript. While script-side documentation and development tools still need improvement, it opens up new possibilities for cross-platform development.

7 Conclusions

In this paper we have discussed a mashup editor for client-side mashups that are well-suited for mobile environment. The system was developed using tools and techniques that the Qt framework provides. Even in its current state, the Qt Creator extensions can be used to successfully create simple applications with a considerable reduction in the lines of code – for basic applications that simply display data, no code has to be written at all.

The comparison between a mashup engine-driven and a vanilla Qt implementation reveals performance issues, but also a promise of better reusability and reduced effort in software development. While the vanilla implementation does not use many lines of code either in absolute terms, it can be argued to be a considerable difference to a developer new to C++ development, which is typically considered difficult.

Interesting future directions also include better integration to device-specific facilities such as a cell phone contact book and data engines provided by the operating system, such as Akonadi (http://pim.kde.org/akonadi/) available on the K Desktop Environment (http://www.kde.org/). Moreover, provided with a declarative fashion to compose user interfaces in the form of Qt Quick, the developer is offered an extended toolset for the development of interesting, interactive mobile mashups.

References

1. Mikkonen, T., Taivalsaari, A., Terho, M.: Lively for Qt: A Platform for Mobile Web Applications. In: Proceedings of the Sixth ACM Mobility Conference, Nice, France, September 2-4 (2009)
2. Nyrhinen, F., Salminen, A., Mikkonen, T., Taivalsaari, A.: Lively mashups for mobile devices. In: Proceedings of the MobiCase 2009 Conference, San Diego, CA, USA, October 26-29 (2009)
3. ECMA Standard 262: ECMAScript Language Specification, 3rd edn. (December 1999)

4. Yahoo Inc. Yahoo! Maps Web Services – Geocoding API (2010)
5. Nokia Corporation. Qt 4.6: View Classes (2010),
   http://qt.nokia.com/doc/4.6/model-view-view.html (reviewed January 18, 2010)
6. Nokia Corporation. Qt 4.6: An Introduction to Model/View Programming (2010),
   http://qt.nokia.com/doc/4.6/model-view-introduction.html (reviewed January 18, 2010)
7. Nokia Corporation. Qt 4.6: QPicture Class Reference (2010),
   http://qt.nokia.com/doc/4.6/qpicture.html (reviewed January 18, 2010)
8. Lassila, O.: Enabling Semantic Web Programming by Integrating RDF and Common Lisp. In: Proceedings of the First Semantic Web Working Symposium. Stanford University (July 2001)
9. World Wide Web Consortium. Resource Description Framework (RDF) Model and Syntax Specification. World Wide Web Consortium (1999)
10. Marincu, C., McMullin, B.: A comparative assessment of Web accessibility and technical standards conformance in four EU states. First Monday 9(7-5) (2004)
11. Riggs, R., Taivalsaari, A., Van Peursem, J., Huopaniemi, J., Patel, M., Uotila, A.: Programming Wireless Devices with the Java 2 Platform, Micro Edition, 2nd edn. Java Series. Addison-Wesley (2003)
12. Lahtinen, E., Ala-Mutka, K., Järvinen, H.-M.: A Study of the Difficulties of Novice Programmers. In: Proceedings of the 10th Annual SIGCSE Conference on Innovation and Technology in Computer Science Education, Capariga, Portugal, pp. 14–18 (2005)

Cuckoo: A Computation Offloading Framework for Smartphones

Roelof Kemp, Nicholas Palmer, Thilo Kielmann, and Henri Bal

Vrije Universiteit, De Boelelaan 1081A, Amsterdam, The Netherlands
{rkemp,palmer,kielmann,bal}@cs.vu.nl

Abstract. Offloading computation from smartphones to remote cloud resources has recently been rediscovered as a technique to enhance the performance of smartphone applications, while reducing the energy usage.

In this paper we present the first practical implementation of this idea for Android: the Cuckoo framework, which simplifies the development of smartphone applications that benefit from computation offloading and provides a dynamic runtime system, that can, at runtime, decide whether a part of an application will be executed locally or remotely. We evaluate the framework using two real life applications.

Keywords: Mobile Computing, Computation Offloading.

1 Introduction

In the last decade we have seen, and continue to see, a wide adoption of advanced mobile phones, called *smartphones*. These smartphones typically have a rich set of sensors and radios, a relatively powerful mobile processor as well as a substantial amount of internal and external memory. A wide variety of operating systems [1,2,3,4] have been developed to manage these resources, allowing programmers to build custom applications.

Centralized market places, like the Apple App Store [5] and the Android Market [6], have eased the publishing of applications. Hence, the number of applications has exploded over the last several years – much like the number of webpages did during the early days of the World Wide Web – and has resulted in a wide variety of applications, ranging from advanced 3D games [7], to social networking integration applications [8], navigation applications [9], health applications [10] and many more.

Not only has the number of third-party applications available for these mobile platforms grown rapidly – from 500 [12] to 200,000+ [13] applications within two years for the Apple App Store –, but also the smartphones' processor speed increased along with its memory size (see Figure 1), the screen resolution and the quality of the available sensors. Furthermore, the cell networking technology grew from GSM networks allowing for 14.4 kbit/s to the current 4G networks that will provide around 100 Mbit/s, while simultaneously the local wireless networks increased in bandwidth [14,15].

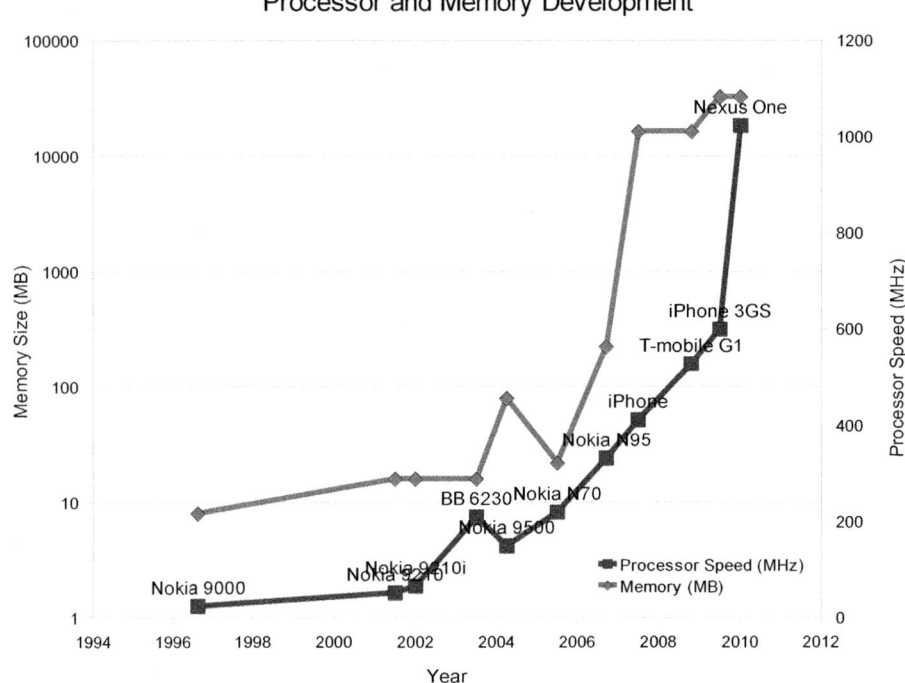

Fig. 1. Starting from the earliest Nokia 9000 Communicator, the processor speed as well as the memory size have grown enormously. In this graph, we show the increase in processor speed (linear scale) and the increase of the memory size (logarithmic scale) of some of the top model smartphones over the last 15 years. Data gathered from http://www.gsm-arena.com.

Todays' smartphones offer users more applications, more communication bandwidth and more processing, which together put an increasingly heavier burden on its energy usage, while advances in battery capacity do not keep up with the requirements of the modern user.

The original idea of offloading computation from a thin client to a rich server is well known from a speed performance point of view. Only recently, it has been recognized that offloading computation using the available communication channels to remote cloud resources can also help to reduce the pressure on the energy usage [16,17,18].

In this paper we elaborate on the idea of computation offloading and present a practical system, called *Cuckoo*, that can be used to easily write and efficiently run applications that can offload computation. Cuckoo is targeted at the Android platform, since Android provides an application model that fits well for computation offloading, in contrast with other popular platforms, such as the iPhone.

Cuckoo offers a very simple programming model that is prepared for mobile environments, such as those where connectivity with remote resources suddenly disappears. It supports local and remote execution and it bundles both local and remote code in a single package. Cuckoo integrates with existing development tools that are familiar to developers and automates large parts of the development process. Furthermore, it offers a simple way for the application user to collect remote resources, including laptops, home servers and cloud resources.

The contributions of this paper are:

- We show that computation offloading can be implemented elegantly, if the underlying operating system architecture differentiates between interactive user-oriented activities and computational background services.
- We present Cuckoo[1]: a complete framework for computation offloading for Android, including a runtime system, a resource manager application for smartphone users and a programming model for developers, which integrates into the Eclipse build system. The runtime system supports custom remote implementations optimized for remote resources and can be configured to maximize computation speed or minimize energy usage.
- We evaluate the framework with two real life example applications.

The outline of this paper is as follows. In Section 2 we will discuss the Android operating system, which is the platform that we have selected for our research. Section 3 then discusses the design of the Cuckoo framework and Section 4 presents important details about the implementation. In Section 5 we evaluate our system with two example applications. The related work is covered in Section 6 and we conclude with presenting our future work and conclusions in Sections 7 and 8.

2 Background: Android

First, we will turn our attention to the Android platform, as we need to understand how mobile applications on Android are composed internally, before detailing the design and implementation of Cuckoo.

Android is an open source platform including an operating system, middleware and key applications and is targeted at smartphones and other devices with limited resources. Android has been developed by the Open Handset Alliance, in which Google is one of the key participants. Android applications are written in Java and then compiled to Dalvik bytecode and run on the Dalvik Virtual Machine.

2.1 Android Application Components

The main components of Android applications can be categorized into *Activities*, *Services*, *Content Providers*, and *Broadcast Receivers*, which all have their own specific lifecycle within the system.

---

[1] The framework is named after the Cuckoo bird which offloads egg brooding to other birds.

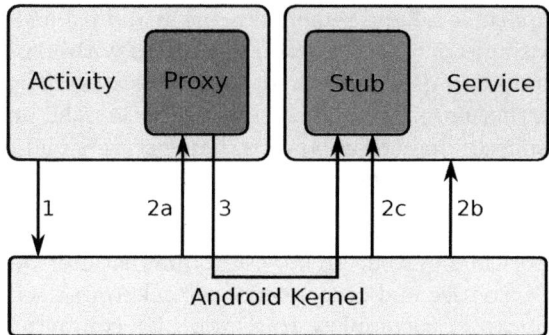

Fig. 2. A schematic overview of the Android IPC mechanism. An activity binds to a service (1), then it gets a proxy object back from the kernel (2a), while the kernel sets up the service (2b) containing the stub of the service (2c). Subsequent method invocations by the activity on the proxy (3) will be routed by the kernel to the stub, which contains the actual implementation of the methods.

Activities are components that interact with the user, they contain the user interface and do basic computing. Services should be used for CPU or network intensive operations and will run in the background, they do not have graphical user interfaces. Content Providers are used for data access and data sharing among applications. Finally, Broadcast Receivers are small applications that are triggered by events which are broadcasted by the other components in the system.

For computation offloading, we focus on activities and services, because the separation between the large computational tasks in the services and the user interface tasks in the activities form a natural basis for the Cuckoo framework. We will now have a closer look at how activities and services communicate in Android.

2.2 Android IPC

When a user launches an application on a device running the Android operating system, it starts an activity. This activity presents a graphical user interface to the user, and is able to bind to services. It can bind to running services or start a new service. Services can be shared between multiple activities. Once the activity is bound to the running service, it will communicate with the service through inter process communication, using a predefined interface by the programmer and a stub/proxy pair generated by the Android pre compiler (see Figure 2). Service interfaces are defined in an interface definition language called AIDL [19]. Service methods are invoked by calling the proxy methods. These proxy methods can take primitive type arguments as well as *Parcelable* arguments. Parcelable arguments can be serialized to Parcels and created from Parcels, much like Java Serialization serializes and deserializes Objects from byte arrays. The Android

IPC also supports callbacks, so that the service can invoke a method on the activity, allowing for asynchronous interfaces between activities and services.

2.3 Android Application Development

Android applications have to be written in the Java language and can be written in any editor. However, the recommended and most used development environment for Android applications is Eclipse [20], for which an Android specific plugin is available [21].

Eclipse provides a rich development environment, which includes syntax highlighting, code completion, a graphical user interface, a debugging environment and much more convenient functionality for application developers.

The build process of an Android application will be automatically triggered after each change in the code, or explicitly by the developer. The build process will invoke the following builders in order:

- **Android Resource Manager**: which generates a Java file to ease the access of resources, such as images, sounds and layout definitions in code.
- **Android Pre Compiler**: which generates Java files from AIDL files
- **Java Builder**: which compiles the Java source code and the generated Java code
- **Package Builder**: which bundles the resources, the compiled code and the application manifest into a single file

After a successful build, an Android package file (.apk) is created, which can be installed on a device running the Android operating system.

3 Cuckoo Design

After initial research on the topic of computation offloading in [18], we found that computation offloading as a technique is very useful to enhance the performance and reduce the battery usage of applications with heavy weight computation. However, adding offloading to such applications requires additional effort and skills of application developers. We focus on minimizing this effort by:

- offering a very simple programming model that is prepared for connectivity drops, supports local and remote execution and bundles all code in a single package.
- integrating with existing development tools that are familiar to developers.
- automating large parts of the development process.
- offering a simple way for the application user to collect remote resources, including laptops, home servers and other cloud resources.

In this section we show which parts of the development process we automated, how we integrated our system into the existing development tools and what the programming model looks like. We will also discuss how the user can collect remote cloud resources. In Section 4, we will present the details of the implementation.

3.1 Programming Model

One of the key contributions of Cuckoo is the programming model that is offered to application developers. This programming model acts as the interface of the system to the developers and therefore should be easy to use and to understand.

Secondly, since smartphones are not always connected to networks, making cloud resources sometimes unreachable, the programming model must support both local and remote method implementations.

Furthermore, the programming model must specifically support remote implementations to be different from the local implementation, so that, for instance, parallelization can be used at the remote implementation to get the full performance out of a remote multiprocessor resource.

As a last requirement, the programming model must bundle all local and remote code to be together, so that the user will always have a compatible remote implementation.

To make the programming model easy to use and to understand, we decided to use the existing *activity/service model* [22] in Android that makes a separation between compute intensive parts (services) and interactive parts of the application (activities), through an interface defined by the developer in an interface definition language (AIDL). For Android applications that contains compute intensive operations, there will be already such an interface available if it is well programmed. Otherwise an interface can easily be extracted from the code. This interface will be implemented as a local service that has, when used, a proxy object at the activity.

Next to this local service implemented by the programmer, the Cuckoo framework generates an implementation of the same interface for a remote service. Initially this implementation will contain dummy method implementations, which the programmer has to replace with real method implementations that can be executed at the remote location. The real methods can be identical to the local service implementation, but may also be completely different, since the remote implementation can run a different algorithm, use a different library, run in parallel, etc.

3.2 Integration into the Build Process

The Cuckoo framework provides two Eclipse builders and an Ant [23] build file that can be inserted into an Android project's build configuration in Eclipse.

The first Cuckoo builder is called the *Cuckoo Service Rewriter* and has to be invoked after the Android Pre Compiler, but before the Java Builder. The Cuckoo Service Rewriter will rewrite the generated Stub for each AIDL interface, so that at runtime Cuckoo can decide whether a method will be invoked locally or remote.

The second Cuckoo builder is called the *Cuckoo Remote Service Deriver* and derives a dummy remote implementation from the available AIDL interface. This remote interface has to be implemented by the programmer. Next to generating the dummy remote implementation, the Cuckoo Remote Service Deriver also generates an Ant build file, which will be used to build a Java Archive File (jar) that contains the remote implementation, which is installable on cloud resources. The Cuckoo Remote Service Deriver and the resulting Ant file have to be invoked after the Java Builder, but before the Package Builder, so that the jar will be part of the Android Package file that results from the build process. A schematic overview of how the Cuckoo components integrate into the default Android build process is shown in Figure 3, whereas Table 1 shows the order of the development process.

3.3 Integration into the Runtime

At runtime method invocations to services are intercepted at the service side of the Stub/Proxy pair, which is at the Stub. The Cuckoo framework will then decide whether or not this method should be invoked locally or remotely. The Cuckoo framework will query the Cuckoo Resource Manager for any available resources to see whether these resources are reachable and, if so, use them.

3.4 Cloud Resources

An application which uses Cuckoo for computation offloading, can offload its computation to any resource running a Java Virtual machine, either being machines in a commercial cloud such as Amazon EC2 [24] or private mini clouds such as laptops, desktops, home servers or local clusters. The user runs a simple Java application, the *server*, on such a resource to enable it to be used for computation offloading. The server that runs on such a resource does nothing by itself, however, services available on a phone can be installed onto such a server.

A part of the Cuckoo framework is a Resource Manager application that runs on the smartphone. In order to make a remote resource known to a phone, the remote resource has to register its address to this Resource Manager using a side channel. If a resource has a display, starting a server will result in showing a two dimensional barcode – a QR code [25] – on the resources' screen. This QR code contains the address of the server. Smartphones are typically equiped

Table 1. Application Development Process. An overview of what steps need to be performed during the development process of an Android application that supports computation offloading.

step	actor	action
1	developer	creates project, writes source code
2	developer	defines interface for compute intensive service
3	build system	generate a stub/proxy pair for the interface and a remote service with dummy implementation
4	developer	writes local service implementation, overwrites remote service dummy implementation
5	build system	compiles the code and generates an apk file
6	user	installs the apk file on its smartphone

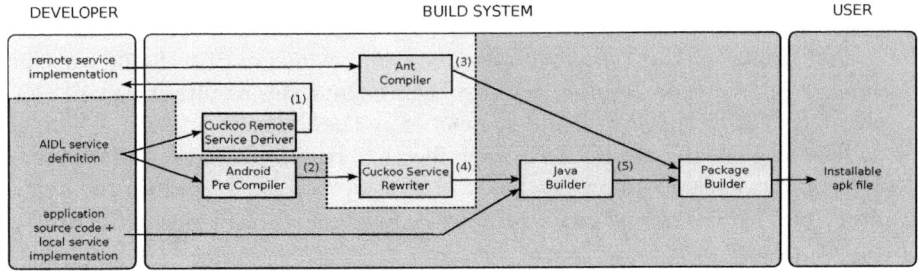

Fig. 3. Schematic overview of the build process of an Android application, the area within the dashed line shows the extensions by Cuckoo for computation offloading. The figure shows the process of building an Android application from source code to a user installable apk file. The developer has to write application source code and if applicable AIDL service definitions and implementations for these services. Then the build system will use these to generate Java Source files from the AIDL files (2), compile the source files into Dalvik bytecode (5) and bundle them into an installable apk file. To enable an application for computation offloading, the only thing the developer has to do is to overwrite remote service dummy implementations (1) generated by the Cuckoo Remote Service Deriver. The Cuckoo Service Rewriter will insert additional code into the generated Java Source Files and these rewritten Source Files (4) will be compiled and subsequently bundled with the compiled remote implementation (3) to again an installable apk file.

with a camera and can scan this QR code using a special resource manager application (see Figure 4). If the resource does not have a visual output, a resource description file can be copied from the resource to the phone to register the resource. When the resource is known to the Resource Manager application, it can be used repeatedly for any application that uses the Cuckoo computation offloading framework.

3.5 Limitations

Cuckoo does not yet support callbacks, although they are supported by the Android interface definition language. Implementing asynchronous callback communication for remote services is challenging and is part of our future work.

Method arguments can only be used as input parameters and cannot be used in a C-style way as output parameter. With Cuckoo, only the return object of a method will be available to the activity. Currently, Cuckoo does not support any form of security, which means that the remote resources can be accessed by untrusted phones, which in turn can install any code onto the system. However, setting up a security infrastructure can be realized and will be part of our future work.

Cuckoo supports only stateless services. Although the programming model does not forbid a service to maintain internal state, Cuckoo can arbitrarily change from local to remote execution or from one remote resource to another without transfering state. We do not support such state transfers, because at the

Fig. 4. A screenshot of the resouce manager collecting the address of a remote resource. From now on this resource is known to the smartphone and will be used by any application that uses Cuckoo for computation offloading.

moment the state needs to be transfered, it is generally too late to access the state, because the connection to the running service has been lost. Nonetheless, this limitation is acceptable, since application developers understand well how to transfer state between stateless services by using the Representational State Transfer (REST) architectural style [26].

4 Implementation

In this section we will highlight the important implementation details of the Cuckoo framework. We will describe the Ibis communication library that we use to communicate with the remote resources and the protocol between the phone and the remote resources. Finally we show how Cuckoo can be used in different configurations.

4.1 Offloading Decision

When an activity invokes a method of a service, the Android IPC mechanism will direct this call through the proxy and the kernel to the stub. Normally, the stub will invoke the local implementation of the method and then return the result to the proxy. The Cuckoo system intercepts all method calls and then decides whether it is benificial to offload the method invocations or not. We plan to use a combination of heuristics, context information and history to evaluate an offloading decision. For now, we use the very simple heuristic to always prefer remote execution. The only context information we use is to check whether the remote resource is reachable. We will incorporate more advanced intelligence in Cuckoo in our future work (see Section 7).

4.2 Communication: Ibis

In order to execute methods on a remote resource, the phone has to communicate with the remote resource. We use the Ibis communication middleware [27] for this purpose, because it offers a simple and effective interface that abstracts from the actual used network, being either WiFi, Cellular or Bluetooth.

The Ibis communication middleware is an open source scientific software package developed for high performance distributed computing in Java. Since it is written in Java, it can also run on Android devices. While targeted at high performance distributed applications that typically run on large supercomputer clusters, the Ibis middleware has proven to be useful on mobile devices too [18,28].

The Ibis middleware consists of two orthogonal subsystems, the Ibis Distributed Deployment System, which deploys applications onto remote resources and the Ibis High-Performance Programming System, which handles the communication between the individual parts of a distributed application.

Cuckoo has been implemented on top of the Ibis High-Performance Programming System, which offers an interface for distributed applications in the form of several high-level programming models and a low-level communication library. The low-level communication library in turn abstracts from the actual implementations of this library. Currently the library has implementations for TCP, UDP, SmartSockets, Bluetooth and Myrinet. The Cuckoo framework is built directly on top of the communication library and offers a programming model that is at the same abstraction level as other existing programming models, such as Remote Method Invocation (RMI).

The communication library offers unidirectional communication channels between so called *ports* and allows for messages to be sent between multiple ports. Send Ports can be connected to Receive Ports in one-to-one, many-to-one, one-to-many and many-to-many mode. Receive Ports receive messages either explicitly or implicit with upcalls. Every port has a unique identifier, which include a program-wide Ibis identifier and the name of the port.

4.3 Client / Server Protocol

In the Cuckoo system we set up Receive Ports at the server to handle requests from clients. The clients can find a server using its Ibis identifier and bind a Send Port to the matching Receive Port at the server to exchange messages with the server.

Normally, the client will request the server to execute a particular method with particular arguments and return the result of the method execution to a Receive Port that the client has set up especially for this method invocation. However, if the service is not available on the server, the client has to install the service onto the server. It does so, by sending a message to the install Receive Port. This message contains the jar file created in the build process (see Section 3.2), all the required external jars, and the package name of the service.

After a successful installation, the client can invoke methods on the remote service. Although the installation of a service introduces the overhead of sending the jars to the remote server, this occurs only once per service, while many method invocations can be done.

Since the client runs on a mobile phone in a dynamic networking environment, it will sooner or later experience disconnections with resources. These disconnection can be due to network switching or limited network coverage. Cuckoo will react on disconnection by switching to another remote resource. As a final fallback, it can use the always available, but less preferred, local implementation.

A Cuckoo server can be shared with multiple clients. For instance a family home server can be used to enhance the computation of each of the family's mobile device. Since the supported services are stateless, multiple invocations cannot interfer with each other. Users can exchange remote resources by simply presenting the QR-code containing the Ibis identifier on their phone, while another person uses the camera of the phone to scan the QR-code.

4.4 Configuration of Cuckoo: Trade Offs

Now that we have explained how programmers can write remote implementations and how these implementations eventually will be executed on the remote server, we will turn our attention to the different configuration options of Cuckoo. We implemented several configuration points in the run time system of Cuckoo to make it flexible and therefore valuable across a large number of applications with different requirements.

Cuckoo can do both early and late binding to remote resources. When speed is of key importance to an application, early binding will give the best results: Cuckoo will try to find and, if needed, install a remote resource at the moment the activity binds to the service. Then, when service methods are invoked, the resource discovery process has already been done and the remote methods can be invoked right away. In contrast with early binding, using late binding the resource discovery process will be delayed until a method will be invoked. This will avoid energy overhead in the form of unnecessary reconnections, which can occur during the time between the binding of the activity to the service and the moment a method of the service is invoked. During this time the established connection may already be lost, making a reconnection to a new resource necessary. Although early binding optimizes for speed, late binding optimizes for energy usage. A developer can specify per service, whether the service is optimized for speed or for energy usage.

Another property that the developer can configure is the maximum number of remote resources that Cuckoo will try before giving up and use the local implementation. A higher number of resources will give a bigger chance to find one that can be used, but can also introduce more overhead in terms of time and energy if Cuckoo cannot connect to any of the resources. Determining whether a resource can be used involves a simple ping-pong message exchange between the client and the resource. If the cost of multiple ping-pong messages is neglectable compared to the benefit of a potential remote execution, the developer should specify a high maximum number of remote resources that will be tried, to increase the chance of remote execution. Ideally, the developer should be able to

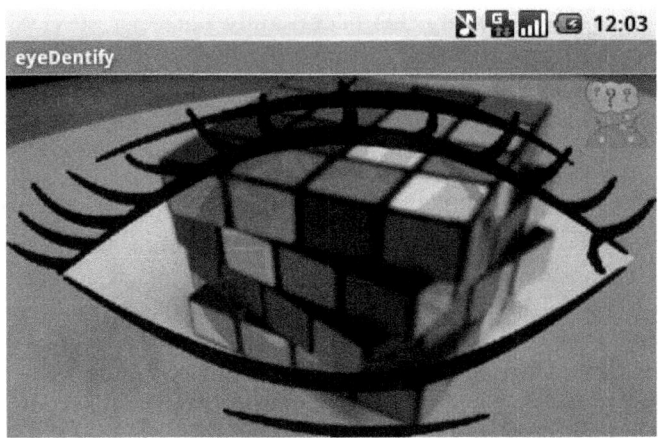

Fig. 5. This figure shows a screenshot of the object recognition application *eyeDentify*. The application can recognize objects in the center of the eye, which shows the camera preview. Users can also teach the name of new objects to the application.

specify a budget in time and/or energy that Cuckoo might spend in searching for a remote resource. We plan to include this in our future work.

To improve responsiveness of an application – that is execution speed – the developer can also specify a method of the service to be executed in parallel on multiple resources, including the local implementation. The result of the first implementation that returns, will then be forwarded to the proxy in the activity. Note that this method is energy hungry, because there always will be some execution of code that will be discarded.

Finally, the developer can configure individual methods of a service to be not offloadable at all, which is useful for instance if methods interact with sensors on the smartphone.

5 Example Applications

In this section we will evaluate the Cuckoo computation offloading framework with two smartphone applications that contain heavy weight computation. We will show which parts of the applications are offloaded and how the computation offloading impacts the performance of the applications in both speed and functionality.

5.1 eyeDentify

Our first example application is *eyeDentify* [18], a multimedia content analysis application that performs object recognition of images captured by the camera of a smartphone (see Figure 5). The idea of the application is similar to the Google Goggles application which can recognize contact info, places, logos, landmarks,

Table 2. eyeDentify parameters. This table shows the parameters used for the local and the remote implementation of the service that converts an image into a feature vector. The quality of the feature vector is higher if the image size is larger, the number of receptive fields in the algorithm is higher, if there are more color models and if there are more bins. See [18] for an explanation of the *-marked terms.

	image size	receptive fields*	color models*	bins*
local	64 x 48	19	1	500
remote	512 x 384	37	3	1000

artworks, and books [29]. In addition to the recognition of objects, eyeDentify allows the user to teach new objects to the application.

The key algorithm in eyeDentify is an algorithm that converts an image into a *feature vector*, a mathematical representation of the image that can be compared to other feature vectors. This algorithm is both compute and memory intensive and therefore suitable for computation offloading. There are several parameters of the algorithm that influence the computation time and the memory footprint. For instance, a larger input image will result in a bigger memory footprint and a larger computation time. Other parameters can be tuned to reduce the memory and computation requirements, while also reducing the quality of the feature vector and thereby the quality of the object recognition.

In [18] we show that it is possible to perform object recognition on a T-Mobile G-1 smartphone, with a 528 MHz processor and 16 MB available application memory within a reasonable time, however, the algorithm used can take at most 128 x 96 pixel images as input, with the algorithm parameters set to poor quality object recognition. By offloading the computation needed for this algorithm, we have shown that we can speed up the computation with a factor of 60, reduce the battery consumption with a factor of 40 and increase the quality of the recognition [18].

We have rewritten eyeDentify to use the Cuckoo computation offloading framework. First, we specified an interface in AIDL for a service that hosts the algorithm that converts an image into a feature vector (see Figure 6). Then we implemented the local service with parameters that are suitable for local computation (see Table 2, local). Cuckoo generated a dummy remote implementation

```
package interdroid.eyedentify;

import ibis.dog.shared.FeatureVector;

interface FeatureVectorServiceInterface {
    FeatureVector getFeatureVector(in byte[] jpegData);
}
```

Fig. 6. The AIDL interface definition of the service that converts an image into a feature vector. The eyeDentify application has a local and a remote implementation of this interface, where the only differences between the implementation are the value of accuracy parameters of the algorithm.

Fig. 7. A screenshot of PhotoShoot - The Duel, a distributed augmented reality smartphone game in which two players fight a virtual duel in the real world. Both players have 60 seconds and six bullets to shoot at each other. They shoot with virtual bullets, photographs of the smartphones' camera, and use face detection to evaluate whether shots are hit. The area covered in the crosshairs will be used for face detection.

and we replaced the dummy implementation simply with the same implementation as for the local one and then changed the algorithm parameters (see Table 2, remote), so that the remote implementation would perform high quality object recognition on large images. Cuckoo then generated the jar file for the remote service and the application was ready to be deployed on a smartphone.

At runtime this application can always perform object recognition, even when no network connection is available, in contrast to, for instance, Google Goggles that only works when connected to the cloud. However, if a network connection is available, then eyeDentify can use remote resources to speed up the recognition and reduce the energy consumption, while providing higher quality object recognition. Another advantage is that the client and server code of eyeDentify are bundled and will remain compatible with each other.

Compared to our earlier work on computation offloading, using Cuckoo results in similar performance improvements.

5.2 PhotoShoot

The second example that we will consider is a distributed augmented reality game, called *PhotoShoot* [28], with which we participated in the second worldwide Android Developers Challenge [30] and finished at the 6th place in the category 'Games: Arcade & Action'. This innovative game is a two-player duel that takes place in the real world. Players have 60 seconds and 6 virtual bullets to shoot at each other with the camera on their smartphone (see Figure 7). Face detection will determine whether a shot is a hit or not. The first player that hits the other player will win the duel.

The major compute intensive operation in this game is the face detection. The Android framework comes with a face detection algorithm, so it is possible to create a local implementation to detect faces in an image. However, due to memory limitations this algorithm works only for images up to 3.2 MegaPixels – which is rather low for a modern smartphone – and consumes about 9 seconds of compute time on the Google Nexus One with a 1.0 GHz processor or 44 seconds on the older T-Mobile G-1 with a 528 MHz processor. Thus, without offloading, the slower the processor of the smartphone, the longer it takes for the shot to be analyzed, which gives the user of a slow smartphone a significant disadvantage. Offloading can, however, be used to make the game fair again.

We have modified PhotoShoot, by refactoring the face detection into a service (see Figure 8), so that it can use computation offloading using the Cuckoo framework. Using offloading, slower phones can get fast face detection and phones with very high resolution cameras will not experience memory shortage.

We used a different face detection algorithm for the remote implementation to demonstrate that offloading does not only result in faster execution, but gives also more precise results. The remote implementation is based on the Open Source Computer Vision library (OpenCV [32]) and can, next to frontal faces, also detect profile faces, and will therefore give more accurate results (see Figure 9).

A face *recognition* algorithm would further improve the application so that only pictures containing the face of the opponent would count as valid hits. This would open up possibilities for team-based augmented reality multiplayer games. But to our knowledge, existing face recognition algorithm libraries for Java are not available in the public domain.

6 Related Work

Computation offloading is a technique that dates back to the era of dumb clients that used mainframes for computation. With the introduction of *personal computers*, the need for computation offloading decreased, but with the introduction of todays' portable devices, a new need for remote compute power emerged.

```
package interdroid.photoshoot.market;

import interdroid.photoshoot.market.Face;

interface FaceDetectorInterface {
  List<Face> findFaces(in byte[] jpegData, int maxFaces);
}
```

Fig. 8. The interface of the face detection service in AIDL. The services will search for a maximum number of faces in the provided image and returns a list of Face objects that it has found. The local implementation will use the face detection library available on Android, which can only detect frontal faces, while the remote implementation uses the OpenCV library, which can also detect profile faces.

Fig. 9. Comparison of the local and remote implementation of the face detection service on several test images from [31]. The local version (top row) uses the face detection algorithm provided by the Android framework, while the remote version (bottom row) is able to use a much more powerful algorithm, which for instance can also detect profile images. Since the remote implementation runs multiple detectors over the image, some faces are detected several times.

In this section we give an overview of what has been proposed by others with respect to computation offloading for smartphones and how that relates to the Cuckoo framework.

In the early days of the portable handheld devices and before the popularity of the commercial cloud, Satyanarayanan [33] proposed a computation offloading model called *cyber foraging*, in which portable resource constrained devices can offload computation to more powerful remote resources called *surrogates*. Important in cyber foraging are the discovery of surrogates, the trust relation between the client and the surrogates, load balancing on the surrogates, scalability and seamlessness.

Cuckoo provides resource discovery through QR-codes as explained in Section 3.4 and addresses scalability by allowing for both commercial and private cloud resources. In our future work we will investigate the security aspects of computation offloading and load balancing on the surrogates.

Yang et al [34] describe an offloading system based on cyber foraging that, like Cuckoo, uses a Java stub/proxy model. Using profiling, surrogate matching and graph partitioning algorithms their system decides what part of the computation can be offloaded. They provide an evaluation of the system using an automatic translation application that uses OCR and Machine Translation. The system runs on the HP iPaq platform.

From the field of security Chun et al [17] propose an architecture, called *CloneCloud*, that can offload computation, such as virus scans and taint checking to a clone of the smartphone in the cloud. An implementation of their proposed architecture should support the following types of augmented execution:

- primary functionality outsourcing: offloading of expensive computation
- background augmentation: offloading of background processes
- mainline augmentation: offloading light weight computation, for heavy weight analysis (taint checking, debugging, profiling).
- hardware augmentation: offloading of computation because of hardware limitations of the smartphone
- augmentation through multiplicity: parallel execution of offloaded computation, for computation speedup or speculative execution.

The Cuckoo framework implements all the proposed execution types in its single programming model. It does not run a complete clone of the smartphone at the remote cloud resource, as proposed by Chun et al, but rather runs a temporary clone limited to only the service that the application is using, thereby avoiding the costly process of keeping the smartphone synchronized with an application clone in the cloud.

Another offloading model, called AlfredO, which is based on the modular software framework OSGi, is proposed in [35]. They contribute partitioning algorithms to optimize the distribution of OSGi software modules between the phone and remote resources and evaluate their system with a home interior design application. Since AlfredO will use the same modules on the phone as on the remote resource, it is not possible to differentiate between local and remote implementations, as is possible with Cuckoo.

In [16], Kumar et al describe a cost/benefit model for offloading with respect to energy usage. This model takes into account the speed of both the smartphone and the remote cloud resource, the number of bytes that need to be transfered, the network bandwidth, the energy consumption of the smartphone in idle, computing and communicating state. They assume that the number of local instructions is identical to the number of remote instructions and the number of bytes that need to be communicated are known beforehand. The assumptions do not completely hold for Cuckoo, since the remote implementation might be different from the local implementation and since the number of bytes that will be received by the smartphone after a remote method execution can be unknown beforehand. However, we plan to extend the model and incorporate it into Cuckoo.

Next to computation offloading systems, where computation components are transfered from a smartphone to the remote resource, there exist many applications that are separated into a light weight client and a heavy weight server hosted in the cloud. Examples of such applications are the music search service Shazam [36] and image search service Goggles [29]. The remote parts of these services typically have to interact with very large databases and therefore are not suitable to be bundled with the client application. The drawback of unbundled services is that the application provider has to provide the remote service, which will include hosting and maintanance costs. Futhermore, since these services are typically commercial, users might not want to hand over input data, because of privacy concerns. With an offloading system, like Cuckoo, users will run their heavy weight computations on their private machines, or machines that they trust.

In case studies about computation offloading we found that the applications that benefit from computation offloading exist in the following domains:

- **image processing**: object recognition [18], OCR [34], face detection [28], barcode analysis [37]
- **audio processing**: speech recognition [38]
- **text processing**: machine translation [34]
- **artificial intelligence for games**: chess [16]
- **3D rendering**: 3D home interior design [35]
- **security**: taint analysis and virus scans [17,39]

The Cuckoo framework supports the development of applications from these domains in a simple and developer friendly environment.

Several other solutions, complementery to computation offloading, have been proposed at different abstraction levels to reduce the pressure on the energy usage. At the hardware level, manufacturers build processors that can switch to lower frequencies to save energy. Techniques to harvest energy from surrounding sources, such as movement, light and wireless signals, have been proposed to charge the battery during operation [40]. Furthermore, some modern smartphone operating systems have been tuned to be more energy efficient. Finally, the users of smartphones have developed habits to turn off sensors or radios when they are not needed.

7 Future Work

In this section we present our research agenda, we describe what parts of Cuckoo will be enhanced and which parts we feel that are missing in Cuckoo.

In our future work we will improve the intelligence of the offloading by improving the heuristics and the addition of more context information, such that the estimation whether or not offloading is going to save energy or increase the computation speed will be more accurate. Although computation offloading can speed up computation and save energy, it is not guaranteed that it does. Offloading introduces additional communication, which consumes energy and takes time. To take this overhead into account we plan to incorporate the energy analysis model for computation offloading described by [16].

We also want to add context information of the remote resources to the system, such as processor speed, number of processors, available memory, etc., which could be added to the address encoded in the QR-code. In addition we want to use context information available on the phone, such as location, network status, etc.

Another direction of our future work is to investigate which security measures need to be taken to secure the communication between the smartphone and the remote cloud resources. We also have to pay attention to the security implications of multiple users using a single remote resource, running foreign code on the remote resources, and making sure that remote services cannot disturb the working of other remote services.

Furthermore, we will extend the programming model to support callbacks from the remote resource to the smartphone and investigate whether we can support method parameters to be used as return values, like the AIDL specification supports.

We will improve the Eclipse integration, by integrating the builders with the graphical user interface of Eclipse.

8 Conclusions

In this paper we have presented Cuckoo, a framework for computation offloading for smartphones, a recently rediscovered technique, which can be used to reduce the energy consumption on smartphones and increase the speed of compute intensive operations.

Cuckoo integrates with the popular open source Android framework and the Eclipse development tool. It provides a simple programming model, familiar to developers, that allows for a single interface with a local and a remote implementation. Cuckoo will decide at runtime where the computation will take place. Furthermore, the Cuckoo framework comes with a generic remote server, which can host the remote implementations of compute intensive services. A smartphone application to collect the addresses of the remote servers is also included.

We have evaluated the Cuckoo framework with two real world smartphone applications, an object recognition application and a distributed augmented reality smartphone game and showed that little work was required to enable computation offloading for these applications using the Cuckoo framework.

References

1. Android, http://developer.android.com/
2. iPhone OS, http://developer.apple.com/iphone
3. Symbian, http://developer.symbian.org/
4. Windows Phone 7 Series, http://www.windowsphone7.com/
5. iPhone App Store, http://www.apple.com/iphone/appstore
6. Android Market, http://www.android.com/market/
7. Raging Thunder, http://www.polarbit.com/our-games/raging-thunder-2/
8. Motorola MOTOBLUR, http://www.motorola.com/Consumers/US-EN/Consumer-Product-and-Services/MOTOBLUR/Meet-MOTOBLUR
9. Google Maps Navigation, http://www.google.com/mobile/navigation/
10. Health to Go, http://www.healthymagination.com/
11. Data gathered from: http://www.gsmarena.com
12. iPhone 3G on Sale Tomorrow, http://www.apple.com/pr/library/2008/07/10iphone.html
13. Steve Jobs. Thoughts on Flash, http://www.apple.com/hotnews/thoughts-on-flash/

14. IEEE Standard for Information technology–Telecommunications and information exchange between systems–Local and metropolitan area networks–Specific requirements Part 11: Wireless LAN Medium Access Control (MAC) and Physical Layer (PHY) Specifications Amendment 5: Enhancements for Higher Throughput. IEEE Std 802.11n-2009 (Amendment to IEEE Std 802.11-2007 as amended by IEEE Std 802.11k-2008, IEEE Std 802.11r-2008, IEEE Std 802.11y-2008, and IEEE Std 802.11w-2009), 29 (2009)
15. Bluetooth High Speed. Product Zone, Bluetooth High Speed Technology, http://bluetooth.com/
16. Kumar, K., Lu, Y.-H.: Cloud computing for mobile users. Computer 99 (2010)
17. Chun, B.-G., Maniatis, P.: Augmented smart phone applications through clone cloud execution. In: Proceedings of the 12th Workshop on Hot Topics in Operating Systems, HotOS XII (2009)
18. Kemp, R., Palmer, N., Kielmann, T., Seinstra, F., Drost, N., Maassen, J., Bal, H.E.: eyeDentify: Multimedia Cyber Foraging from a Smartphone. In: IEEE International Symposium on Multimedia (2009)
19. AIDL, http://developer.android.com/guide/developing/tools/aidl.html
20. Eclipse, http://www.eclipse.org/
21. ADT Eclipse plugin, http://developer.android.com/sdk/eclipse-adt.html
22. AIDL, http://developer.android.com/guide/topics/fundamentals.html
23. Apach Ant, http://ant.apache.org/
24. Amazon Elastic Computing, http://aws.amazon.com/ec2/
25. Denso Wave's QR website, http://www.denso-wave.com/qrcode/index-e.html
26. Fielding, R.T., Taylor, R.N.: Principled design of the modern web architecture. ACM Transactions on Internet Technology 2(2), 115–150 (2002)
27. van Nieuwpoort, R.V., Maassen, J., Wrzesińska, G., Hofman, R.F.H., Jacobs, C.J.H., Kielmann, T., Bal, H.E.: Ibis: a flexible and efficient Java-based Grid programming environment. Concurrency and Computation: Practice and Experience 17(7-8), 1079–1107 (2005)
28. Kemp, R., Palmer, N., Kielmann, T., Bal, H.: Opportunistic Communication for Multiplayer Mobile Gaming: Lessons Learned from PhotoShoot. In: MobiOpp 2010: Proceedings of the Second International Workshop on Mobile Opportunistic Networking, pp. 182–184. ACM (2010)
29. Google Goggles, http://www.google.com/mobile/goggles/
30. Android Developer Challenge 2, http://code.google.com/android/adc
31. Schneiderman, H., Kanade, T.: A Statistical Model for 3D Object Detection Applied to Faces and Cars. In: IEEE Conference on Computer Vision and Pattern Recognition. IEEE (June 2000)
32. Open Source Computer Vision library, http://opencv.willowgarage.com/wiki/
33. Satyanarayanan, M.: Pervasive computing: Vision and challenges. IEEE Personal Communications 8, 10–17 (2001)
34. Yang, K., Ou, S., Chen, H.H.: On Effective Offloading Services for Resource-Constrained Mobile Devices Running Heavier Mobile Internet Applications. IEEE Communications 46, 56–63 (2008)
35. Giurgiu, I., Riva, O., Juric, D., Krivulev, I., Alonso, G.: Calling the Cloud: Enabling Mobile Phones as Interfaces to Cloud Applications. In: Bacon, J.M., Cooper, B.F. (eds.) Middleware 2009. LNCS, vol. 5896, pp. 83–102. Springer, Heidelberg (2009)
36. Shazam website, http://www.shazam.com
37. Kallonen, T., Porras, J.: Use of distributed resources in mobile environment. In: International Conference on Software in Telecommunications and Computer Networks, pp. 281–285 (2006)

38. Goyal, S., Carter, J.: A lightweight secure cyber foraging infrastructure for resource-constrained devices. In: WMCSA 2004: Proceedings of the Sixth IEEE Workshop on Mobile Computing Systems and Applications, pp. 186–195. IEEE Computer Society (2004)
39. Portokalidis, G.: Using Virtualisation to Protect Against Zero-Day Attacks. PhD thesis, Vrije Universiteit, Amsterdam, The Netherlands (February 2010)
40. Paradiso, J.A., Starner, T.: Energy scavenging for mobile and wireless electronics. IEEE Pervasive Computing 4, 18–27 (2005)

Debugging Tools for MIDP Java Devices

Olli Kallioinen[1] and Tommi Mikkonen[2]

[1] Sasken Finland, Tampere, Finland
olli.kallioinen@sasken.com
[2] Tampere University of Technology, Tampere, Finland
tommi.mikkonen@tut.fi

Abstract. Mobile Java development using CLDC and MIDP can be very restricting, not only because of the more restricted libraries and older Java language, but also because some very basic development tools are not available in many situations. One of the biggest problems when debugging a midlet – a CLDC/MIDP application – is that when running a mobile Java application in a real device, stack traces are not available. Also other tools, like profiling tools, only work in certain emulators. In this paper, a set of improved tools for mobile Java development is introduced. Instrumentation, a well-known technique is used to work around the restrictions of the Java sandbox. Consequently no special support is required from the platform.

Keywords: Mobile Java, debugging.

1 Introduction

Sun's mobile Java platform (Java Micro Edition, Java ME) has not been a very hot topic lately. New, advanced mobile devices with platforms like Android, Maemo, and iPhone have been stealing most of the media attention. However, when comparing mobile devices that are able to run third party software, Java ME is still by far the most common mobile platform at the moment.

One of the most frustrating problems when developing application targeted to Java ME devices, is that stack traces are often unavailable. Stack traces are normally used to locate the cause of run-time errors. The traces are printed to the standard system out stream that is usually not available when running a program on a real device. In Java SE it is possible to redirect the stream to any desired destination [1], but this redirection possibility was left out from Java ME [7]. It would not be a big problem if it would be possible to access the traces programmatically, like in Java 1.4, but this is not available either. This problem describes well how the combination of small shortcomings in Java ME prevent some very basic functionality that most developers take for granted in modern environments. Also other tools like profilers usually work only in an emulator.

This paper proposes better tools for Java ME development. The practice of manipulating Java binaries after compilation is used to implement the tools. This method allows all the tools to be used regardless of the environment where the application is running. The tools can be used to find the cause of a problem in situations where the emulator cannot be used. For example a problem could only occur on certain devices, or in some cases an application cannot be run on an emulator at all.

The rest of the paper is structured as follows. Section 2 discusses mobile Java and Java language internal workings.Section 3 describes improvements and the principles that have been used to implement them. Section 4 evaluates the implemented tools and describes how they can be used. Section 5 summarizes the presented improvements..

2 Java Micro Edition

As the Java platform was growing it was split into multiple editions. In addition to the Standard Edition (Java SE) two other editions were created: Enterprise Edition (Java EE) for application servers, and Micro Edition (Java ME) for devices with limited resources. Java ME is further divided to smaller parts to be able to support a very heterogeneous set of devices. Each Java ME runtime environment consists of three parts: (1) a *configuration* that defines the set of basic libraries and virtual machine capabilities, (2) a *profile* that defines a common set of APIs for a smaller group of similar devices, (3) a set of *optional packages* for other specific technologies like Bluetooth or SMS.

Two different configurations are currently available: *Connected Limited Device Configuration* (CLDC), and *Connected Device Configuration* (CDC). Multiple profiles exist, but in general *Mobile Independent Device Profile* (MIDP) is the most commonly used one. The combination of CLDC and MIDP, which is targeted for low-end devices like feature phones, is one of the most common runtime environments in existence, with billions deployed in different mobile phones. A Java ME application built using CLDC and MIDP is called a *MIDlet*.

2.1 Mobile Java Development Using CLDC and MIDP

There is a variety of different kinds of tools available for a Java developer. Multiple different Java specific tools can be used for writing code, compiling, packaging, debugging, testing, static analysis of source code, and so on. Some of these tools can also be used to develop mobile Java applications, but the restrictions of CLDC also prevent the use of many tools.

The build process for building a MIDP application has some differences compared to building a normal Java SE application. In addition to the normal Java Development Kit (JDK), the Java Platform Micro Edition Software Development Kit (Java ME SDK) is required. The standard Java compiler can be used, but the Java source code version and target class file version need to be set appropriately, as the newest class file format supported by CLDC is the JDK 1.4 class format. The Java standard libraries must also be replaced with the CLDC ones. The standard Java compiler supports cross-compiling, meaning that classes can be compiled against bootstrap libraries instead of the normal JDK class libraries and the target class file format can be set to other JDK versions [4].

In comparison to the class format of Java 1.3 that CLDC is based on, an extra step is added to the build process [7]. The classes need to be *preverified* before the classes can be used by the CLDC virtual machine. The CLDC virtual machine uses a simplified process to verify the correctness of classes to make loading faster and to save memory, and consequently classes must be preverified before they can be loaded. The Java ME SDK provides a preverification tool that is usually run after compiling the classes.

Finally, compiled and preverified classes need to packaged into a JAR package before the MIDlets in them can be installed and run. Information about the MIDlet must be included in the manifest file inside the JAR file. Furthermore, the JAR file is usually accompanied by a JAD file that describes the MIDlet properties [10].

2.2 Restrictions When Developing MIDlets

When developing for mobile devices many special considerations must be taken into account. Devices usually have less resources like memory and storage space than in a full Java SE desktop environment and the CPU of the device might not be very powerful. One further speciality is the user interface. Typically only one MIDlet can be running at a time and there is no standard support for inter-MIDlet communication. Also, in many platforms MIDlets cannot be running on the background while the phone is running its native applications.

From developers point of view, the most notable difference between Java SE and Java ME is that a lot of classes have been dropped from the basic Java library and some functionalities have been replaced with Java ME specific classes. In MIDP the AWT and Swing UI libraries are replaced with the LCDUI library [8]. In addition, many language features have been dropped from CLDC. Furthermore, all the newer features of the Java language are missing.

Same debugging tools can be used in Java ME applications as in Java SE. When using an emulator, all that needs to be done is to launch the emulator in debugging mode and attach it to a debugger. Debugging with a real Java ME device (on-device debugging) requires specific support from the device. Device specific differences are common, and on many platforms no support is provided. In reality, using a debugger on an actual device is usually cumbersome, slow or not possible at all. Often, logging to a COM port or to log files is the best available solution for finding the cause for issues in the developed MIDlet.

Exceptions and stack traces are good tools for finding issues in Java applications. The restrictions of CLDC limit the use of stack traces, however. When an exception is thrown in Java SE, it is easy to locate the cause of the exception by following the stack trace attached to the exception. In CLDC it is only possible to print the traces to standard system output using *Throwable.printStackTrace()* [7]. This makes it hard to get the traces when using a real device as the standard output may not be accessible. Some devices provide a way to read the standard output prints, but such features are device and platform specific.

In Java SE it is possible to access the stack trace information programmatically [3], but the API is not available in CLDC. Another Java SE option that is not available in CLDC is to reassign the *PrintStream* used by *System.out* using *System.setOut()* [2]. Finally, setting a default exception handler for threads is not possible in CLDC.

Depending on the emulator that is used, it is usually possible to configure the emulator to show exceptions implicitly. Some vendor specific tools may be used to get the traces even on actual devices but the tools are in many cases not publicly available. In some cases a MIDlet can use proprietary functionality that is not available on the emulator and the only possibility is to run it in a real device. In these cases the exception traces provided by the emulator are again unavailable. Even if the developer can access the stack traces, they may not include the source code line numbers like Java SE traces do. In fact, the debug information is not used by the reference CLDC virtual machine implementation even if the source line debug information is available in the classes.

The lack of stack traces when running an application in a real device becomes even a bigger issue because of another Java ME related problem: *fragmentation*, caused by the huge number of different Java ME devices that have different kind of hardware, operating system, and virtual machine implementation [5]. It is common to have an application that runs on an emulator and on some devices, but crashes in some other phone models. Finding out what is wrong without proper stack traces can be a time consuming operation. Usually debugging prints need to be added to pinpoint the problem source and each time the application needs to be built, packaged, installed, and restarted on the device. All the information needed to locate the exception source are in stack traces.

The sandbox of Java ME is much more restricted in comparison to Java SE. Due to Java ME restrictions, it is generally not possible to extend the functionalities that are offered by the platform.

2.3 Java Virtual Machine and Bytecode

The virtual machine and its instruction set resemble real hardware and the instruction set of a real hardware machine. Especially the instructions that are used to manipulate memory and for performing arithmetic operations are similar. In addition, some higher level concepts have been also introduced. For example, the virtual machine directly supports objects and exception handling [6].

When Java source code is compiled, it is compiled to a hardware-independent binary format called the *class file format*. Although it is called a file format, and usually files are used, the binaries do not necessarily need to be stored to files. Each class file represents the Java class or interface it was compiled from. The file format also addresses details like byte ordering to achieve platform independence. The instruction set of the virtual machine is used to represent the Java source code. In the class format, each instruction is denoted by one byte, allowing 256 different instructions. That is also why term *bytecode* is often used to describe the compiled instructions. The bytecode instructions take only a small part of typical class files, as other information, like the symbol table containing literals, need to be also included in each class file. The file format and the instruction set are designed to be used with the Java language, so the supported features match closely to the features of the Java language [6].

The class files preserve the basic structure of the source code classes. The names of packages, classes, methods and fields are also preserved in the class file. All the information to reference a class is available in the compiled Java class so it is possible to reference the class from Java source code, even if the source code for the referenced class is not available. This enables decompilation of Java class binaries so that the produced decompiled source code is close to the original source code. Even local variables names can be preserved if the information was included in the class files when it was compiled [6].

The virtual machine specification only specifies the abstract virtual machine. Many implementation details can be defined by the virtual machine implementation. For example, the used garbage collection algorithm is virtual machine specific. Similarly, the bytecode could be interpreted at runtime, compiled to native code, or even run directly on hardware. When interpreting bytecode, just-in-time (JIT) compilation is often used to improve performance. This means that when the bytecode is being executed, some often called classes may be further compiled to native machine code.

3 Improving the Java ME Toolset

When custom class loaders are available, class bytecode can be modified at run-time. A custom class loader can be defined that alters the bytecode of the class when it is being loaded into memory. Java 5.0 has a specialized *java.lang.instrument* package that provides built-in support for modifying the bytecode of classes when they are being loaded [3]. Modifying an application to gather information about the execution of the application is often called *instrumenting* the bytecode or just *instrumentation* [9]. This kind of dynamic runtime modification is not possible in CLDC. There is no support for custom class loaders either [7]. The only available possibility to modify the bytecode, is to do it statically after compiling, before creating the final MIDlet JAR package. The modified classes also need to be preverified again before packaging.

Modifying bytecode introduces the possibility of adding new errors to otherwise working code. Modifications can be checked already on build-time and the instrumentation libraries have support for verifying that the modification produces valid bytecode. Such problems are usually found when loading and verifying the class, or when preverifying a class in CLDC. If the aim of the bytecode modifications is to change the functionality of the application, possibly introduced problems can be very cryptic and hard to debug. Even if the application is just instrumented to gather information, errors in the instrumentation can cause the functionality of the application to change.

The basic idea of all the improvements this paper is modifying the compiled Java bytecode before creating the final distributables. As the Java class format has no direct connection to Java language, also constructs that would not be possible in normal Java language can be used in valid Java class files.

3.1 Tracing Method Calls

The basic idea of tracing method calls is that a piece of Java bytecode is added to the beginning and end of each method. This makes it possible to receive a callback every time a method is entered or exited, which in turn acts as the basis for other more complex features and is not as useful by itself. However, these automatically added callbacks can be used to eliminate the manual labor of adding similar tracking code.

Tracing in General. This first improvement makes it possible to track the execution of Java methods and to write the information to the standard system output or a log file. This kind of tracking of code execution by log messages is generally known as tracing. The trace messages are usually disabled during normal development and in final builds to prevent excessive amount of log messages. Then, if more information is needed by the developer, the trace level messages can be enabled.

Automated Tracing. Using the implemented tool, the developer can automatically get trace messages without the need to add source code individually to each traced method. All that the developer needs to do, is to enable the method traces option in the instrumentation phase. No modifications to the existing code are needed. This option only outputs the traces in a simple predefined format to standard system output. The developer can implement specific callback methods himself if more control over the output is wanted. No references to any external Java libraries are needed, since callbacks to the method will be generated in the instrumentation phase and there is, for example, no need to register a listener. The location and name of the callback method name can be defined in the instrumentation phase. This callback can then be used to print information about each method invocation in any way wanted. A separate callback is available for both entering a method and exiting a method.

Modifying the Bytecode. Adding code to the beginning of a method is a straightforward operation. On the other hand determining when a method exits is a bit more complicated because there can be multiple return statements and thus multiple possible exit points. In addition, an exception can cause the method to exit at any point. The method tracing only generates the exit callback if the method exits normally. The more complicated special cases caused by exceptions are ignored for now. The improved stack traces explained in the following can be used to detect the thrown exceptions. When exceptions are ignored, all that is needed to detect exiting from a method is to find all the return statements in the method and to add additional bytecode for handling the callbacks just before them. The end of a void method is also considered as a normal return statement in bytecode. The instrumentation phase can be demonstrated with an example Java method. Listing 1 defines a method that calculates the mean for an array of integers. From the instrumentation point of view it is irrelevant what the method actually does, but it was chosen because it is short and meaningful but it still has enough code to demonstrate some special cases.

Listing 1. Java method before it has been instrumented.

```
public int calculateMean(int[] values) {
   if (values == null) {return 0;}
   int sum = 0;
   for (int i = 0; i < values.length; i++) {sum += values[i];}
   int mean = sum / values.length;
   return mean;
}
```

The Java source code in Listing 2. demonstrates how instrumenting the class with method tracing enabled affects the class. A call to a static method *Trace.methodEntered* has been added to the beginning of the method. This is a call to the method that handles the tracing for entering a method. This method will invoke a callback method that has been defined in the instrumentation phase. The *Trace* class is a Java class that is a part of an implemented run-time library. This and some other utility classes need to be included to the final application JAR package for the tracing to work. Also, a call to *Trace.methodExited* has been added before each return statement. This call will relay the information about exiting the method to the user defined callback method. The actual insertion is done as bytecode but if the resulting class would be decompiled it would result in source code that would be similar.

Listing 2. Same method after instrumentation.

```
public int calculateMean(int[] values) {
   Trace.methodEntered("MyMath", "calculateMean", 1);
   if (values == null) {
      Trace.methodExited("MyMath", "calculateMean", 2);
      return 0;
   }
   int sum = 0;
   for (int i = 0; i < values.length; i++) {sum += values[i];}
   int mean = sum / values.length;
   Trace.methodExited("MyMath", "calculateMean", 6);
   return mean;
}
```

More detailed information regarding the method that is being executed is provided in the callback method parameters, including the name of the class and the name of the method. The last parameter is the source code line number. It is added to make it possible to later distinguish multiple methods with same name, and to help finding the method in a source file. The class name and the method name are known in the instrumentation phase and the line number is also available if debug information has been included in the class file.

3.2 Improved Stack Traces

Stack trace is a structure representing the call path of a thread at some specific time of execution. It lists the method invocations that lead to the state that the trace represents. The position in source code where each method was invoked can also be stored to the structure. Stack traces are very useful when trying to locate and fix errors in the source code. The trace can be displayed when an exception has happened to easily find the exact place in source code what caused the exception. Unfortunately getting stack traces with full source line number information can be a problem when using CLDC.

Tracking the Execution of an Java Application. It is not possible to access the stack trace information programmatically in CLDC, but it is possible to solve the problem of not being able to get the stack traces with another approach. The solution is to modify the bytecode of the application itself, so that the application keeps track of its own execution. Each method needs to be modified to detect when the execution of a thread enters the method and when it leaves the method. In addition, each thread needs to be associated with a data structure that defines the current method call path of that thread. As the application itself will be keeping track of the execution of its code, it is possible to get the current stack trace at any point regardless of the execution environment.

The structure used to define the invocation path is a stack, where each element represents a method invocation that was done from the previous element. This stack mimics the actual call stack that keeps track of the method execution in the virtual machine [6]. Similar kind of structure has been available for the programmer in Java SE version 1.4 and in newer versions, where it is possible to call *throwable.getStackTrace()* method to receive an array of *StackTraceElement* objects. Java version 1.5 (or 5.0) takes this even further and provides multiple method in the *Thread* class for accessing the current stack trace programmatically. As CLDC is based on Java 1.3, neither *getStackTrace()* method nor *StackTraceElement* class are available [2,3,7].

An element representing the current method is pushed to the stack each time a thread enters a method and popped when the method is exited. It is also necessary to catch possible exceptions and pop the method if one is detected to keep the stack correctly up-to-date. Each time a method is entered and the thread is not known from before, a new stack trace is created and associated with that thread. Similarly, as the last item is popped from a stack, we know that the stack trace related to that thread is not needed anymore.

At the same time it is also possible to solve the problem of not getting source code lines in stack traces. This is however only possible if the required debug information has been enabled when compiling the classes that are being instrumented. If this line numbers debug information is enabled, the compiler adds a special bytecode instruction defining the source line number before each group of bytecode instructions that were generated from the same source code line. When instrumenting a method every time this kind of bytecode instruction is encountered instructions need to be added to update the current stack trace. The added instructions need to update the top element in the stack of method elements to point to the correct source line.

Adding Bytecode to Track the Execution. The instrumentation phase is demonstrated with the same example Java method that was used previously in Listing 1. The Java source code in Listing 3 demonstrates the effect of instrumenting with improved stack traces enabled. The *Trace* and *StackTrace* classes are normal Java classes that are also part of the implemented run-time library. Most of the functionality related to the stack traces is implemented as normal Java code in the run-time library to make the instrumentation simple. Amongst other things, the included utility classes keep track of the threads and map them to the right stack traces.

Listing 3. The same method after instrumentation with stack tracing enabled.

```
public int calculateMean(int[] values) {
    StackTrace trace = Trace.push("MyMath", "calculateMean", 1);
    try {
        trace.setSourceLine(2);
        if (values == null) {
            trace.pop();
            return 0;
        }
        trace.setSourceLine(3);
        int sum = 0;
        trace.setSourceLine(4);
        for (int i = 0; i < values.length; i++) {
            sum += values[i];
        }
        trace.setSourceLine(5);
        int mean = sum / values.length;
        trace.pop();
        return mean;
    } catch (Throwable t) {
        trace.exception(t);
        throw t;
    }
}
```

A call to *Trace.push()* is added to the beginning of the method. This method pushes a new element, representing this method, to the call stack of the current thread. The name of the class, the name of the method, and the method beginning line number are passed as parameters. The parameters are needed later to print a stack trace that looks like a normal Java stack traces. For convenience the current stack trace element is saved to a local variable called *trace*.

The *trace.pop()* method pops the top element from the stack. A call to it needs to be added before each return statement to keep the stack state up-to-date. This is done similarly as with the method tracing before.

The whole original method is enclosed inside a try-catch block to catch any exceptions that would otherwise escape. Any exception that escapes the original method will be caught and thrown again, so that the functionality of the method does not change. Before throwing the exception, *trace.exception()* is invoked to update the stack trace. Calling *trace.exception()* pops this method from the stack and records the thrown exception. In normal Java source code it is actually not possible to throw a *Throwable* object without declaring it to be thrown in the method declaration, but it is possible on the bytecode level.

A call to *trace.setSourceLine()* needs to be added before each original line of code, so that the line number that is being executed is updated to the stack trace. If an exception will be thrown when executing the following original line, the trace will now point to the same line. In the example code, the execution can jump back when the for loop is being executed and the source line will not be set correctly. In the actual bytecode implementation this is not a problem as the call to the *trace.setSourceLine()* method is placed just after the bytecode instruction defining the line number debug information so that it is set correctly regardless of any jump instructions.

For the tracking to work, each thread needs to have its own stack trace. Every time a *Trace.push()* is invoked we need to find out which thread invoked the method. The current thread is resolved inside the *Trace.push()* method using *Thread.currentThread()*. A separate mapping is needed to find the correct stack trace for the current thread as the thread object itself is part of the standard Java library and cannot be modified.

Logging Exceptions Traces. Now that the application itself keeps track of the execution of its threads, it is possible to get a up-to-date stack trace at any given point in the code. Similarly to Java SE, the tools provide the user with an API to receive the trace explicitly as an array of method elements.

In exception situations it is convenient to log the stack trace implicitly without any code written by the developer. The tool provides an build time option that can be used to determine how the exceptions are logged. The options are to log all exceptions, to log only uncaught exceptions, or to not log exceptions implicitly at all. All that the developer would need to do is to define a callback method that will be invoked each time an exception has occurred.

It is a common practice to use *throwable.printStackTrace()* to print the stack trace to standard output whenever an caught exception could be interesting when debugging. As stated before, the standard output is often not available and the stack trace should be redirected to the logging system that is in use. This can be achieved by detecting each call to *throwable.printStackTrace()* in the instrumentation phase and modifying each call so that the stack trace of the exception will also be logged.

Listing 4 provides an example method that will be used to demonstrate how the exception logging works. This method invokes another method called *con-*

nection.open() that is known to possibly throw an *IOException*. If an exception is thrown, it will be caught and the exception stack trace will be printed.

Listing 4. A Java method catching a possible exception and printing the stack trace of the exception to the standard output stream.

```
public void open() {
   try {
      connection.open();
   } catch {IOException ioe} {
      ioe.printStackTrace();
   }
}
```

Now we want the exception stack trace to appear in the logging system that is used by the application. As it is necessary to detect exceptions to keep the stack trace correctly up-to-date, it is easy to add a callback to some logging code each time an exception is detected. The stack trace can be given as a parameter. In addition, we need to detect exceptions that are caught by the original method. Otherwise it is not possible to detect exceptions that never leak outside the method they were thrown in. Exceptions that are caught inside the original method need different kind of processing as they should not pop the method from the call stack.

Listing 5 shows how the instrumentation modifies the bytecode of the method to enable detecting all exceptions. All the same modifications are done as before when adding the general stack tracing (Listing 3). In addition, code is added to detect caught exceptions and to enable their logging.

A call to *trace.exceptionCaught()* is added to be the first instruction in the catch block of the original code (after line 4). This call needs to be added to the beginning of every catch block in the instrumented bytecode. If the developer has chosen to print all exceptions, a callback to the logging method will be done each time *trace.exceptionCaught()* is invoked. Otherwise the current stack trace is just saved so that the callback can be called later with correct stack trace if necessary. The stack trace needs to be saved before any code is executed inside the catch block to be able to get the correct source line information later. Any possible finally blocks do not need changes as they do not catch exceptions.

Finally, a call to our own *Trace.printStacktrace()* needs to be added after every standard *throwable.printStackTrace()* method. The added method call logs the stack trace that was saved when the exception object that is given as a parameter was detected. It uses the callback method defined by the developer to log the trace.

If the developer has chosen to implicitly print all uncaught exceptions, all we need to do is to add an extra check inside the final *trace.exception()* method that is invoked when an exception escapes the original method. If the call stack

Listing 5. The method after instrumentation for logging exception traces.

```
public void open() {
   StackTrace trace = Trace.push("MyClass", "close", 1);
   try {
      try {
         trace.setSourceLine(3);
         connection.open();
      } catch {IOException ioe} {
         trace.exceptionCaught(t);
         trace.setSourceLine(5);
         ioe.printStackTrace();
         Trace.printStacktrace(throwable);
      }
   } catch (Throwable t) {
      trace.exception(t);
      throw t;
   }
}
```

is empty after popping the current method we know that the exception was not caught by any instrumented code and it should be logged. An extra warning message is shown when a thrown exception escapes uncaught.

3.3 Deadlock Detection

Synchronization is required to keep data from corrupting when multiple threads are using shared resources. Synchronization restricts the access to certain memory areas so that only one thread may access the data at a time. In Java, synchronization is implemented on the language level and also the Java virtual machine has its own specialized instructions for it [4,6].

Adding synchronization adds the possibility of deadlocking however. A deadlock happens when multiple threads are accessing synchronized resources so that a thread has to wait for another thread that is directly or indirectly waiting for the original thread.

Errors that cause deadlocks are usually hard to detect and hard to debug. The deadlock may only happen in some special case with certain specific timing. There are many techniques that can be applied to prevent deadlocks from happening and it is usually best to apply some well known strategy to rule out the possibility of a deadlock. One commonly applied strategy is called *resource ordering*. When using resource ordering deadlocks can be avoided by accessing the resources always in the same order.

Even if preventing deadlocks has been taken into account when developing an application, it is still possible to miss a possible deadlock situation. If a deadlock is already occurring and it can be reproduced, a good way to find the cause for the deadlock is trying to detect it on run-time and print the stack trace for each of the deadlocking threads. Java SE provides a built-in method for finding deadlocked threads when the application is running. The *findMonitorDeadlockedThreads()* method in *java.lang.management.ThreadMXBean* class can be used to find threads that are currently deadlocked [3]. The same functionality is not available in Java ME, but similar functionality can be implemented by instrumenting the applications bytecode.

Bytecode Manipulation to Detect Deadlocks. The application needs to be modified to track when each of the running threads enters a synchronized block and thus acquires a lock. Every synchronized method and synchronized block must be modified to be able to detect deadlocks when they happen on runtime. Before entering a synchronized code block the current thread is marked to be waiting for the lock object defined in the synchronization block. Just after entering the synchronized area the thread needs to marked as the one owning the lock. Similarly the lock is marked as released when the method or block is exited.

A data structure containing the locks acquired by each thread needs to be maintained. Each time a new lock is requested by a thread, the data structure must be checked to see if a deadlock situation has occured. When a deadlock is detected the stack trace of all the threads that are causing the deadlock can be printed using the same method as earlier when printing normal exception stack traces.

For synchronized member methods the lock object is the *this* reference and for static methods it is the *Class* object for the class in question [4]. Using the synchronized keyword in a method declaration has the same effect as enclosing the whole method in a synchronized block. As, an example, let us consider the following Java method:

```
public synchronized void deadlockTest() {connection.open();}
```

When a synchronized method is instrumented, the synchronized keyword is removed, and a separate synchronized block is added to achieve the same effect. This needs to be done because we need to mark the thread as waiting for a lock before it enters the synchronized block. Listing 6 shows the instrumented version of the method from Listing 6. The resulting code for Listing 6 would be exactly the same, but the line number have been aligned to match the former example. The added instructions for tracking stack traces have been excluded to make the example clearer. In addition, the current stack trace must be passed to the deadlock detector so that the current thread can be determined and the stack trace printed upon a deadlock.

The call to *DeadLockDetector.waitingForLock()* method is added to mark the thread as waiting for a lock. Waiting is not necessary if the thread already owns the lock. A call to *DeadLockDetector.locked()* is added right after entering a synchronized block to mark the current thread as the owner of the lock. A counter must be increased to be able to mark the lock as released when the

Listing 6. The method after it has been instrumented for deadlock detection.

```
public void deadlockTest() {
   DeadLockDetector.waitingForLock(this);
   synchronized (this) {
      DeadLockDetector.locked(this);
      connection.open();
      DeadLockDetector.released(this);
   }
}
```

outermost synchronized block using the same lock is exited. Before the end of the block a call to *DeadLockDetector.released()* method is added. This call decreases the counter for the used lock and marks the lock as released if this was the outermost block using this lock. The lock object reference that is used for synchronization is given to all the method calls as a parameter

The code shown in the example is not enough to correctly handle exception situations. Fortunately in Java bytecode the instruction for releasing a locks is automatically added to each exception catch block when a class is compiled. The *DeadLockDetector.released()* method call just needs to be added before each instruction releasing a lock. The case when a method is exited with an exception must be however handled separately and all the acquired locks must be released. The exception tracking described earlier can to be used for that.

The *DeadLockDetector* class handles the logic for detecting deadlock situations. The class keeps track of the locks that are owned by each thread and each time before entering a synchronized block a check is made if a deadlock situation has occurred. A deadlock occurs if the lock that the thread is waiting for is locked by another thread that is again waiting for the current thread. Each time when trying to acquire a lock a check needs to be made for these kind of cyclic locking situations.

In addition to synchronized blocks and methods, also *object.wait()* methods affect the ownership of locks and can cause deadlocks. A wait method invocation stops the thread and releases the lock until a notification wakes up the thread again or a timeout happens. When the thread is woken up it tries to acquire the lock again and can cause a deadlock.

Listing 7 shows how each call to a *object.wait()* method in the original class is modified. before the call a call to *DeadLockDetector.wait()* is added. This method call marks the lock as released and marks this thread as waiting for the lock again. Even though this thread might sleep for a long time and is not technically waiting for the lock, it is close enough for the purpose of deadlock detection, as the next executed instruction cannot be reached before the lock is acquired. After the wait, a call to *DeadLockDetector.locked()* is added as the lock is now again acquired by this thread.

Listing 7. Instrumenting wait calls to enable deadlock detection.

```
...
   DeadLockDetector.wait(this);
   this.wait();
   DeadLockDetector.locked(this);
...
```

3.4 Simple Profiling

The performance of an application may be found to be be less than satisfactory at some point of development. The process of determining which parts of the application are using the most resources is called *profiling*. Optimization in these so called hot spots will also give the most benefit. If optimizations are done without knowing what the actual performance bottlenecks are, a lot of time can be wasted without much improvement in performance. Profiling can be used to find problems in memory use and to find execution speed bottlenecks [11].

Bytecode Manipulation to Enable Profiling. The basic idea of the implemented profiling tool is to provide information about the time that is spend executing in each of the methods of the application. This can be achieved by getting the system clock time before executing a method and the time after the execution. The difference in the time can be used as an estimate of the amount of time spent in the method. Also the number of calls to each method is stored. The memory usage of the application is not analyzed by the implemented tool. Let us next consider the following Java method that is to be instrumented with profiling information:

```
public void profilingTest() {connection.open();}
```

Listing 8 shows the example method after instrumentation. A line is added to the beginning of the method that stores the starting time of the execution of the method to a local variable. Code for determining how long the execution took is added to the end of the method. The difference between the start time and end time is calculated and added to the total time used in the method. After that the counter for the number of method calls is incremented.

An array is used to store the profiling information. Reference to the array is kept in a static field called *$methods*. This field needs to be added to each instrumented class. The array contains *ProfiledMethod* objects. Each object holds the profiling data for one specific method: the total time spent in the method and the call count for the method. Each method of a class is associated with an index number, which is used to access the correct object in the array.

Listing 8. The same method after it has been instrumented for profiling.

```
public void profilingTest () {
  long start = System.currentTimeMillis ();
  connection.open ();
  $methods [0].totalTime +=
      System.currentTimeMillis () - start;
  $methods [0].callCount++;
}

static final ProfiledMethod [] $methods = new ProfiledMethod [1];

static {
  $methods [0] = new ProfiledMethod ("profilingTest ()");
  Profiler.addClass ("MyClass", $methods);
}
```

The profiling information array needs to be initialized in the static initialization block so that every method contained in the class is added to the array. Also the name and signature of the method is saved so that the data for different methods can be identified later. The array is also registered to the *Profiler* class with the name of the current class. The profiling information can be later accesses using methods that are available in the *Profiler* class.

4 Evaluation

Using Java ME and SE simultaneously made the old and limited CLDC libraries feel even more limited compared to the less restricted Java SE. The full Java library of Java SE and the new Java language features will still be missed after the implemented improvements.

4.1 Limitations and Potential Problems

All the standard Java libraries and other libraries that are part of the platform cannot be instrumented. Fortunately, the interesting parts of the code that is being debugged are usually the ones that are being developed and therefore it is possible to instrument them. For performance reasons it might make sense to instrument only part of the whole project.

Tracing is not possible in classes that have not been instrumented. For example if an instrumented method calls another uninstrumented method that again eventually calls instrumented code, we have no way of knowing what happened in the uninstrumented method and if there were other method calls before calling the instrumented code again. In these situations the stack trace will just contain an unknown element to mark the uninstrumented code. Similar problems exist with deadlock detection where some deadlocks may not be detected if all of the code is not instrumented.

To include correct line numbers in stack trace, the instrumented class files are compiled with debug information. If the application uses some third party library whose source code is not available, some parts of the application might not include the necessary debug information. Source file name and line number information is needed to map the byte code instructions in the compiled classes to source code lines. Stack traces will work even if the information is missing but the stack traces will not contain the line number and source file name.

When the optimizations are enabled in the instrumentation phase, the instructions for catching exceptions are not added to some very simple methods. This has the side effect that some exceptions that can happen at any point of execution may be reported to have happened on a wrong line. Normally this is not a problem and the optimizations can be disabled if necessary.

Instrumenting modifies the Java bytecode and can also cause issues with code that was previously working. Many errors are detected immediately when the preverification fails but some errors may only occur on run-time. Extensive testing has been done to make sure as many as possible different kind of Java language constructs work without problems.

Biggest problem with the implemented deadlock detection is that the instrumentation affects the timing of the code execution. As the deadlock may only happen on some specific timing it is possible that the deadlock cannot be reproduced when the detection is enabled. Similar problem also exists in profiling: The profiling results are affected by the code that is added to measure performance.

4.2 Performance and Size Impact

The instrumentation phase adds bytecode instructions to every instrumented method. The added instructions make each class bigger and affect the execution speed of the program. The added instructions also include calls to the runtime library part of the debugging tool. These runtime library classes need to be included in the final JAR package and require about 10 kilobytes of space. Also the memory usage of the application is slightly increased as each thrown Exception and the current stack trace for each thread needs to be kept in memory when running an instrumented MIDlet. The overall size and speed overhead of the instrumentation was tested using some existing Java ME benchmarking software and using an actual Java ME application. The memory usage impact of the implemented tools was not measured as the increase in memory usage is very small compared to the amount of memory that is usually available.

JBenchmark (*http://www.jbenchmark.com/*) is a set of Java ME benchmarking tools that are meant to measure the performance of Java ME enabled devices. The differences between the performace of different devices is not interesting from point of view of this paper. However, the same benchmarking MIDlets can be used to measure how the performance changes after instrumenting the benchmarking MIDlet to enable the improved debugging features. Multiple versions of the JBencmark MIDlet were used to measure both the impact on the MIDlet size and the impact on the MIDlet execution speed.

Table 1. The size impact of instrumentation

Test MIDlet	Original size	Size after instrumentation	Size increase
JBenchmark	26236 bytes	28747 bytes	9.6%
JBenchmark 2	63994 bytes	67794 bytes	5.9%
JBenchmark Pro	207103 bytes	238981 bytes	15.4%
Dromo client	513167 bytes	671686 bytes	30.9%

When instrumenting, the size of the original uninstrumented classes is increased approximately 5–30%, depending on the application and the instrumentation parameters. Table 1 shows the size impact of instrumentation on some existing Java ME applications. The constant increase of approximately 10 kilobytes caused by the runtime library classes is ignored in these calculations so that it will not skew the size increase percentage. The overhead is smaller in the JBenchmark MIDlets as they do not include debug information and thus the information to track line numbers in exception will not be added to them. JBenchmark and JBenchmark 2 were also already obfuscated before instrumentation was done. This reduces the size increase as the string literals that need to be added when instrumenting are much shorter. The *Dromo client* (*http://sourceforge.net/projects/dromo/*) is an open source MIDlet for remotely controlling a video recording server. It was chosen to represent a real life Java ME application. The used version was compiled to include all the necessary debug information and it was not obfuscated.

The maximum allowed size of a MIDlet JAR package can be very restricted depending on the target device. The limit can be as little as 100 kilobytes in some older mobile devices. In current devices the limit is considerably higher, up to megabytes. A 30% increase in MIDlet size can become an issue in some cases. The MIDlet can be obfuscated after the instrumentation to reduce the size if necessary. Obfuscation renames all the classes and normally this would also make the stack traces unreadable. If the instrumentation is done before obfuscation, the original class and method names will be preserved even after obfuscation in the stack traces. This is possible because the method and class names are literals in the code and will not be changed by the obfuscator. The reduction in JAR size is not as much as it would be without instrumenting. Obfuscation also reduces the size of the runtime classes that are added to the JAR by the implemented tools.

Each method call is instrumented with multiple new calls to methods in the runtime debugging classes and the current source line number needs to be updated on each new line of code. If an instrumented method is called repeatedly in a loop, this can add up to a significant amount of time. The speed impact of instrumentation is shown in Table 2. Each benchmark was first run uninstrumented and then instrumented. The shown percentage shows the performance of the instrumented benchmark relative to the uninstrumented run. All the measurements were done using the default emulator of Sun's Wireless Toolkit and a Nokia X3 device. The table shows an average of three runs of each benchmark.

Table 2. The speed impact of instrumentation

Benchmark	Speed after instrumentation	
	WTK emulator	Nokia X3
JBenchmark	97.3%	96.3%
JBenchmark 2	84.9%	97.5%
JB Pro: Business math	89.2%	99.8%
JB Pro: Chess AI	23.1%	9.4%
JB Pro: Game physics	28.8%	20.6%
JB Pro: Image processing	22.9%	17.1%
JB Pro: Shortest route	46.6%	15.6%
JB Pro: XML parsing	85.7%	87.8%
JB Pro: ZIP compression	12.9%	7.6%

The speed impact varies a lot depending on the code that is being instrumented. The JBenchmark and JBenchmark 2 tests mostly consist of different kind of graphics operations where the majority of execution time is not spent in the Java code and the impact on MIDlet performance is very small. JBechmark Pro was used to run some specific processing intense operations. These operations are closer to the worst case scenario where a very simple method is run a huge number of times in a loop. Also the performance impact is very significant in these kind of tests.

In UI and graphics operations the application speed is not affected very much and even a slowdown to half or one fourth of the original speed is normally acceptable when debugging an application. Instrumentation can however cause the application to slow down up to one tenth of the original speed in some processing intense operations.

The performance degradation and the increase in program size can further be tackled by different options in the instrumenting phase. It is possible to limit the number of instrumented classes by including only interesting classes, or by excluding some classes that are uninteresting or processing intense. Another option is to disable some tracing features. Both of the options compromise the amount of information that will be available when executing the program.

5 Conclusions

This paper presents a set of tools to make Java ME development and debugging faster and easier. Implementing dynamic tools that are used when the application is running is challenging in the very restricted sandbox of Java ME. As a workaround, we instrument the compiled application binary statically before running it. This way the application can itself gather information at run-time.

The developed tools have already been used in Java ME projects being developed at Sasken Finland. The feedback from the developers has been very positive. Mainly the ability to get stack traces has been used, and it has proven its usefulness multiple times already during a short period of two months.

References

1. Java^TM 2 Platform Standard Edition 1.3 API specification, http://java.sun.com/j2se/1.3/docs/api/ (accessed on February 2010)
2. Java^TM 2 Platform Standard Edition 1.4 API specification, http://java.sun.com/j2se/1.4.2/docs/api/ (accessed on February 2010)
3. Java^TM 2 Platform Standard Edition 5.0 API specification, http://java.sun.com/j2se/1.5.0/docs/api/ (accessed on February 2010)
4. Arnold, K., Gosling, J., Holmes, D.: The Java^TM Programming Language, 4th edn. Addison-Wesley (April 2008)
5. Lau, A.: The fragmentation effect. JavaWorld (May 2004), http://www.javaworld.com/javaworld/jw-05-2004/jw-0524-fragment.html
6. Lindholm, T., Yellin, F.: The Java^TM Virtual Machine Specification, 2nd edn. Prentice-Hall (April 1999), http://java.sun.com/docs/books/jvms/
7. Sun Microsystems, Inc. Connected Limited Device Configuration (CLDC) Specification 1.1 (March 2003), http://jcp.org/aboutJava/communityprocess/final/jsr139/
8. Sun Microsystems, Inc. Mobile Information Device Profile (MIDP) Specification 2.1 (June 2006), http://jcp.org/aboutJava/communityprocess/mrel/jsr118/
9. Tanter, É., Ségura-Devillechaise, M., Noyé, J., Piquer, J.: Altering Java Semantics via Bytecode Manipulation. In: Batory, D., Blum, A., Taha, W. (eds.) GPCE 2002. LNCS, vol. 2487, pp. 283–298. Springer, Heidelberg (2002)
10. Topley, K.: J2ME in a Nutshell. O'Reilly (March 2002)
11. Wilson, S., Kesselman, J.: Java^TM Platform Performance: Strategies and Tactics. Addison-Wesley (June 2000)

Dynamic Reduction of Rollbacks in Wireless Multi-user Virtual Environments

Abdul Malik Khan[1], Sophie Chabridon[1], and Antoine Beugnard[2]

[1] Institut TELECOM, TELECOM SudParis
CNRS UMR SAMOVAR
9 rue Charles Fourier
91011 Evry cedex, France
{Abdul_malik.Khan,Sophie.Chabridon}@institut-telecom.fr
[2] Institut TELECOM, TELECOM Bretagne
Computer Science Department, CS83818
29238 Brest cedex 3, France
Antoine.Beugnard@institut-telecom.fr

Abstract. In distributed virtual environments such as multiplayer games, where many users interact in real time while communicating through a network, the users may have an inconsistent view of the game world because of the communication delays across the network. Consistency maintenance algorithms must be used to have a uniform view of the game world. The majority of these algorithms use rollback mechanisms to correct the inconsistencies that occur because of the disorder of the arrival of update messages. These rollbacks are very costly, especially when playing a game, using high-latency wireless networks, on mobile terminals which have limited memory and processing speed. In this paper, we present a dynamic and adaptive approach for reducing the number of rollbacks in distributed virtual environments on wireless mobile devices. This approach takes into account the underlying network latency and the semantics of the game virtual world to dynamically decide whether a rollback is needed in case inconsistencies have occurred or can be possibly avoided. We evaluate our approach on a simplified version of a Football game on hand-held devices and show that this dynamic rollbacks' reduction approach improves the responsiveness of the game and maintains consistency of the game state while limiting the use of processing power and memory space.

Keywords: Multiplayer Mobile Games, Latency Hiding, Data Synchronization, Consistency Algorithm.

1 Introduction

Networked Multiplayer Games are becoming more and more popular with the advances in hardware technologies and enriched game design improving immersive feelings. However one of the main issues hindering the real-time interaction in these games is network delays. To enforce strong consistency, events transfer should satisfy total order [2], however this would degrade [4] the performance of the game. Because of the unordered delivery of events and the network delays induced by unreliable protocols, the game state can be inconsistent at different points in time.

To address this issue, synchronization algorithms are used to reach a consistent state at all the players. There are two approaches to state consistency maintenance. In the conservative approach, all the participants must wait for the acknowledgement of their update messages and reach a global consistent state before advancing. Unfortunately, this approach cannot be applied to continuous applications such as multiplayer games where the state of the simulation changes not only as the result of user actions but also with the passage of time. Also, waiting for acknowledgements would violate the interactivity property. In the optimistic approach, the participants process events without waiting for the arrival of late events and repair any potential inconsistency when it actually occurs. This approach is suitable for mobile continuous interactive applications as it preserves the game experience of the players even with a high network latency and variations in the connectivity as in wireless networks.

Different mechanisms [10,3,5,6,13] can be used to repair inconsistencies in the optimistic approach. Among the proposed solutions, rollback mechanisms are a popular solution to remove the inconsistencies that occur because of late arriving messages when using unreliable network protocols. However, these rollback mechanisms use static approaches to solve inconsistencies without keeping in view the changes in the network and game environments. In wireless networks, the communication delays are higher as compared to wired networks and also jitters can occur impacting the network delays. Also, we believe that in a multiplayer game, different objects and regions in the game have different consistency requirements that vary during the game play as we have already discussed in [9]. In this paper, we propose an optimistic approach based on a dynamic rollback mechanism which adapts its behaviour according to the changes in the network conditions and the consistency requirements of the game at a particular time. The aim is to considerably reduce the number of rollbacks which occur because of variations in the wireless network latency and to improve the playability of the game on mobile devices. Furthermore, we introduce the idea of critical actions to denote highly sensitive events in the game which impact the outcome of the game results. The update messages representing these actions cannot be discarded and must be eventually delivered. Also the rolling back of these messages can have an adverse effect on the result and the usability of game. Our approach tries to avoid, whenever possible, the rollbacks of such critical actions (or events) by relaxing the consistency requirement during the game play.

This paper is organized as follows. In section 2, we discuss some related works. In section 3, we discuss the need for a more dynamic approach to consistency maintenance. In section 4, we discuss our proposed solution for dynamic rollback reduction. In section 4.1, we propose the idea of critical actions and then in section 4.2 we use this concept to define relations between game events that can help discard late arriving events without performing any rollback. In section

5, we present an example scenario to explain our approach. Based upon the concepts discussed in sections 3 and 4, we give in section 6 an algorithm for the reduction of the number of dynamic rollbacks. In section 7, we discuss the trade-off between interactivity and consistency. In section 8, we give the results of our experimentation on the proposed approach. Section 9 concludes this paper and discusses some future work.

2 Related Works

Time Warp [10] is an optimistic synchronization algorithm that allows remote participants to execute their events without the guarantee of a causally consistent execution. Time Warp takes a snapshot of the state at the reception of a command and issues a rollback to an earlier state if a command older than the last executed command is received. With an unreliable transmission protocol and high delays in wireless networks, the number of late arriving commands can greatly increase the number of rollbacks which can be very costly for users of mobile terminals keeping in view their limited memory and processing capacity. Moreover, frequent rollbacks can affect the playability of the game. The local lag approach can be used with Time Warp to force the local events to wait for the late arriving events before displaying them. This allows to reduce the number of inconsistencies that can occur, but at the cost of a decreased responsiveness of the game.
In [5,6], the authors present an event correlation algorithm for mirrored server architectures to decrease the number of rollbacks. The events which are not correlated to earlier events and arrive late can be discarded without rolling back the previously executed commands. These papers, however, do not present any definition or approach for finding the correlation between events.
[13] uses a correlation algorithm for P2P Massively Multiplayer Online Games (MMOGs). According to their definition, two events are correlated if "they both update the same state variables associated to a given game element". Apart from repairing inconsistencies, an approach called dead-reckoning [1] can be used to predict the future positions of objects in the game world for increasing the interactivity of the virtual environment and hiding the network latency. Additionally, the local lag [10] allows to delay the display of local command(s) so that update messages from remote players can arrive, thereby avoiding a causal disorder of messages. We have already proposed in [9] the idea of dynamically changing the parameters for dead-reckoning and local lag according to the network delays and the positions and speeds of the game's objects. In this paper, we combine this idea with our dynamic rollback reduction approach to achieve consistency in multiplayer games according to the need of the situation.

3 A Dynamic Approach to Consistency Maintenance

In the previous section, we discussed different synchronization algorithms used for consistency maintenance in multiplayer games. Because of the rollback and re-processing of commands, Time Warp is very costly in terms of memory and processor usage and hence is not very suitable for mobile games. The local lag approach, combined with dead-reckoning, is suitable for high latency networks. But because of changing delays and jitters in wireless networks, a fixed local lag can cause inconsistencies in the game state across different nodes. Also, we believe that the message discarding approach presented in [7] can be interesting to combine with local-lag. This is because in wireless networks, there are messages which arrive late due to the network jitter and hence must be discarded as these late arriving messages may cause inconsistencies. In this section, we present a dynamic approach in which the consistency maintenance algorithm changes its parameters according to different factors of the environment: e.g. the network load, the type of objects, the location of an object in the virtual world etc. In this section, we first discuss the conditions under which these different parameters are dynamically changed and then we combine these different approaches into the form of an algorithm.

3.1 Observations

While playing a multiplayer game, some inconsistencies may occur due to the communication delays across the network. The game programmers estimate these delays in order to compensate for the late arriving messages from remote users. Because of the jitters, especially in wireless networks, these delays may vary greatly. Therefore, it is necessary to observe the network load during the game play and to compensate for these delays accordingly. In a rich multiplayer game, there are many different types of objects in the virtual environment. The velocities and directions of these objects vary according to their nature. For example, in a tennis game, the speed of the tennis ball is greater than the speed of the players. Also, a player has to react sharply to the movement of a ball. Therefore, these different types of objects in the game world have different consistency requirements. The algorithm responsible for consistency management has to react not only to the varying latencies of the underlying network infrastructure, but also has to deal appropriately with the various types of objects of the game. We also observe that the consistency requirement for an object not only depends upon its speed, but also upon its position in the virtual world. For example, in a car racing game, we need strict consistency management when two cars are very close to each other and they both are approaching the finishing line. Based upon these observations, we believe that a consistency maintenance algorithm must take into account the context of the game along with the variations in the network communication delays.

3.2 An Adaptable Local Lag

When a message about an object is received from a remote user, this object has a certain distance, possibly zero, from the other objects, destination, or any other important entity in the game world, called pivot. We therefore change the value of the local lag according to three factors:

1. If the object concerned by the message we have received from the remote user is coming closer towards the pivot, we reduce the value of the local lag. If the object is going away from the pivot, we increase the value of the local lag upto a certain limit called *Local-lag$_u$*. This increase can be continuous with respect to the motion of the object, or can be discrete based on zones as in [12]. The rate of the change of the value of the local lag is application dependent, and the programmer must specify it during the development of the game. In the next section, we present an approach to help the programmer to specify these values to a component responsible for consistency management.
2. The value of the local lag also changes according to the network load. When the number of messages arriving later than a certain waiting time reaches a certain level N_d, we increase the value of the local lag. This increase in the number of messages arriving late can be due to jitters in the network. Note that the value of the local lag is proportional to the network latency, but we cannot increase it more than a certain limit because it would have a bad effect on the responsiveness. This upper limit can be dependent on the pace of the game. If the game or an object in the game demand high responsiveness, we should not increase the local lag value above a certain limit. This limit can be specified by the game developer.
3. We can set different local lag values for different types of objects according to their importance in the game. For example, we can have a smaller value of local lag for the ball and a higher value for the players in a tennis game. Indeed, the responsiveness of the ball must be high to satisfy the interaction requirement as in [14]. Again, we need an interface provided by the component implementing the algorithm so that game developer can specify the relative values of local lags for different objects in the virtual environment.

3.3 Adaptable Dead Reckoning

The dead-reckoning approach relies on a threshold represented the maximum distance tolerated between the real object trajectory and the predicted one. This threshold value should be situation and environment dependent. Different objects have different consistency requirements at different regions in the game and hence the dead-reckoning should adapt dynamically according to the situation. For this purpose, we propose the idea of critical regions to denote those regions where strong consistency is required.

Critical Regions. We define a Critical Region as a region in the game where we need strict consistency so that all the players have a consistent view of the game in that region. A critical region is one in which inconsistencies can violate the fairness of the game or can annoy a player because his expectations are violated. Therefore, we propose to use real update messages instead of predictions in those regions. For example, in the case of a tennis match, if the ball hit by one player is touching the ground near the base line on the other side of the court, and the opponent player

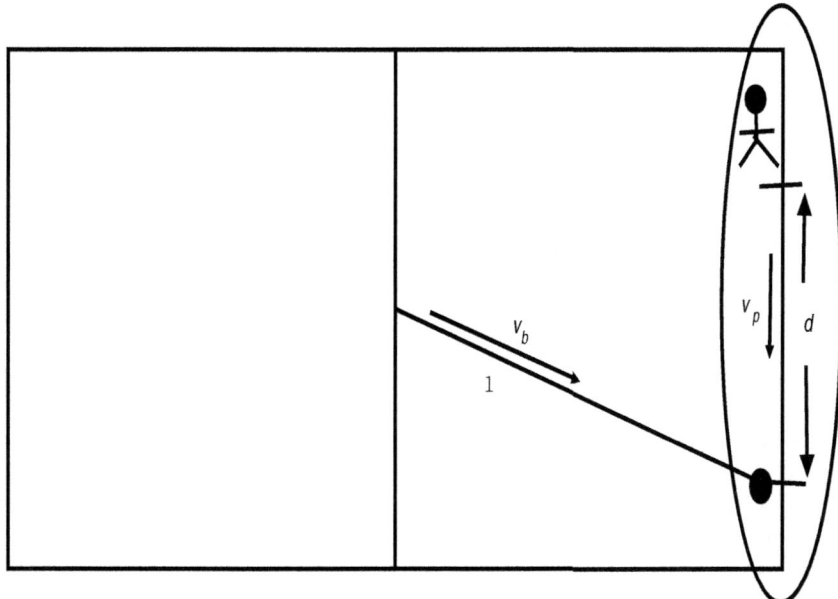

Fig. 1. Area around the base line constitute a critical region in Tennis game

is quite far from the ball, we stop the dead-reckoning so as to increase the fairness between the players. The ball will touch the ground only when the original message is received and hence the result of the score will be correct.

The calculation of the decision whether to use dead-reckoning or not can normally be done through some easy arithmetics. For example, in Figure 1, let v_b and v_p be the current speed of the ball and the maximum speed of the opponent player respectively. Let l be the distance covered by the ball from the centre of the court to the point where it touches the ground and d be the distance of the player from that point. The ball will reach the ground in l/v_b time units and during that time the player can cover a distance of $v_p * l/v_b$ distance units. So if $d > v_p * l/v_b$, then the player cannot reach the ball, and we can stop dead-reckoning from near the net where we did the calculation. Of course, the ball will stop and jump into the air for a short period of time near the centre of the court. This will not affect the playability and the fairness of the game because we know that the opponent could not reach the ball in any case. Note that, if we do not use the idea of critical region and continue with dead-reckoning, we may predict a wrong position for the ball touching the ground near the base line and that wrong decision may cause one player to loose a point which he otherwise would have won.

Apart from critical regions, the dead-reckoning mechanisms is also dependent upon the network load and the nature of the object. For example, we can have different dead-reckoning thresholds for different objects according to their movement and importance in the game world. This threshold must be specified by the game developer to the component responsible for consistency maintenance.

4 Dynamic Reduction of the Number of Rollbacks

In this section, we present the concepts that can help to dynamically reduce the number of rollbacks in high latency wireless networks.

4.1 Critical Actions

We first introduce the concept of *critical action* (or *event*). We define *critical actions* as commands in the game, that unlike position update messages, affect the output of the game for other objects. For example, a shoot command can increase the points of one object and can kill or reduce the moving capability of another object. Hence, we consider the shooting command as a critical action. The delivery of critical actions is highly essential, however, as we will see in the next section, there are times when dropping such events does not violate the outcome of the game. Moreover, the rollback of a critical action can cause the users to quit the game because their expectations were violated [14]. This happens when there is difference between the state that they could achieve because of their own actions and the resultant state after the rollback. Therefore, efforts must be made to deliver the critical actions before users' expectations are violated and rollbacks of critical actions should be avoided whenever possible. We argue that in certain cases, such as with a very high latency in wireless networks and when the action is not taking place in a critical region, the critical action message can eventually be dropped if arriving *very* late. The definition of *very* depends upon the nature of the game and the value of the network latency and the local lag at that time.

4.2 Weak and Critical Correlation

Introduced in [5], the concept of obsolescence states that an event that arrives late at a recipient while some event that was issued earlier than this event (i.e. having greater time stamp) has already been processed, is obsolete and must be discarded. Additionally, the concept of correlation states that if this obsolete message is correlated to an earlier processed event, then all the events till that correlated event must be rollbacked and reprocessed along with this new arriving message. [5], however, does not define any mechanism to calculate the correlation between any two events. [13] defines two events to be correlated if they are associated with a single object.

We introduce here the new concepts of *weak correlation* and *critical correlation* by incorporating the notions of *critical regions* and *critical actions*. We propose that any late arriving event should be considered obsolete if it is neither in a critical region nor it is a critical action. A critical action in a critical region should never be considered obsolete and must be guaranteed to be eventually delivered.

We now give the definitions [1] of the properties of weak and critical correlations based upon the concept of correlation.

Correlation χ,
Given two events e_{ci} and $e_{cj} \in E_c$, where E_c is the set of all events that belong to o_i and $o_j \in O$, the set of all objects, with time stamps T_i and T_j, such that $T_i < T_j$, then,
$e_{ci} \chi e_{cj}$ iff
$(e_{ci}; e_{cj}, s) \to s1 \wedge (e_{cj}; e_{ci}, s) \to s2 \wedge s1 \neq s2$. $s1, s2 \in S$, the set of all states.

It means that the ordered execution of correlated events is necessary to reach a consistent state. We now define the two properties of weak and critical correlations.

Property 1: *weak correlation* χ_w Two events are weakly correlated if they are correlated to each other, but are either non-critical actions or do not occur in any critical zone.

Property 2: *critical correlation* χ_c Two events are critically correlated if they are correlated to each other, and in addition, those two events belong to the set of critical actions and they occur in a critical zone.

All correlated events that are non-critical events are weakly correlated.

4.3 Relaxed Consistency

In our definition of relaxed consistency, when the number of late arriving messages increases above a threshold value, thereby indicating a high network latency, we rollback only those events that are critically correlated and we discard weakly correlated events. We also increase the local lag value for critical events so as to further decrease the number of rollbacks and better satisfy the user expectations with regard to the game.

Whenever the number of late arriving messages decreases below the threshold value, then we stop the discarding of weakly correlated events. This is because the number of rollbacks has already decreased following the decrease in the network latency, so processing weakly correlated events will not increase the number of rollbacks beyond a certain limit. Also we decrease the value of the local lag to increase interactivity for the users. For all other events that are not critical, we discard them when arriving late, because it will not affect the playability of the game and will unnecessarily increase the number of rollbacks thereby wasting the limited processing power and memory space of mobile terminals.

---

[1] We adopt a simplified version of a syntax based on a Plotkin-style operational semantics (Plotkin 1981 [11]). In particular, $(e_i; e_j, s) \to s*$ denotes an initial game state s from which the final game state $s*$ is reached through two subsequent events, namely e_i and e_j.

5 An Example Scenario

In figure 2, we show an example game scenario in which a character hits a moving target (an animal in the figure) with their gun shots. To hit the target, the shooter first sends a warning (or ready) sign before shooting it. Without a warning sign, it cannot shoot the target. In the figure, a circle around the animal denotes a critical region where a gun shot can hit the target.

Suppose that the warning sign $w2$ arrives later than the actual shot $s2$. Since $s2$ hits an area which is outside of the critical region, we do not need to perform any rollback when $w2$ arrives since it will not affect the outcome of the game. We say that $s2$ and $w2$ are weakly correlated. In case $w1$ arrives later than $s1$, then we have to apply a rollback because we cannot discard W1 as it will violate the game rule of warning before shooting. We say that $w1$ and $s1$ are critically correlated. If we have already experienced a large number of rollbacks thus suggesting a high network latency, we increase the value of the local lag so as to give more time to late arriving messages and hence decrease the number of rollbacks in the critical region.

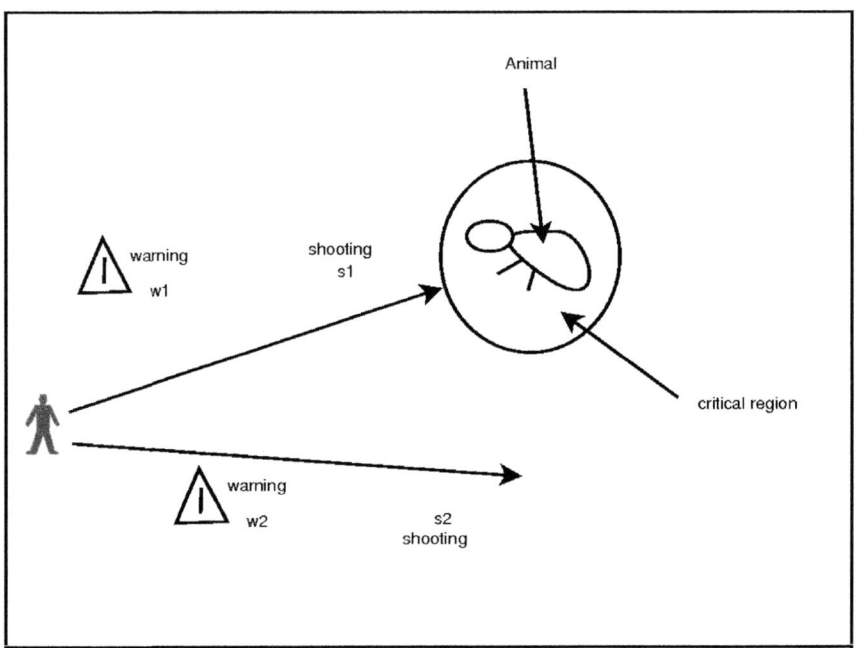

Fig. 2. A game with two players

6 Dynamic Rollbacks Reduction Algorithm

In this section, we present a dynamically adaptive synchronization algorithm based on the concepts of obsolescence and correlation using the optimistic approach as already discussed in the previous sections. In our algorithm, we propose to minimize the number of rollbacks with our dynamic approach. The purpose of reducing the number of rollbacks is to avoid unexpected behaviour during the game and to allow a smooth flow of the game. Also, in case of a game played on mobile phones, rollbacks are very costly in terms of their use of memory and processing resources which are limited on small portable devices.

In our approach, we combine the use of local lag and the notions of critical region and critical action. A local lag is the artificial introduction of a delay (both at emitter and receiver nodes) so as to allow an ordered processing of those messages which were isssued earlier but arrived later than other messages because of the network delay. If the number of rollbacks oversteps a certain threshold, we increase the value of the local lag, so that more and more messages could arrive on time. However, when a player enters a critical region and the latency is not very high, we reduce the value of local lag and increase the frequency of message sending to achieve high interactivity. When the number of required rollbacks increases in the critical region, then, instead of doing rollbacks for all arriving events, we discard weakly correlated events and increase the frequency of update messages to minimize the dependency, i.e. correlation, on a single late arriving message.

The pseudo code for the algorithm is given in Algorithm 1. In the given algorithm, lines 10 through 14 are the most important as far as the concept of correlation and obsolescence and the avoidance of rollbacks are concerned.

We discard any message that arrives late (and became obsolete) and is not correlated with any previous message (lines 7 and 8). Otherwise, if a message arrives late but is correlated to some previous message e_c and the number of rollbacks is less than a certain threshold, then we apply a rollback on all the messages processed before e_c including e_c itself and re-process them in the correct order (line 11). In line 14, we rollback only critically correlated events as now the latency is very high and the number of rollbacks has reached a certain limit. In lines 16 to 18, we increase the value of the local lag if the number of rollbacks oversteps some limit so as to avoid processing related messages in an incorrect order. This will also decrease the number of obsolete discarded messages which could be related to any future message(s). We keep a different value for rollbacks threshold in critical regions and change the values of the local lag and Dead-Reckoning thresholds in these regions if the number of rollbacks reaches a certain level (lines 24 and 25). Lines 2 to 25 are repeated in a loop throughout the execution of a game session.

There is no doubt that the interactivity will be decreased at the cost of consistency and correctness of results. We discuss the issue of interactivity in the next section.

Algorithm 1. Correlation and Obsolescence based adaptive algorithm

1: Calculate the *local lag* at the beginning of the game for each class of objects according to network latency and responsiveness requirement of the object(s).
2: Change the *local lag* if the network load has changed or the location of an object is changed
3: **if** the message arrives during its local-lag specified time **then**
4:     Buffer the message according to its *local lag* value before playing out
5:     GOTO line 20 (apply DR)
6: **end if**
7: **if** the message is obsolete (not arriving in its speicified local lag time) and not correlated with events already processed during the local lag time interval **then**
8:     Discard the message
9: **else**
10:    **if** the number of rollbacks is less than a certain threshold **then**
11:       Rollback the messages, process this message and then all others
12:    **end if**
13: **else**
14:    Rollback only critically correlated events and discard all others
15: **end if**
16: Calculate the number of rollbacks
17: **if** the number of rollbacks reaches a certain limit **then**
18:    Increase the value of *local lag*.
19: **end if**
20: Apply Dead Reckoning
21: **if** the object has entered the critical region and/or the network load has changed **then**
22:    Change the threshold value for DR for that object
23: **end if**
24: **if** the number of rollbacks reaches a certain limit in the critical region **then**
25:    decrease the value of DR threshold and increase local lag value
26: **end if**

7 Responsiveness vs Consistency

By waiting for late arriving messages with an increase of the local lag value, the consistency is improved. However, it means that even local actions (and those remote events that arrived earlier because of low latency) must be delayed before playing out. Thus, it decreases the responsiveness (or interactivity) of the game. If a game requires high responsiveness, then we need to reduce the value of the local lag which can disturb the causal order of events (and increase the need for rollbacks), thus compromising the consistency of the system. Hence there is a trade-off between these two properties in a distributed interactive application.

We propose to apply different degrees of interactivity and responsiveness for different situations and/or regions of a game. For example, if a player has to shoot a static target, we need high consistency but low responsiveness since the target is static and during the time the bullet hits (or misses) the target, we do not need high interactivity from the system, but we need a fair result.

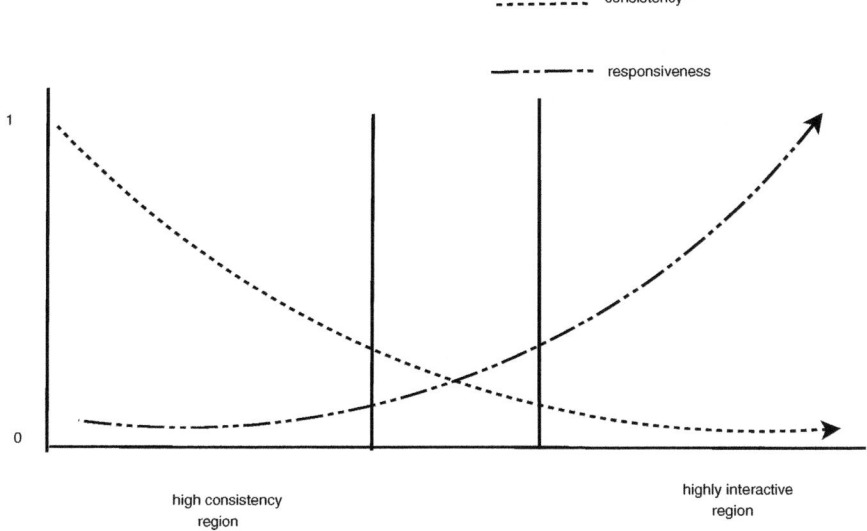

Fig. 3. Trade-off between consistency and responsiveness in different game regions

By applying a suitable value of local lag, the shooting player will observe that his shoot action has taken place at a slow pace, but will get the true results. However, we need high responsiveness from the system in some other cases, such as hitting a ball coming towards a player in a baseball game. The trade-off between responsiveness and consistency is shown in figure 3. Although we have shown consistency (and responsiveness) on a scale from 0 to 1, where 1 means absolute consistency, absolute consistency is never achieved in distributed virtual environments and hence consistency should be compromised for the sake of a high system responsiveness.

8 Evaluation

In the evaluation of the proposed approach, we are interested primarily in the measurement of two parameters.

1. the number of rollbacks;
2. the amount of dropped events;

For experimenting our proposed solution, we have developed a simple Football (Soccer) game using J2ME and Java servlet technologies. The game logic resides on a mobile phone and different players interact with each other via a server which is a simple message queue servlet. In the game, we have characters representing the different players and a goal post, which corresponds to a critical region. Each player has a circle of specific radius representing the critical region around them. A player can earn points by either hitting another player or the

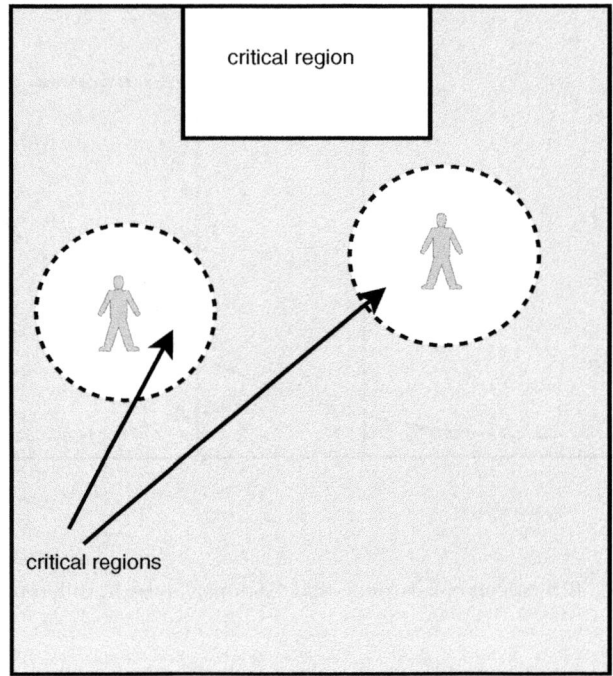

Fig. 4. A simple Football game

goal post with a bullet. The rules of the game allow the player to be hit only in the critical region shown by the rectangular shape. The game is shown on Figure 4.

The server randomly selects messages to be delayed deliberately by storing them in a server side queue before clients can receive these messages. When a delayed message arrives at the client, it calculates whether this message is correlated with another message or not. In our game semantics, a message is correlated with another message if it belongs to a bullet (critical action) or if a player entered a critical region. In case of a very high consistency, we can drop even the *hit* message because it will have no effect upon the result since our game rules allow a player to be hit only in a critical region. If the message is not correlated, we simply drop it, otherwise we apply a rollback mechanism and reprocess all the already processed messages. We continuously calculate the number of rollbacks and apply our dynamically adaptive algorithms by changing dead-Reckoning and local lag thresholds to control the number of rollbacks. In essence, we measure the number of rollbacks in the critical regions and outside the critical regions, and the number of messages discarded while the players are in a critical region and outside it. We also compare these results with the fixed-correlation based approach and the Time Warp algorithm, where there is no concept of obsolescence and correlation. The results are shown in figure 5.

From the figure, it is clear that the number of rollbacks required for Time Warp is higher than for the other two approaches. This was expected, as Time

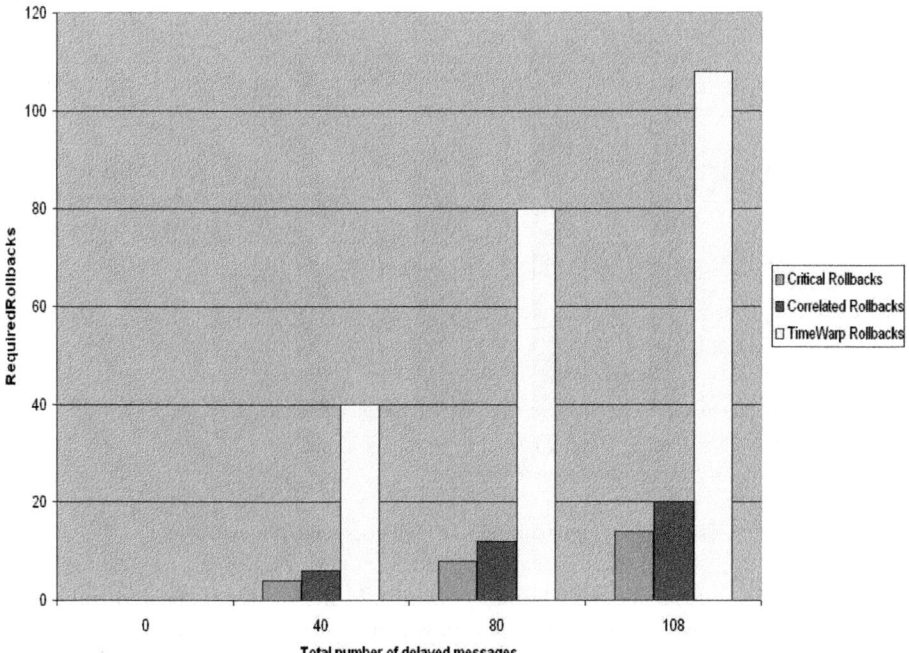

Fig. 5. Rollbacks comparison in three different approaches

Warp does not drop any message and applies rollbacks for all late arriving messages. The result of our approach is better than the result for fixed-correlation based approach, because we rollback only those messages which are critically important, and discard non-critical messages in case of high delays.

Figure 6 compares the number of messages re-processed per rollback as a function of their correlation probability. In case of a low probability of correlation (high No-Correlation probability on the x-axis), our Critical Correlation based approach reprocesses a smaller number of messages as compared to the Time Warp and Simple Correlation based approaches. This is because in case of lesser correlation probability, there are even lesser chances that the messages will be critically correlated and hence they are discarded without affecting the outcome of the game. However, in the case of a high correlation (low No-Correlation probability percentage), our result approaches that of the Time Warp and Simple Correlation based approaches. This happens only when almost all messages are critically correlated which is very rare in a real game scenario, where players enter and exit the critical regions and only a few of their actions are critical.

Figure 7 compares the increase in the total number of rollbacks as a function of the time elapsed during the game play.

Note that on the Y-axis, we show the number of rollbacks required as a whole i.e. when the rollback mechanism has to be applied, and not the number of

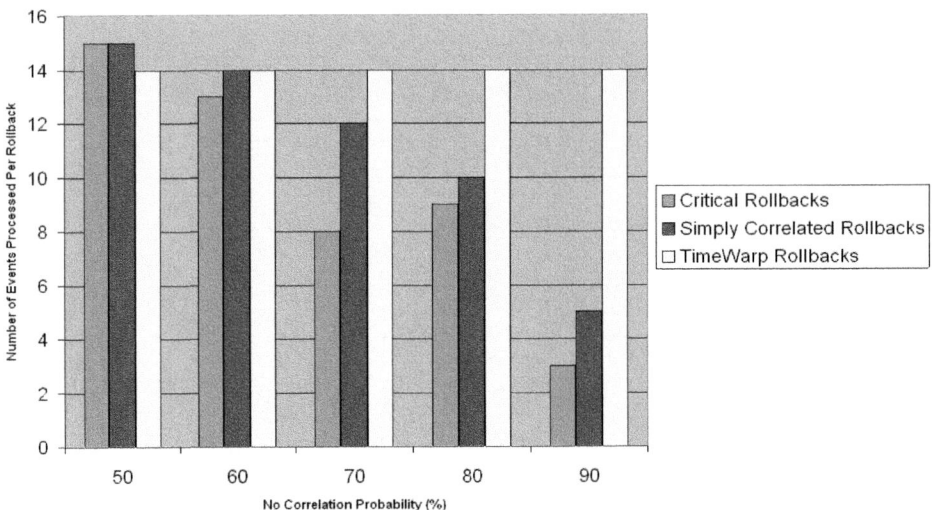

Fig. 6. Comparison of events processed per rollback

messages to be rollbacked which can be manifold higher than these numbers. For the reasons discussed in the previous paragraph, our approach performs better than simple correlation. The increase in the number of required rollbacks is linear in case of simple correlation because we periodically delay only correlated update messages at regular intervals. In the figure, for the critical correlation-based approach, the increase in the number of rollbacks is not uniform (the graph is not straight). This is because even if we delay messages periodically at regular intervals, they may not be critically correlated at that junction of time and hence are discarded without the need for applying a rollback mechanism. It must be noted here that our approach is dependent upon the strategy of the players of the game as when and where they send critically correlated messages. In the worst case scenario, when all the received messages are critically correlated, our mechanism is equal to that of the simple correlation mechanism.

9 Conclusions and Perspectives

In this paper, we introduced a dynamic and adaptive approach for consistency maintenance. We proposed a dynamic mechanism for the reduction of the number of rollbacks in multiplayer games on mobile phones. We introduced the concepts of critical actions and critical regions, and with the help of these two notions, we showed that we can relax the consistency requirements whenever the game rules allow it, thus reducing the number of rollbacks considerably. We showed, through our experimentation on a simple Football game, that our approach performs better than the Time Warp and the simple correlation approaches.

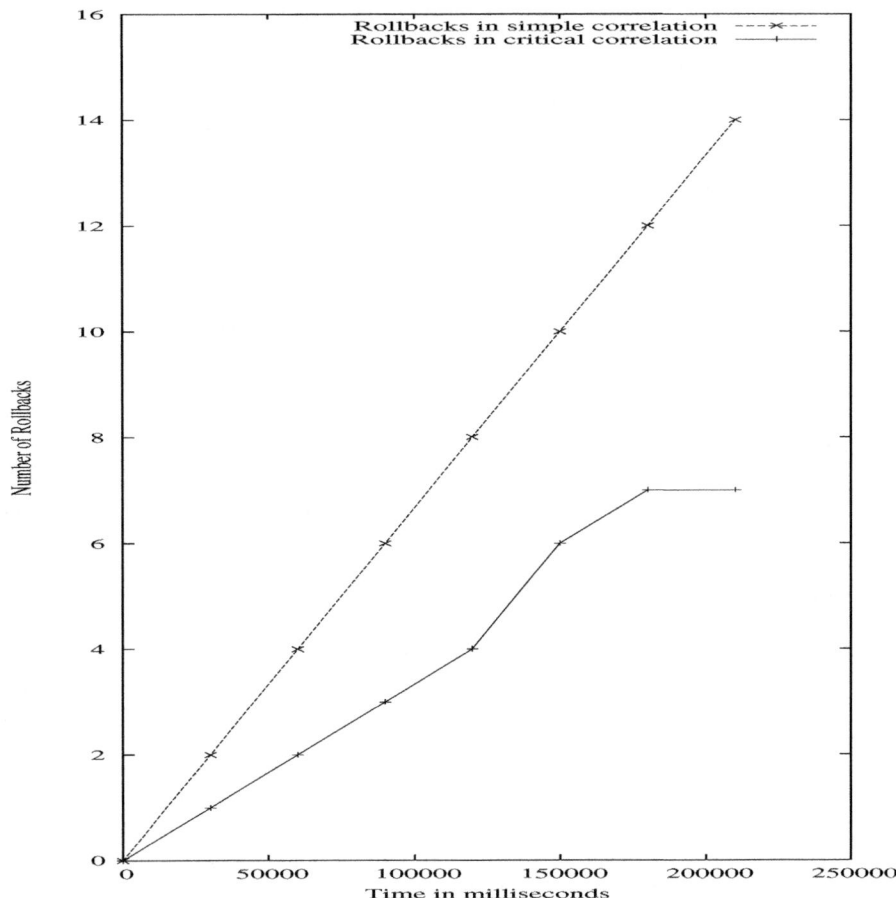

Fig. 7. Rollbacks comparison as a function of elapsed time

In the future, we would like to insert this algorithm as a part of our synchronization medium, a distributed component responsible for consistency maintenance as proposed in [8]. This way, the game developer(s) will be able to reuse these complex consistency maintenance algorithms without changing the game logic code.

Acknowledgments. This work is in parts supported by the Higher Education Commission of Pakistan (www.hec.gov.pk).

References

1. Application protocols. In: IEEE Standard for Distributed interactive Simulation. IEEE Std. 1278.1-1995 (1995)

2. Cheriton, D.R., Skeen, D.: Understanding the limitations of causally and totally ordered communication. SIGOPS Oper. Syst. Rev. 27(5), 44–57 (1993)
3. Cronin, E., Filstrup, B., Kurc, A.R., Jamin, S.: An efficient synchronization mechanism for mirrored game architectures. In: NetGames 2002: Proceedings of the 1st Workshop on Network and System Support for Games, pp. 67–73. ACM Press, New York (2002)
4. Défago, X., Schiper, A., Urbán, P.: Total order broadcast and multicast algorithms: Taxonomy and survey. ACM Computing Surveys 36(4), 372–421 (2004)
5. Ferretti, S., Roccetti, M.: A novel obsolescence-based approach to event delivery synchronization in multiplayer games. Int. J. Intell. Games & Simulation 3(1), 7–19 (2004)
6. Ferretti, S., Roccetti, M.: Fast delivery of game events with an optimistic synchronization mechanism in massive multiplayer online games. In: ACE 2005: Proceedings of the 2005 ACM SIGCHI International Conference on Advances in Computer Entertainment Technology, pp. 405–412. ACM, New York (2005)
7. Ishibashi, Y., Tasaka, S., Tachibana, Y.: Adaptive causality and media synchronization control for networked multimedia applications. In: IEEE International Conference on Communications (ICC), pp. 232–241. IEEE Computer Society (2001)
8. Khan, A.M., Chabridon, S., Beugnard, A.: Synchronization medium: a consistency maintenance component for mobile multiplayer games. In: NetGames 2007: Proceedings of the 6th ACM SIGCOMM Workshop on Network and System Support for Games, pp. 99–104. ACM, New York (2007)
9. Malik Khan, A., Chabridon, S., Beugnard, A.: A dynamic approach to consistency management for mobile multiplayer games. In: NOTERE 2008: Proceedings of the 8th International Conference on New Technologies in Distributed Systems, pp. 1–6. ACM, New York (2008)
10. Mauve, M., Vogel, J., Hilt, V., Effelsberg, W.: Local-lag and Timewarp: Providing Consistency for Replicated Continuous Applications. IEEE Transactions on Multimedia 6(1), 47–57 (2004)
11. Plotkin, G.D.: A Structural Approach to Operational Semantics. Technical Report DAIMI FN-19, University of Aarhus (1981)
12. Santos, N., Veiga, L., Ferreira, P.: Vector-Field Consistency for Ad-Hoc Gaming. In: Cerqueira, R., Campbell, R.H. (eds.) Middleware 2007. LNCS, vol. 4834, pp. 80–100. Springer, Heidelberg (2007)
13. Xiang-bin, S., Fang, L., Ling, D., Xing-hai, Z.: An event correlation synchronization algorithm for mmog. In: Eighth ACIS International Conference on Software Engineering, Artificial Intelligence, Networking, and Parallel/Distributed Computing, SNPD 2007, June 30-August 1, vol. 1, pp. 746–751 (2007)
14. Zhou, S., Shen, H.: A consistency model for highly interactive multi-player online games. In: ANSS 2007: Proceedings of the 40th Annual Simulation Symposium, pp. 318–323. IEEE Computer Society, Washington, DC (2007)

Handling the M in MANet: An Algorithm to Identify Stable Groups of Peers Using Cross-Layering Information

Hoa Dung Ha Duong and Isabelle Demeure

Institut Telecom, Telecom ParisTech, CNRS, UMR 5141,
46 rue Barrault, 75013, Paris, France
{hoa.haduong,isabelle.demeure}@telecom-paristech.fr

Abstract. This paper proposes an algorithm to identify groups of users connected to a mobile ad hoc network that remain stable over time. Several similar algorithms have been proposed to manage mobility either by predicting disconnections or by identifying groups of peers stable over time. They all rely on information such as GPS, signal strength or routes and result in message overhead. The algorithm proposed here uses information from the routing layer to detect groups that are stable over time. This algorithm is fully distributed and creates no message overhead as the result of using cross-layer information.

Keywords: MANet, Group mobility, Cross-layering.

1 Introduction

A MANet, or Mobile Ad Hoc Network, is a network established spontaneously between mobile terminals with wireless capacities, that do not require preexisting network infrastructures [5]. It therefore allows people in geographical proximity to collaborate without paying for network infrastructures, or when infrastructures are absent or down.

In this paper, we present an algorithm to build groups of peers stable over time in order to enable collaboration over a MANet. The proposed algorithm uses cross-layering information acquired from the routing layer, in our case OLSR [4]. It does not require specific equipment (such as GPS device), and does not generate traffic overhead.

This algorithm can be useful as part of CSCW[1] applications for MANet. When collaborative applications users go out of their usual working environment, such as researchers meeting at a conference, or kids on a field trip, they want to be able to work as they usually would. and keep sharing files, doing collaborative edition, editing wiki, etc. This is made easy if they use MANets because they do not require any preexisting network infrastructure.

However, when using MANets, mobility of terminals may cause network partition, when the nodes initially connected split in connected groups isolated from

---
[1] Computer Supported Cooperative Work.

each other. This creates new issues for collaborative users, and especially for collaborative services, such as:

- A peer P has been elected to provide a service such as hosting an index of all the documents available in the network. This way finding a document requires a unique message exchange. What will then happen when the network is partitioned?
- A user wants to read a document stored on a distant terminal T. Should it be accessed once and then discarded, in which case the user would not be able to access it if T disappears, or should it be replicated, creating consistency issues and additional network traffic?

The proposed algorithm can be coupled with other algorithms to maintain data availability and data coherence in a nomadic context: proactive data replication may then be enforced within the stable groups. More about these other algorithms is described in [7].

Note that in this proposal, we focus on the use of MANets by pedestrians gathering in groups to collaborate. This context allows us to make hypothesis about mobility and traffic. Traffic and data accesses are human generated and therefore sporadic (as opposed to sensors). Groups of users are stable over time but may split and merge (for example as researchers go from one conference room to another) and users may come and go.

The proposed stable group building algorithm therefore tries to build stability over time rather than looking for proximity between terminals, although this parameter can also be taken into account.

A key issue is that the target terminals for our algorithm are mobile devices. Such devices are often battery operated and are therefore limited in energy. The two main sources of energy consumptions in a mobile device are the screen and the wifi card. The extent of use before having to reload is therefore correlated to the network load. Hence, any algorithm intended for MANet should try to limit its communication needs. This is addressed in our proposed algorithm that does not create message overhead as a result of using cross-layer information.

Several similar algorithms have been proposed to manage mobility either by predicting disconnections or by identifying groups of peers stable in time. Such propositions will be surveyed in next section.They all rely on information such as GPS, signal strength or routes and result in message overhead. The algorithm proposed here uses information from the routing layer to detect groups that are stable over time.

The remainder of this paper is organized as follows: we first present existing solutions and their context of use. We then present our proposal, illustrate it with a simple example and discuss the choice of a few parameters. In section 4 we evaluate our algorithm; a validation protocol is presented, and several scenarios are tested.

2 Existing Solutions

Several solutions have already been proposed for dealing with mobility. They can be classified in three categories: those that predict when a partition will occur; those that detect groups of terminals moving along; and finally those that create clusters of terminals in dense networks, with no management of terminals volatility.

These solutions rely on different technologies and techniques: the first one is the use of absolute coordinates, such as those used by GPS. This is the most precise assessment but it necessitates that peers exchange information; another technique is to use relative distances, for example based on signal strength (this is less precise and allows only to evaluate distance between peers within reach); another possibility is the use of higher level information, such as the routing graph or capacities of the terminals, for example the battery level. Some solutions, such as [18] aim at creating clusters of terminals in dense networks. This is a context where a lot of peers are connected and nodes are grouped by proximity, for example to produce a hierarchical routing algorithm. While an interesting approach, this does not consider the partitioning problem and is therefore of little interest in our context.

Older solutions aimed at creating groups to achieve hierarchical routing and therefore limit network traffic. In [13], Lin proposes a simple algorithm to build 2-hop wide clusters. Peers broadcast a list of their one-hop neighbours, on one hop, and the cluster ID is set to the ID of the peer with the smallest identifier among its neighbours. The stability of the algorithm is evaluated by counting the number of nodes changing cluster in a 100 ms interval. This solution is clearly not aimed at detecting human user groups, working together over periods of several minutes. In [3], Basu proposed a distributed 2-hops clustering algorithm by computing relative mobility between terminals based on the frequency of a beacon signal. It aggregates terminals by 2 hops wide clusters, where the clusterhead has the lower mean relative mobility.

Other works try to predict partition in order to adequately replicate services. In [15], Su proposes such an algorithm, based on GPS and an absolute dating system (it comes with GPS devices). Velocity vectors are exchanged and the duration of a link is computed based on communication range, velocity and positions, which is then used to predict partitioning. In [8], Hauspie proposes to detect partitions by computing the 'robustness' of a path between two nodes. This is done by checking for redundancy in the routing graph. When the robustness decreases, the communication may be interrupted and a partition occurs. This solution, if distributed, requires a full reconstruction of the graph, and would likely cause high traffic overhead. In [6], De Rosa, proposes to predict partitions and to prevent them: a coordinator centralizes distance information between nodes, computed by using signal strength. It then predicts future nodes position, and therefore possible partitions. The coordinator asks nodes to fill the possible gaps to ensure full connectivity. While the idea of moving peers is interesting and acceptable in some context such as rescue or military operations, it does not suit our problem.

Finally, some proposals sort the nodes in mobility groups. In [16] and [17], Wang proposes to build groups based on mobility to replicate services in each group. In [16], two algorithms are proposed to build groups of peers with similar velocity. The first one, detailed in [17], is centralized, and relies on a server to collect velocity vectors and cluster them. The second algorithm is fully distributed: every node computes a mean distance to each of its neighbours, and a standard deviation over time, and uses those criteria to build groups. In [11], Huang proposes to cluster peers based on GPS information. Each terminal broadcasts to n-hops a list of positions with a timestamp. The peer with the smallest IP address among the messages received becomes a zone-master, and organizes nodes based on the position vector, by computing velocities. Since it is fully distributed, this solution generates traffic each time the clusters are updated. In [19], Zheng proposed a distributed clustering algorithm using a positioning device (GPS). Nodes exchange their positions, and α-stable clusters are constructed, where α is a probability of keeping the connectivity at t+1 between 2 peers, knowing the velocity vector and the position of each peer at t. In an α-stable cluster, every pair of peers is α-stable. While presenting good stability results, this solution does not evaluate the network overload.

The validation of [3] and [15] assumes a maximum speed of 72km/h. These solutions are therefore probably meant for vehicular networks. The proposals [15], [11], [19], and possibly [16] rely on GPS.

Our solution is closer to this third class of algorithm: we postulate that users are moving in groups and we want to create groups of terminals stable over time. Our proposal uses cross-layering information, obtained from the routing tables, and therefore creates no network overhead. We only require a proactive routing algorithm that maintains the routes, so that we always know which peers are reachable at any time. In our evaluation we use the OLSR routing algorithm [4]. Our algorithm is fully distributed and does not require the use of a GPS device. Hence, it should be compared to proposal such as [3], [16] and [8].

3 Proposal for Stable Groups

In this section we introduce our proposal. First of all we present our working hypotheses, and we propose a definition for the notion of stable group. We then present the proposed algorithm, both with pseudocode and an example. Our algorithm relies on several parameters: the refreshing rate, the stability threshold, and the number of tolerated sporadic absences. We see how these values have been chosen.

3.1 Assumptions

In our proposal, we make the following hypotheses. First of all, we assume pedestrian users, moving in groups, that can be modeled with the Reference Point Group Mobility Model (see section 4.3) [10]. We also assume that communication are symmetrical, with a maximal range of communication of 100m (the maximal range of 802.11b outdoor). Finally, the routing algorithm is proactive and we use OLSR.

3.2 Stable Group: Definition

We define a stable group as a group of peers able to continuously communicate over time. In other words, to become stable neighbours at time t, A and B must have been in contact for at least δ_p seconds. If they cannot establish contact for δ_f seconds, they stop being stable neighbours.

Therefore, we consider that peer A is able to continually communicate with peer B around time t if:

- B receives all broadcast messages sent by A since $t - \delta_p$.
- B receives all broadcast message sent by A between t. and $t + \delta_f$

If these conditions are true, B is a stable neighbour of A. Since communications are symmetrical by hypothesis, if B is a stable neighbour of A, A is a stable neighbour of B.

We define a stable group around peer P as the set G of peers where each pair of peers in G are stable neighbours of P.

3.3 OLSR

Our work is part of the Transhumance [14] research project, a middleware for MANet. As designing a routing protocol was not part of the project, we wanted a solid routing protocol implementation. We settled on OLSR because of its active community and the availability of the uniK implementation [2].

In our algorithm, we acquire information about which peers are in view by looking at OLSR routing tables.

OLSR is a proactive routing algorithm for mobile ad hoc networks. Proactive routing algorithms maintains routes by periodically sending messages to check if routes still exist, and offers low latency, to the cost of maintaining the routes. Reactive routing algorithms create routes on demand, by flooding the network. Both approaches have their strong points and drawbacks and [12] shows that for sporadic traffic, proactive algorithms are better suited. This clustering algorithm is a building block for human manned collaborative applications, thus generating sporadic traffic.

3.4 The Algorithm

Figure 1 presents the pseudocode of our algorithm.

In this algorithm we use a function called *getRoutesFromRoutingLayer*. It returns an associative array including all the nodes in view and the number of hops to reach it.

In a nutshell, our algorithm is a periodic algorithm that behaves in two phases, in this fashion:

- **Observation phase**: when a peer P comes into view, we create a counter, called *Presence Counter*, or PC, and set it to 1. Each period, we check if P is present. If it is, PC is increased (line 28) and when PC gets over a stability threshold, P rejoins the stable group (l. 34). Else, PC is decreased (l. 25), and if it reaches 0, we stop observing P (l. 23).
  To become part of a stable group, P must be present for at least (stability threshold) periods, plus the number of periods when it was absent.
- **Stability phase**: when P is in our stable group, we check at each period if P is present. If it is, PC increases, up to a second threshold (l. 28), else, PC is decreased. When PC gets under the stability threshold (l. 35), P gets out of the stable group.
  A number of consecutive sporadic absences are thus tolerated before it is withdrawn from the group.

3.5 An Example

Before explaining the way our algorithm is parameterized, we present its functioning with an example, where of stable threshold = 5 and maxAb=3. We consider the value associated to the name of a peer in the associative array *heardOf* to be its presence counter. In this figure we see the evolution of the presence counter, PC, associated to peer P by peer A over time and the consequences on the stable group.

1. at t=0, A had never heard of P.
2. from t=0 to t=2, PC increases.
3. at t=3, P is absent so PC is decreased.
4. from t=4 to t=6 PC increases ; at t=6, P becomes part of peer A's stable group. We can see that to become part of a stable group, a peer has to be more present than absent.
5. from t=6 to t=7, PC increases.
6. at t=8, P is not reachable so its PC is decreased; it stays in the stable group: a sporadic absence can be tolerated.
7. from t=9 to t= 13, PC is increased to *stableThreshold+maxAbs*, and then stabilizes.
8. at t=15, P disappears; its PC is decreased but it stays in the stable group.
9. from t=14 to t=16, PC decreases.
10. at t=18, PC goes below *stableThreshold*, and so after *maxAbs*, it is withdrawn from the stable group.

3.6 Algorithm Parameters

In this algorithm, three parameters can be adapted: The refreshing rate, represented by the PERIOD parameter (lines 1, 42) ; The length of the observation phase, represented by the stableThreshold parameter (lines 28,34) ; The number of absences tolerated in the stability phase, represented by the parameter maxAbs (line 34).

In this section we discuss their significance, and how we chose their values.

```
1   PERIOD=2
2   stableThreshold = 120
3   maxAbs = 60
4
5   class Peer :
6      def __init__(self, group, filename):
7          self.group = group
8          self.heardOf =dict()
9          self.stableNeighbourhood=[]
10
11  def updateHeardOf(self, routes) :
12     heardOfSet= set(self.heardOf.keys())
13     routesSet= set(routes.keys())
14     absent = heardOfSet-routesSet
15     stillpresent= routesSet & heardOfSet
16     newcomers = routesSet-heardOfSet
17
18     for p in newcomers :
19         self.heardOf[p]= 1
20
21     for p in absent :
22         if self.heardOf[p] == 1 :
23             del self.heardOf[p]
24         else:
25             self.heardOf[p]= self.heardOf[p] - 1
26
27     for p in stillpresent :
28         if self.heardOf[p]<stableThreshold+maxAbs :
29             self.heardOf[p]=self.heardOf[p]+1
30
31  def buildStableGroup(self):
32     self.stableNeighbourhood=[]
33     for p in self.heardOf.keys():
34         if self.heardOf[p]>stableThreshold :
35             self.stableNeighbourhood.append(p)
36
37  def buildStableGroup(self) :
38     while true:
39         routes = getRoutesFromRoutingLayer()
40         self.updateHeardOf(routes)
41         self.buildStableGroup()
42         sleep(PERIOD)
43
```

Fig. 1. An algorithm to build stable groups

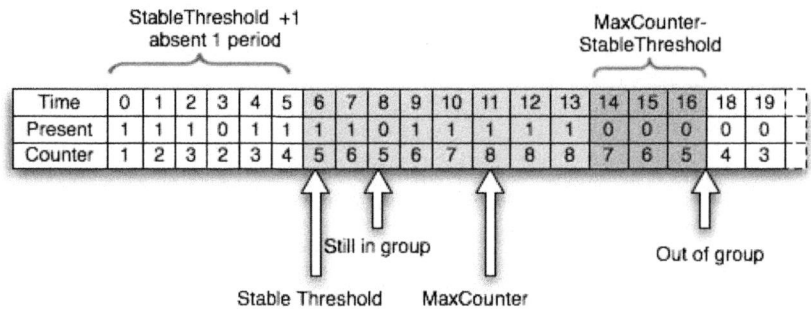

Fig. 2. An example

Stable Group Refresh Period. Our algorithm periodically builds a stable group: it tries to predict which peers will be present between the present time, and the next time it is executed, based on past experience. This must be done frequently enough that the peers in the stable groups are always in reach between two refresh.

Since HELLO messages are sent every 2s, as proposed in the OLRS RFC [4], the routing tables are refreshed every 2 seconds. Therefore, a smaller period would not yield more information than a period of 2s. Therefore we decided on PERIOD_SEC=2s, and in this section, we evaluate this choice.

Consider a peer A in our stable group at time t. We would like to bound the probability that until the next refresh of the group, the peer is actually reachable. At t+2, the peer must still be seen.

We consider a communication range between $r_{min} = 30m$, and $r_{max} = 100m$. The peer velocity is v= 1ms. If two peers are walking in opposite directions, the distance between them will increase by $2 * period * v = 4m$. Therefore, as illustrated by fig. 3, if the peer is in the grey area at t, we are certain that we will still see the peer at t+Period. Therefore, to bound the probability that the peer is still seen at t+period, we compute the probability for the peer to be in the grey area at t, knowing that it can be seen.

$$p_{see\ at\ t+2} = \frac{grey\ area}{whole\ area} = \frac{\pi * (R - 4v)^2}{\pi * R^2} = (\frac{R - 4v}{R})^2$$

For r_{max}, $p_{see\ at\ t+2} \approx 92\%$. For r_{min}, $p_{see\ at\ t+2} \approx 75\%$. For a mean value of r=65, $p_{see\ at\ t+2} \approx 88\%$.

Note that those values are lower bounds. Even if a peer is in the grey area at t, it will not automatically be out of reach at t+1 : it depends on its direction and speed.

Stability Threshold. The stability threshold parameter indicates for how many periods we expect a peer to be seen before it becomes part of our stable neighbourhood.

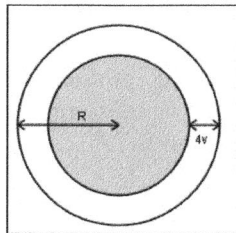 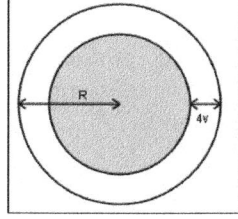

Fig. 3. Bounding the algorithm period

Fig. 4. Two groups crossing at angle a

In the group definition given in section 3.2, this is an approximation of δ_p, the number of seconds peers must continuously communicate before becoming stable neighbours. With our algorithm, two peers become stable neighbours after *(stableThreshold+the number of communication drops)* seconds. Therefore, the time between two peers making contact, and two peers becoming stable neighbours varies and is equal or superior to stableThreshold.

We want to set this threshold low enough that groups can be quickly formed, but high enough that our algorithm is able to discriminate between two groups crossing path and groups merging.

The time two groups crossing path stay in touch depends on the angle a between their trajectories, as illustrated in fig. 4. In the best case, groups come from opposite direction. In the worst case, groups cross at a very small angle and their trajectories seem aligned.

We choose this threshold so that groups crossing at right angle (90°, or $\frac{\pi}{2}$ rad) will not be mixed.

We want to compute the time span T during which two groups crossing at angle α, with a maximum radius of d (this will be explain further in section 4.3, where we present our mobility model), will stay in contact. As illustrated by figure 4, T is the time taken to walk D.

Hence, T can be computed as:

$$\frac{0.5*d}{0.5*D} = sin(0,5*\alpha) \Rightarrow D = \frac{d}{sin(0,5*\alpha)}$$

As v=1m/s, $T = \frac{d}{sin(0,5*\alpha)}$

For $\alpha = 90°$, d=200, D = 282m. Since the velocity is 1m/s, it takes 282s between the time the two groups see each other and the time the two groups split. Since the algorithm is executed every 2 seconds, we set the stableThreshold to 140. Therefore, we consider a peer as part of our group if it is seen for at least 4 minutes 40.

With the stability threshold set to 140, the algorithm can distinguish between two groups crossing at an angle $\alpha = 90°$ or less. If the angle is smaller, our algorithm will momentarily detect one group instead of two before correcting itself.

For $\alpha < 90°$, we can compute $error\_span_\alpha$, a theoretical value for the time during which our algorithm erroneously detects one group instead of two.

At $t_{deb}=0$, groups make contact and the presence counter PC starts to increase. At t_{grp}=2*stableThreshold = 280, the peers are grouped. At t_{end} =D, PC starts to decrease; at that point, we have two cases.

If $D > 2 * (stableThreshold + maxAbs)$:

- PC =stableThreshold+maxAbs
- Mistake is corrected at $t = t_{end} + 2 * maxAbs$ sec.
- The algorithm was incorrect for:
  $error\_span = t_{end} + 2 * maxAbs - t_{grp}$
  $error\_span = D + 2 * maxAbs - 280$

Else:

- $PC = \frac{D}{2}$
- Mistake corrected at $t = t_{end} + 2 * (\frac{D}{2} - stableThreshold)$
- The algorithm was incorrect for:
  $error\_span = t_{end} + 2 * (\frac{D}{2} - stableThreshold) - t_{grp}$
  $error\_span = 2 * D - 560$

Table 5 presents a few results for different angles.

Angle	80	70	60	50
Time	1m2	2m17	4m40	5m53
Angle	40	30	20	10
Time	7m44	10m52	17min11	36m14

Fig. 5. Time to correct discrepancy, depending on the angle, in degrees

Note that if α is small enough, the groups would appear to have identical trajectories to a human observer, who may think for a while that they, in fact, have merged. Our algorithm is misled for the same reason.

Number of Absences Tolerated. The $maxAbs$ parameter indicates for how many period at most we allow a peer in the stable group to be absent before it is removed from the stable group.

In the group definition given in section 3.2, this is an approximation of δ_f, the number of seconds stable neighbours should be able to communicate. When two stable neighbours lose contact, their presence counter is comprised between *stableThreshold* and *stableThreshold+maxAbs*. It is steadily decreased until it gets under *stableThreshold* and the two peers stop being stable neighbours. Therefore, the time to determine that two peers are not stable neighbours anymore after they stop communicating is less or equal to *maxAbs*.

This parameter is used in two cases :

- When a peer part of our stable group leaves, we want it to be withdrawn from the stable group as soon as possible. The same problem arises when a group splits in new groups.
- When a peer part of our stable group is absent for a small period, we want our algorithm to tolerate this absence and keep it in the stable group, so *maxAbs* should be high enough.

The number or frequency of peers leaving have no influence on the choice of *maxAbs* value, except that it should be low enough.

The absences on the other hand do. For example, if we could establish that peers are never sporadically absent for more than 30 seconds, we could set *maxAbs* to 15. The frequency of absences can be influence by different parameters:

- a temporary obstacle, such as a wall, between two peers.
- the network load creates a loss of information at OLSR level.
- the terminal freezes for a few seconds.

Since we have no control over those situations, and no way to predict those parameters, we chose a value for *maxAbs* based on the group definition: if a peer in my stable group at t is absent for more than 2 minutes, it should be withdrawn from the group. Hence, maxAbs=60.

4 Evaluation

In this section we present our algorithm evaluation.

First of all, we see how it compares to others in term of complexity. Then, we run it on a few typical scenarios, to see how it behaves. Since testing on MANets is difficult, we did this part on a simulator.

4.1 Usual Metrics : Distributed Algorithm Complexity

Since it uses information acquired from the routing tables, our algorithm do not need to exchange overhead messages to establish topological information. It is, in this respect, better than any other propositions.

Since each peer computes its neighbourhood independently, the algorithm is fully distributed and presents no centralization bottleneck. Therefore, our algorithm scales as much as the routing algorithm does.

We need another metric to evaluate the performance of our algorithm.

4.2 Proposition: An Accuracy Metric

Different groups are evolving in an area. Suppose an oracle who can set apart groups based on distance and motions, and creates ideal groups.

To computer the accuracy of our algorithm, we compare the composition of the *ideal* group, to that of the *computed* group:

$$accuracy(t) = \frac{|ideal \cap computed(t)|}{|ideal|}$$

Note that $|ideal \cap computed(t)|$ is never null, since both sets contain the peer itself.

This computes a value in $\mathbb{N}$, and an accurate algorithm keeps this metric to close to 1.

4.3 Evaluation by Simulation

MANets are inconvenient to deploy: since each device is mobile, we would need human or robot operators able to move and reproduce mobility at will. They also are inconvenient for reproductibility: wireless communications may be perturbed by external signals upon which we have no control, and two experimentations with strictly identical mobility patterns may produce different results, unless we have a vast isolated space for experimentation. Hence, we have validated our algorithm by simulation.

Modus Operandi. To validate our algorithm by simulation, we designed and used several tools:

1. We generate mobility traces compliant to the RPGM model (cf section 4.3).
2. These traces are then injected in the ns-3 simulator[9], configured with wifi 802.11b[1], and OLSR.
3. Within ns3, we dump the routing tables every 2 seconds on the disk.
4. We run our algorithm using this tables with different values for each parameter.
5. For each run of our algorithm, we compute an aggregated accuracy overtime.

Mobility Model. To generate mobility traces, we used the Reference Point Group Mobility (RPGM) Model [10].

In RPGM, nodes are organized in groups. Each group has a reference point, that is a logical centre : the reference point (RP) moves with a Random Waypoint pattern and other nodes move to stay within range of the reference point.

Since the upper bound of the communication range of IEEE 802.11b in open space is 100m, and since each peer is placed within communication range of RP, two peers will never be more than 200m apart. Therefore, the maximal radius of an RPGM group is 200m.

Node Density, Network Topology, and Influence on the Accuracy Measurement. In a small enough simulated area, groups will interact even if we try to keep them apart because of the nodes density. Therefore, if the simulated area size is less than the group area size by the number of groups, the node density is too high to distinguish all the groups.

We have to calibrate our tests so that the accuracy is not perturbed by the closeness of the groups.

To do so we need to compare two values:

- S = the surface area of the simulated area;
- NG*GS = the number of groups * the maximal area covered by a group.

The maximal area GS covered by a group depends on its topology. For example, as seen in figure 6, the group covers most ground when it is organized as a line.

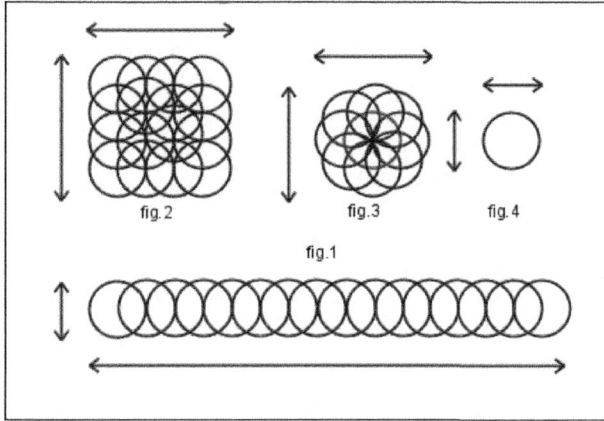

Fig. 6. Possible network configurations

In the RPGM model, peers are organized so that each peer is within reach of the reference point. This corresponds to a network configuration as in fig. 6. A group should therefore occupy a circular area of 200m of radius. In the subsequent scenario, we chose the size of the simulated area so that groups can be isolated.

4.4 Test Scenarios

In this section we examine how our algorithm behaves in characteristic situations. We see how it behave with no groups interfering, and what happens when two groups cross paths, merge, and when one group splits in two subgroups.

For the simulations presented below, the simulated area size is 2000mx2000m, and each group is made up of 10 peers. The mobility model is RPGM.

If not indicated otherwise, the *stable threshold* varies between 100 and 180. The maximum of authorized periods of unavailability varies between 20 and 100.

For each case, we present the scenario, namely which mobility pattern, and which parameters are tested, and the theoretical result. Results are presented as graphs representing the evolution of the algorithm accuracy over time.

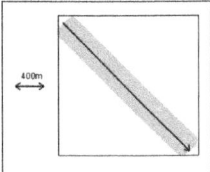
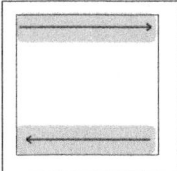

Fig. 7. 1 group   **Fig. 8.** 2 distinct groups

One Group. In this scenario, illustrated by fig.7, we want to verify the behaviour of our algorithm in a simple situation with one group of peers.

Scenario: A group of peers is walking from the upper left corner to the opposite corner. We want to test if our algorithm work when no disruption occurs, and which value of *stableThreshold* maximizes, in that case, the accuracy; *stableThreshold* varies between 100 and 180.

Expected result: We expect the computed group not to be accurate up to *stableThreshold* seconds, to stay accurate after. The lowest value of *stableThreshold* should maximize the accuracy.

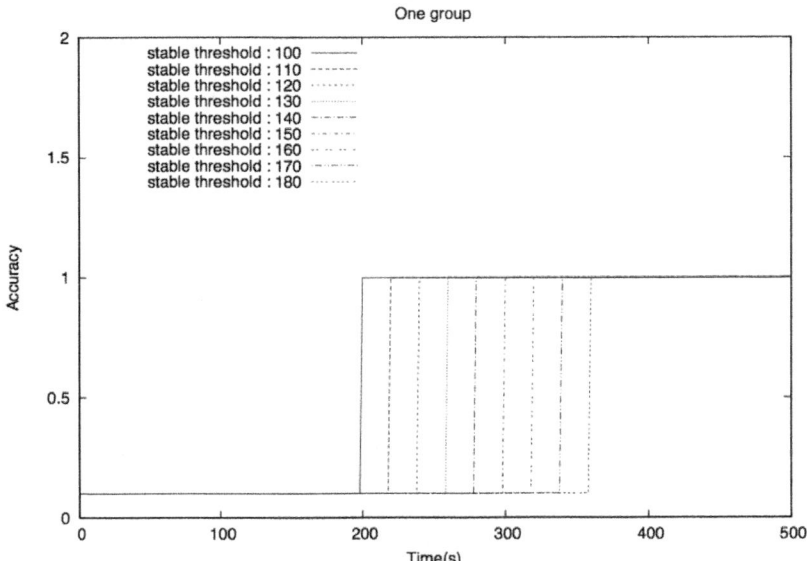

Fig. 9. One group, $\neq$ stable threshold

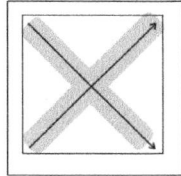
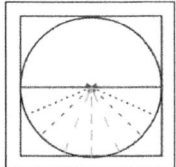

Fig. 10. Paths crossing

Fig. 11. Same distance, different angles

Observation: Graph 9 represents evolution of the accuracy over time. Each curve represents the accuracy for a given value of *stableThreshold*. As we can see, the simulation validates our expected results.

Two Groups with No Interaction. In this scenario, illustrated by fig.8, we want to verify that having two groups with no interaction does not alter the computation accuracy.

Scenario: Two groups follow opposite borders of the area. They never come in contact, and we want to verify that the algorithm accuracy behaves as it does for one group. The *stable threshold* varies between 100 and 180.

Expected result: As groups are not interacting, we expect the same curve as in the previous experiment.

Observation: Results are similar to fig. 9.

Two Groups, Crossing Path. In this scenario, illustrated by figure 10, we examine the behaviour of our algorithm when two distinct groups are crossing paths.

Scenario 1: Two groups start out of reach of each other, at upper left, and bottom left corners; both groups walk diagonally to reach the opposite corner of the area; at t= 1415 seconds, they cross path at right angle, at (0,0). Around t=1415, we want to verify if for the chosen value of *stableThreshold*, our algorithm discriminates between the two groups. The *stable threshold* varies between 100 and 180.

Expected result 1: We examine the behaviour of our algorithm around t= 1415. For a *stable threshold* under 140, no errors should be detected; as the stable threshold decreases, the results should worsen.

Observation 1: Graph 12 show that results are better than expected : the two groups stay distinct for a stable threshold under 120.

Scenario 2: Two groups start out of reach from each other and at t= 1415 sec, they cross path at (0,0); the angle α formed by their trajectories varies between 0 rad and π rad; their starting positions are chosen as illustrated by fig.11 so that all the groups cross paths at time t=1000. We want to verifie that the algorithm discriminates between groups crossing at angle larger than or equal to $\frac{\pi}{2}$ rad. The *stable threshold* is fixed at 140.

Expected result 2: If two groups are walking with a 0 rad angle, they effectively behave as one group, so in this case the accuracy should be 2 from

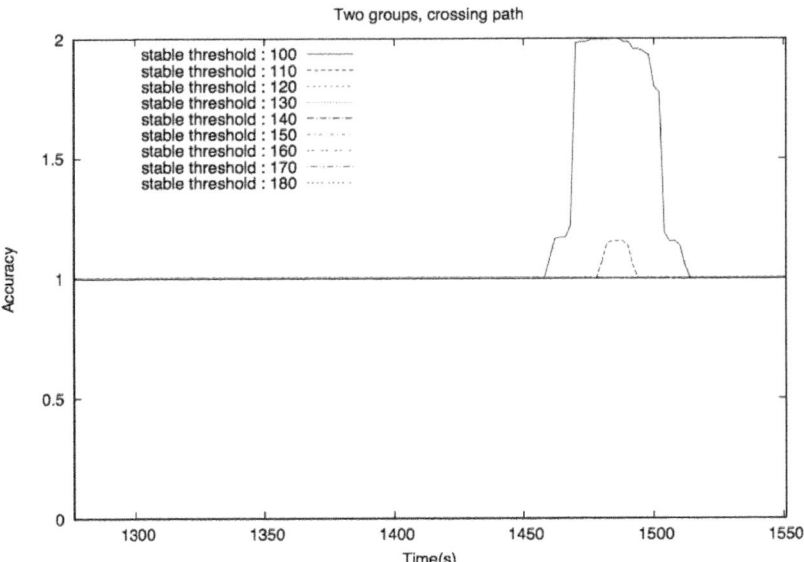

Fig. 12. 2 groups crossing, $\neq$ stable thresholds

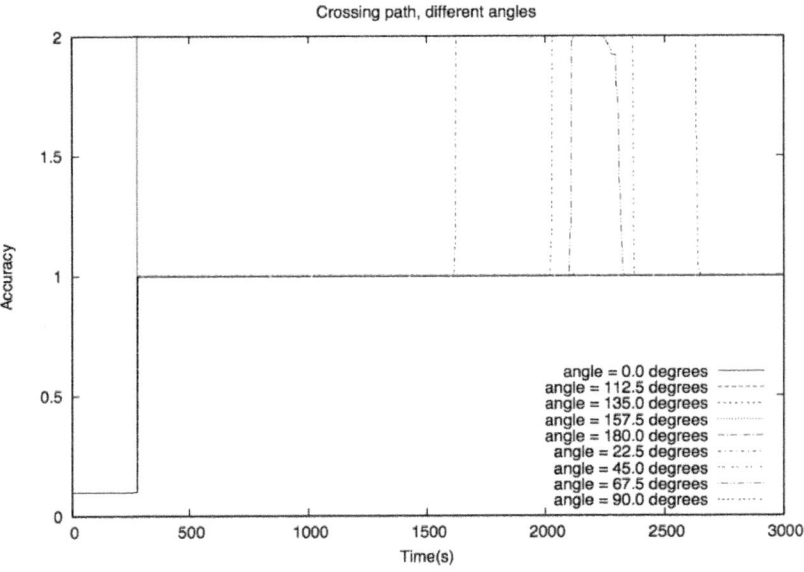

Fig. 13. 2 groups crossing, $\neq$ angles

t=1415 sec and on. For other values of α, the errors should decrease as α grows, and for $\alpha > \frac{\pi}{2}$, no error should be detected.

Observations 2: In figure 4.4, each curve represents the evolution of the accuracy for a given angle. We can see that our prediction for $\alpha = 0$ and $\alpha > \frac{\pi}{2}$ is correct. The correctness of the algorithm is also increasing as α grows. We can also see that the recovery time for $\alpha < 90°$ is lower than expected. For example for $\alpha = 45°$, the error is 4m40, lower than the theoretical time of recovery for $\alpha = 50°$, 5m53, as computed in 3.6.

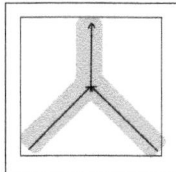
Fig. 14. Merging

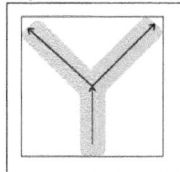
Fig. 15. Split

Two Groups, Fusion. In this scenario, illustrated by figure 14 we measure how our algorithm behaves when two groups merge.

Scenario: Two groups start at the two bottom corners of the area; they both walk at right angle to the middle of the area, taking 1415 seconds; the groups merge and walk straight to the top. We want to verify that the algorithm detects the merging, and the time it takes depending on the *stability threshold*. The *stable threshold* varies between 100 and 180.

Expected result: We examine the behaviour of the grouping algorithm around t=566. The lesser the *stable threshold* is, the faster groups are formed and the better the accuracy.

Observation: As expected, we can see on figure 16, that the algorithm recovers better with a lower *stable threshold*. Note that as trajectories are perpendicular, the peers starts communicating 142 sec before the merge. With a *stable threshold* of 70, no error would occur.

This experiment is also a generalization of the case of a peer joining a group and therefore validates it.

One Group, Splitting. In this scenario, illustrated by figure 4.4, we measure how our algorithm detects a group splitting.

Scenario: The group starts at (0,1000) and walks to the center, taking 1000 sec; there, it splits in two; one half walks to the upper left corner while the other one walks to the upper right corner. We want to verify that the algorithm detects the split, and the time it takes depending on the maximum absences tolerated. The *stable threshold* is fixed to 140 and *maxAbs* varies between 20 and 90.

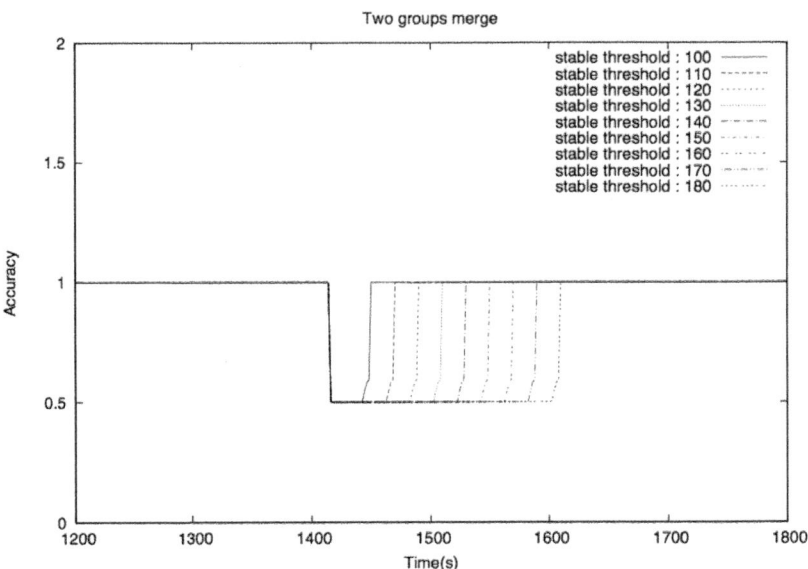

Fig. 16. Two group merge, $\neq$ stable thresholds

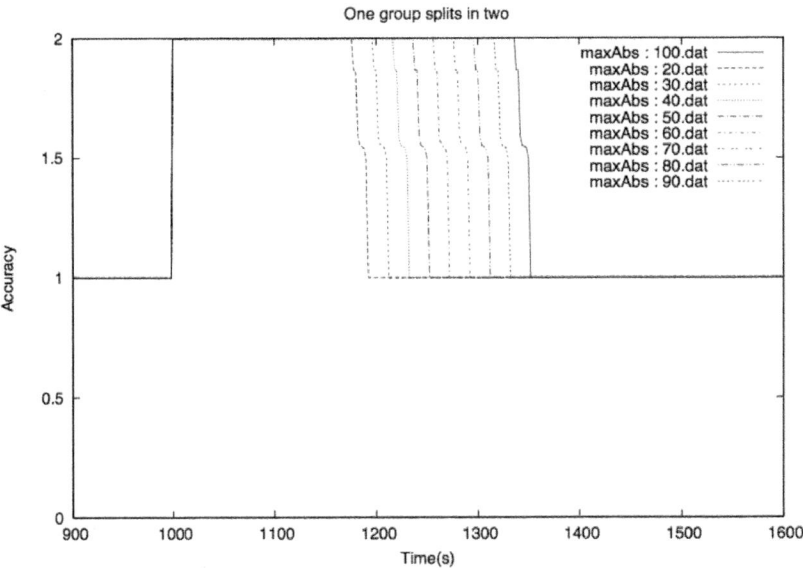

Fig. 17. One group splits, $\neq$ maximum absence

Expected result: We examine the behaviour of the group algorithm around t = 2000. In this scenario, the algorithm cannot always be accurate, even with $maxAbs$=0, since when the groups split at t=1000, they can still communicate for 224 sec. Still, we expect that the lower $maxAbs$ is, the quicker our algorithm will correct the groups.

Observations: In figure 17, each curve represents the evolution of the accuracy over time for a given value of $maxAbs$. It validates the expected result.

Note that this experiment is a generalization for the case of a peer leaving a group and validates it.

Overall Observations. Experiments 4.4, 4.4 and 4.4 show that the lower the *stability threshold* is, the quicker a stable group can be formed, while experiment 4.4 shows that if the *stability threshold* is too low, the algorithm cannot differentiate between a group, and two groups crossing path.

Experiment 4.4 shows that the lowest $maxAbs$ allows a better detection of group splitting. However, while the experiment is not shown here due to the lack of space, $maxAbs$ allows for tolerating transient faults and therefore should not be 0.

Overall, these experiments validate the expected behaviour of our algorithm, and in the case of two groups crossing, shows even better results than expected.

5 Conclusion

In this paper we proposed an algorithm to build groups stable over time. This algorithm relies on cross-layering information, namely routing information maintained by a proactive routing algorithm, to establish which other peers are reachable for long enough to be considered stable neighbours.

The two strong points of this algorithm, compared to existing proposal, are the lack of need for a positioning equipment, such as GPS, and the lack of network overhead. Both are sources of energy use; energy is a critical resource in a MANet context where terminals are mostly handheld and battery-operated.

We presented an evaluation of our algorithm, with a theoretical model and by simulation.

In term of distributed complexity, our algorithm is better than any existing proposal as it does not create network overload. We also validated our theoretical model by simulation.

Further works would be to implement existing proposals to test them again our accuracy metric, in order to provide further comparison.

References

1. http://standards.ieee.org/getieee802/802.11.html
2. http://www.olsr.org/
3. Basu, P., Khan, N., Little, T.D.: A mobility based metric for clustering in mobile ad hoc networks. In: International Workshop on Wireless Networks and Mobile Computing (WNMC 2001), pp. 413–418 (2001)

4. Clausen, T., Jacquet, P.: Optimized Link State Routing Protocol (OLSR). RFC 3626 (Experimental) (October 2003)
5. Corson, S., Macker, J.: Mobile Ad hoc Networking (MANET): Routing Protocol Performance Issues and Evaluation Considerations. RFC 2501 (Informational) (January 1999)
6. De Rosa, F., Malizia, A., Mecella, M.: Disconnection prediction in mobile ad hoc networks for supporting cooperative work. IEEE Pervasive Computing 4(3), 62–70 (2005)
7. Duong, H.H., Demeure, I.M.: Proactive Data Replication Using Semantic Information within Mobility Groups in MANET. In: Bonnin, J.-M., Giannelli, C., Magedanz, T. (eds.) Mobilware 2009. LNICST, vol. 7, pp. 129–143. Springer, Heidelberg (2009)
8. Hauspie, M., Simplot, D., Carle, J.: Partition detection in mobile ad hoc networks (May 20, 2003)
9. Henderson, T.R., Roy, S.: ns-3 project goals
10. Hong, X., Gerla, M., Pei, G., Chiang, C.-C.: A group mobility model for ad hoc wireless networks. In: MSWiM 1999: Proceedings of the 2nd ACM International Workshop on Modeling, Analysis and Simulation of Wireless and Mobile Systems, pp. 53–60. ACM, New York (1999)
11. Huang, J.-L., Chen, M.-S., Peng, W.-C.: Exploring group mobility for replica data allocation in a mobile environment. In: CIKM 2003: Proceedings of the Twelfth International Conference on Information and Knowledge Management, pp. 161–168. ACM, New York (2003)
12. Huhtonen, A.: Comparing AODV and OLSR routing protocols (May 27, 2004)
13. Lin, C.R., Gerla, M.: Adaptive clustering for mobile wireless networks. IEEE Journal on Selected Areas in Communications 15, 1265–1275 (1997)
14. Paroux, G., Demeure, l., Reynaud, L.: A power-aware middleware for mobile adhoc networks. In: NOTERE 2008: Proceedings of the 8th International Conference on New Technologies in Distributed Systems, pp. 1–7. ACM, New York (2008)
15. Su, W., Lee, S.-J., Gerla, M.: Mobility prediction and routing in ad hoc wireless networks. Int. J. Netw. Manag. 11(1), 3–30 (2001)
16. Wang, K., Li, B.: Efficient and guaranteed service coverage in partitionable mobile ad-hoc networks. In: Proceedings of the 21st Annual Joint Conference of the IEEE Computer and Communications Society (INFOCOM 2002), Proceedings IEEE INFOCOM 2002, June 23-27, vol. 2, pp. 1089–1098. IEEE Computer Society, Piscataway (2002)
17. Wang, K.H., Li, B.: Group mobility and partition prediction in wireless ad-hoc networks (May 29, 2002)
18. Yu, H., Hassanein, H., Martin, P.: Cluster-based replication for large-scale mobile ad-hoc networks. In: International Conference on Wireless Networks, Communications in Computing, pp. 552–557 (June 2005)
19. Zheng, J., Su, J., Lu, X.: A Clustering-Based Data Replication Algorithm in Mobile Ad Hoc Networks for Improving Data Availability. In: Cao, J., Yang, L.T., Guo, M., Lau, F. (eds.) ISPA 2004. LNCS, vol. 3358, pp. 399–409. Springer, Heidelberg (2004)

Small World VoIP

Xiaohui Yang[1], Angelos Stavrou[1], Ram Dantu[2], and Duminda Wijesekera[1]

[1] George Mason University, Fairfax VA 22030, USA
{xyang3,astavrou,dwijesek}@gmu.edu
[2] University of North Texas, Denton TX 76203, USA
rdantu@unt.edu

Abstract. We present the analysis and design of a *Small World VoIP system (SW-VoIP)* which is geared towards customers that are communicating with their *Small World* of social contacts. We use the term *Small World* to refer to the Peer-to-Peer (P2P) network of a client and his contacts both incoming and outbound. We reconstruct the *small world* of a user by collecting calling patterns over a configurable period of time. We enable user mobility by using a stepwise *social identity* to an IP address binding propagation model. We propose an efficient algorithm to locate users by electing popular users and leveraging the users closeness. We also introduce a self-stabilized load balancing mechanism to optimize the system performance under heavy network traffic. We evaluate our SW-VoIP system performance by simulating the user's lookup process using real-world telephone logs. Our experimental results show that our SW-VoIP system offers a better performance in optimizing the required routing path and reducing the average lookup delay when compared to traditional, non small-world P2P VoIP systems.

Keywords: small world, VoIP, Peer-to-peer, mobility, electing, popular, closeness, optimize.

1 Introduction

In popular P2P VoIP systems, to make a call, the caller's agent has to first query for the network location information of the called party using the P2P routing. Only after this query is completed, the caller can continue the launch of the SMS, voice, or video stream either using P2P routing directly or other protocols to transmit content. In contrast to traditional client-server VoIP model, the P2P VoIP systems are designed to provide a reliable and cost efficient telecommunication environment, that avoid heavy infrastructure investments, centralized bottlenecks and single points of failures. Unfortunately, due to P2P's decentralized nature of routing and the lack of explicit or implicit trust among peers, P2P VoIP systems are vulnerable to attacks such as Spam, Sybil, Phishing, to name a few. A malicious user with a spurious or unconfirmed network identity can mis-route, eavesdrop, or even resume the identity of a trusted party, leading to call dropping, wiretapping, or unsolicited calls. To make matters worse, even normal user behavior that involves continuous joining and leaving of users and their changing of physical locations lead to higher maintenance cost and lead to communication degradation. Beyond the nuisance that can cause, all of the aforementioned problems can inhibit the widespread adoption of VoIP systems.

To alleviate some these problems, we harness the social nature of the telephone service to design a system that uses social information to optimize the quality of the underlying P2P VoIP network. To that end, we observe that user telephone call graphs exhibit characteristics of a *small-world network* [24]. In such networks most nodes, in our case users, can be reached from every other node by traversing a small number of hops. Inspired by this scale-free property of the telephone call graph, we propose to develop a *small world* call graph based VoIP system. Therefore, we leverage the structural properties of the small world network of the telephone call graph for each user to efficiently route call connection requests, we call our approach SW-VoIP. In SW-VoIP, we cluster users that communicate frequently together. Furthermore, we designate the users that are called by many others as *leaders* for locating users, creating a *hub-and-spoke* type of network. We derive this telephone call graph information from real-world call logs and we developed a prototype system to evaluate the performance benefits of SW-VoIP.

For our performance evaluation, we employed a testbed of 1000+ nodes on 300+ PlanetLab [2] machines using OpenVoIP [1]. We measured our algorithms by comparing user lookup performance on SW-VoIP to OpenVoIP using Kademlia [11]. Based on the call log data set, we explored the performance test on 97 nodes of the SW-VoIP and the OpenVoIP system. We measured the average hop count and system performance by varying the number of and size of user clusters. Our experimental results show that our proposed small world P2P VoIP system offers reduced average routing hop count and minimized average user lookup delays when compared to normal P2P VoIP system.

The rest of this paper is structured as follows: section 2 details the related work. Section 3 explains our design goals and choices. Section 4 describes how we construct the small world overlay using call graphs. Section 5 provides details on the workings of our routing algorithm. Section 6 illustrates our experimental setup and evaluation. We conclude with Section 7.

2 Background and the Related Works

2.1 P2P VoIP System

P2P VoIP traffic differs from P2P file-sharing network traffic and from traditional telecommunication network due to its interactive nature and characteristic resource consumption. Currently, there are several proposals for P2P VoIP system design, and some of them achieved pervasive commercial deployment.

As the most popular commercial P2P VoIP application for Internet Telephony, Skype [3] is deployed to offer free computer-to-computer calls or paid telecommunication services through VoIP-to-PSTN gateways for distributed network users. Skype uses centralized authentication server to establish initial identities, and offers a super node mechanism to manage contacts, relay data flow when necessary, and enable NAT traversal. Skype gives VoIP users advantages by lowering operating costs and facilitating worldwide calls. But deficiencies exist due to the proprietary nature of protocol and the inherent vulnerabilities of P2P system. As a result, lots of general telecommunication Use Cases are unavailable in Skype, and difficulties exist in adding new features as third party solutions to this P2P VoIP system.

On the other hand, work was in progress on developing open standards based P2P VoIP systems for Internet telecommunication. In particular, Session Initiation Protocol (SIP) [17] based P2P VoIP have emerged as a mainstream in system design. In P2P-SIP system, the selection of P2P algorithms and the overlay organizing mode are the two main factors affecting system design. Most current proposals prefer using structured P2P overlay such as Distributed Hash Tables (DHTs) to identify peers and store resources. DHT routing algorithms such as CAN [15], Chord [23], etc, are chosen according to network and user requirements such as tradeoff efficiencies, failure resilience, routing hot spots, node heterogeneity, etc. In addition to intrinsic P2P properties, methods used to combine P2P and SIP protocols are crucial to designing an efficient P2P VoIP system.

Most of current P2P VoIP system design propose to transmit signaling in P2P mode whereas direct media flows end-to-end. For example, the earlier SIPpeer project [19] and SOSIMPLE project [5] use SIP messages to create DHT overlay and convey routing information. Meanwhile, an alternative research [20] at Columbia University uses an external DHT to establish communication sessions among users. Additionally, a few companies claim to offer their own P2P systems for VoIP application, such as Avaya's one-x and Popular Telephony's Peerio.

As commercial P2P VoIP deployments began to merge, the Internet Engineering Task Force (IETF) showed great interest in P2P signaling protocol development. Recently, the REsource Location And Discovery (RELOAD) base protocol [8], a peer-to-peer signaling protocol was proposed to provide clients an abstract storage and messaging service between a set of cooperating peers in the overlay network. RELOAD supports a P2P-SIP network and functions in environments where many nodes are behind NATs or firewalls. The lightweight load on participating peers ensures a high performance system routing.

2.2 Small World Utilization

Watts and Strogatz [24] characterized a small world network as a collection of loosely connected subgraphs with a high clustering coefficient and a low average shortest path. In small world networks, short chains of acquaintances link arbitrary pairs of strangers, connects any user to their *local* and *remote* contacts using a small number of long edge and a large number of short edges, where the length refers to the communication overhead. Additionally, a hub node with a high number of connections serve as a trusted common connection to mediate the short path length between edges of interconnected nodes.

Many researches have investigated properties of small world networks to improve network performance and solve application level complexities. One typical utilization of social network, as shown by Marti [10] is to improve routing performance and to use social links as *paths of trust*. The algorithm proposed by Marti adds a 2-level lookahead *friends links* to Chord [22], and uses the minimum hop distance to efficiently integrate a social network to with a P2P network. Although SPROUT proved to be robust under a large fraction of malicious users, it neglect the differences between a users list of friends and uses a greedy routing algorithm to work with the underlying DHT algorithm without taking trust into consideration during routing.

Hui [6] developed the SWOP protocol, a representative application of small world properties to construct and maintain an structured P2P overlay network. Taking the advantages of high clustering coefficient and low average hop distance, the small world overlay P2P network in SWOP can efficiently lookup resource objects and perform well under heavy traffic. SWOP introduces the concept of head nodes and long links from small world network, but security may be compromised due to choosing inappropriate head nodes during cluster formation.

In order to achieve the shortest routing path in a small world overlay, nodes can be grouped using many criteria. For example, clusters can be constructed using nodes' hashed key value distance to other nodes. Iamnitchi [7] proposes an algorithm to dynamically identify groups with similar interests by using information about data consumed by P2P users. Other criteria such as transmission delay, common interests, etc, are also the popular ways of grouping in current small world P2P networks.

3 Problem Setting and Design Goals

A P2P network to be useful for VoIP has to satisfy some extra conditions in order to accommodate operational realities of telecommunications. Firstly, VoIP customers frequently change their physical location, resulting in changing IP addresses but retaining a fixed telephone number - requiring updating the IP address attributes of the small world social graph. Secondly, telephony requires some centralized control in order to provide services such as E-911 [16], Lawful Interception (CALEA) [18] etc. Thirdly, telephone customers use many customized service packages, and in face of user mobility, these may raise difficulties. Fourthly, telephony communications are delay sensitive - an unwise routing decision or an unexpected node congestion may result in failed calls or inexecutable services. Based on the properties of a small world network and general telephone user's calling behavior, we provide the following three mechanisms as a solution:

- *Optimizing Update:* User mobility makes social links and stored object references obsolete, making many extra transmissions necessary, resulting in call setup delays, and perhaps missing calls. As a solution, we propose continuous IP to telephone number binding updates in a step-wise manner. We propose this mechanism for normal system operations rather than to be used in emergency situations requiring fast connections.

– *Distributed Control:* Administration, services deployment and membership management in telecommunications requires centralized functionality. We propose to distribute these functions to users with high social connections and credibility. These users will play some pivotal roles in controlling message flow (e.g. prioritizing E-911 signal or choosing call dissemination targets) and act as regional auditing center (e.g. checking points for CALEA). They may use different routing policies to send connection requests between friends or unfamiliar users. Messages may also be filtered or differentiated at each distributed center in order to be processed under different services.

– *Delay Protection:* We group users based on the strength of their social relationships and locate called parties using popularly accessed nodes, thereby enhancing message propagation. To avoid single point congestion and reduce message delays, system is optimized to encourage the second best routing choices.

4 Small World Overlay Construction

Telephone call graphs correctly reflect users' calling habits that can be used to measure the *closeness* of their calling behavior, which can be leveraged to enforce identity assurance, reduce call delay, and improve network performance in P2P VoIP systems. The first step is to obtain and update the call graph in order to create the P2P network.

4.1 Distributed PCrawler Design

To acquire current P2P VoIP call graph, we intermittently collect users' call logs, and extract related information such as user ID, number of friends, call frequency, call duration, etc. Because collecting call logs from a large number of peers at one time will lead to a low network performance, using multiple queries and parallel responses are infeasible. Using the idea of web crawlers [9], we introduce a *Peer Crawler (PCrawler)* that crawls a subset of P2P VoIP network to acquire up-to-date copies of call logs. The collected copies of call logs will be downloaded into the cache of the dispatcher with an expiration timer. Meanwhile, important identity information will be extracted and indexed by call log analyzer to provide and validate user identification and accelerate preliminary statistical evaluation.

In addition, the PCrawler automatically maintains the small world overlay. As a user logs in and actively participates in P2P VoIP communication, his/her identity information is stored in a tuple <*SID, IP, Port*> and also harvested by his/her friends using social links. When the user changes the location, his/her telephone number to IP binding will be changed and the PCrawler will find the new binding.

The PCrawler architecture is shown in Figure 1. PCrawler is composed of two components: a call log collector and a consistency checker. PCrawlers are generated by a *Dispatcher*, that is in charge of PCrawlers authentication, call log analysis, and small world network ID generation for the whole P2P VoIP system. PCrawlers are disseminated in each cluster, and the number of PCrawlers is decided by the Dispatcher depending on the size of network. Dispatcher acquires the size of network from enrollment server according to the number of users enrolled. Starting with an list of unexplored peers assigned by the Dispatcher, PCrawlers collect call logs, pick a list of unexplored friends from peers, and crawl those friend's entries recursively while extracting all useful information from the collected call logs and reporting them to the Dispatcher.

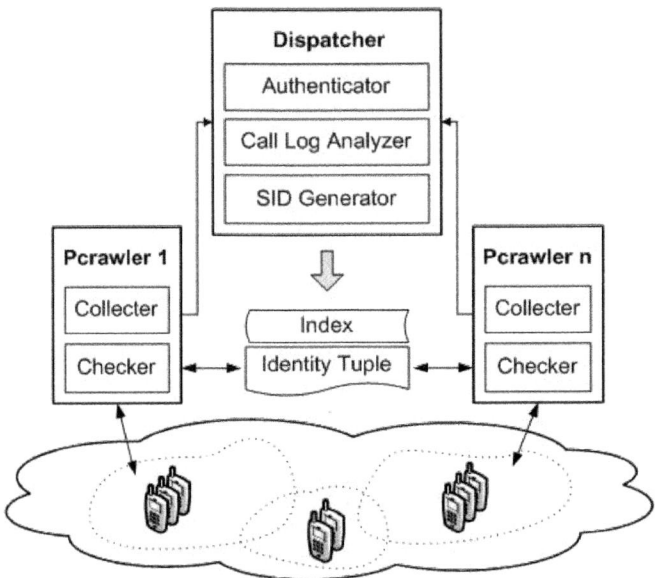

Fig. 1. PCrawler Architecture

Dispatchers are responsible for authenticating PCrawlers before they start a session with P2P peers and request for call logs. Authentication is enforced in order to avoid illegal utilization of PCrawlers to launch malicious attacks. On receiving calling data from PCrawlers, a *Call Log Analyzer* clusters users and generate a small world network ID for each user based on some rules, which are described shortly. *SID* in the identity tuple will thereupon be filled up in the order of call log index.

We implement the Dispatcher in C++ using *Tulip* [4] libraries to construct or update call graphs. We adopt a clustering technique provided by Tulip based on the calculation of the strength of the edges of a small world graph $G = (V, E)$, with a set of vertices V edges E. The PCrawler uses a map $\phi : E \to R$ assigning a real number $\phi(e) \geq 0$ to each edge $e \in E$, and call graphs can be clustered based on removing the edge with the metric ϕ lower than threshold value of $t \in [a, b]$.

In addition to collecting call logs, PCrawlers periodically check the status of users with the help of a Breadth-first search algorithm. If the user's *SID* does not match the *SID* stored in the indexed cache, or if the user's address data is different than that of the identity tuple under the same *SID*, the PCrawler will trigger an update program in the peers to replace the old tuple with the new one. If a user has changed its locality for a sufficient time or have more contact with a different group of users (as measured by the function ϕ on the edges), can be assigned a new *SID* belonging to a new cluster.

4.2 Utilizing Structural Properties of Call Graph

We maintain and leverage the *closeness* of callers in the call graph cluster to enrich cluster our P2P VoIP system. The variable number of clusters m is determined by the number of user n and the closeness of the whole small world network. In each cluster, users are tightly connected, and able to reach others by a small number of hops, which we call *intra-cluster* communications. Clusters are loosely connected, and the users who have connections to users in other clusters provide links for *inter-cluster* communications. Nodes providing interconnections between clusters are called *hub nodes*. The i-th cluster has h_i hubs, where $1 \leq h_i < n$, and every *normal user* in a cluster is directly connected to at least one hub, as shown in Figure 2. These hubs are endowed with three important properties: (1) highly trusted in the whole small world networks, (2) bridge clusters and provide routing path for inter-cluster communications, (3) offers *check points* for the centralized control, making them supervise inter-cluster activities.

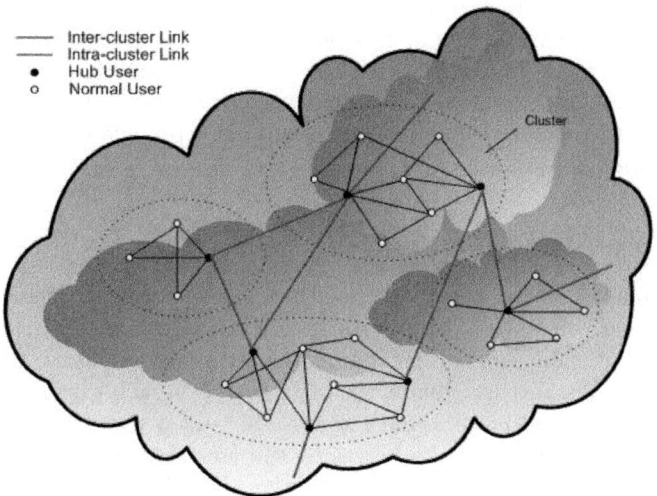

Fig. 2. Clustered Small World P2P VoIP Network

Electing a hub in each cluster relies on many factors such as node degree d, number of inter-cluster links l, calling frequency f, Round Trip Time rtt, etc. We measure a user's popularity based on the assumption that the more friends a user has and the more calls a user made/received over a period of time, the more popular the user is. Popularity p_i is proportional to the average calling frequency $ave(F) = \frac{1}{d}\sum_{k=1}^{d} f_k$ and the contact degree $d^{l/d}$, where the ratio of l to d is calculated to weigh up the importance of a hub for inter-cluster communication. Meanwhile, telecommunications low latency property requires a low average rtt of a hub to its neighborhoods, which is calculated by $ave(RTT) = \frac{1}{d}\sum_{j=1}^{d} rtt_j$ and is reversely proportional to p_i. Given a cluster with y users, hub user election is defined by comparing the popularity p of each user i based on the parameters of d, l, f, and rtt.

$$p_i = \frac{d^{l/d} * ave(F)}{ave(RTT)} \quad (i = 1, 2, \ldots, y) \tag{1}$$

The user with the highest popularity p_{max} will announce its parameter values to all the other users in the same cluster for the normalized popularity $\boldsymbol{p_i}$ calculation.

$$\boldsymbol{p_i} = \frac{p_i}{p_{max}} \quad (i = 1, 2, \ldots, y) \tag{2}$$

Each user will select from its friends list and put forward a hub with $Max(\boldsymbol{p_i})$. As a result, each user is connected to at least on hub which is regarded as the most popular user in its neighborhoods.

5 Algorithms

The incorporation of structural properties [12] of small world network makes user oriented P2P VoIP system more efficiently organized and routing path more intelligently designed. Equipped with the functionalities of quick locating and dynamic updating, our intimacy oriented P2P VoIP system is applicable to any structured or unstructured P2P overlays, and also compatible with any normal P2P system for telecommunications.

5.1 Closeness-Based Routing

We use a numeric *Social Identifier (SID)* chosen from $[0, 2^{160})$ for each social user in the small world call graph. We divide the SID space by the number of social clusters m, and users in the same cluster share the same x-digit prefix, where $m = 2^x$. We further divide the SID space of each cluster according to the number of hubs and users and express the *entities* in the SID space as a binary tree, where users are the leaves of the tree, and the height of the tree is determined by the number of digits in the identifier. We illustrate the SID space binary tree of the above small world network in Figure 3. Thus clusters are differentiated by the identifier sequence along the binary tree, where length of the sequence is determined by x. For clusters with only one hub user, we assign the lowest SID in the cluster to the hub, and assign the subsequent $SIDs$ to normal users. For multi-hub cluster, SID space is divided by the bundles of hub and normal users, and the bundles with higher popularity are assigned lower $SIDs$.

In our setup, a cluster Hub is in charge of collecting and disseminating information and is considered a highly credible routing pivot. In order for a caller's agent to generate a call, it needs to get the address information of the callee. As in any DHT system, all the users information is stored in the small world overlay in the form of $< key, value >$ pairs. We use a hash of the mapping $< IP, Port >$ as the key, and use the same procedure to locate an user object as to locate a closest user to the key.

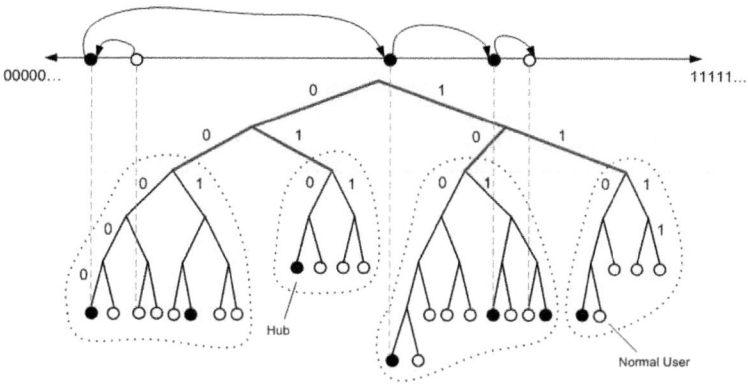

Fig. 3. Routing on the Clustered Small World Network

We illustrate the user lookup process in Figure 3 when $user\_a$ calls $user\_b$ in our structured small world network in Algorithm 1.

The intra-cluster routing process contacts only $O(\log \frac{n}{m})$ users, where n is the total number of user and m is the number of clusters. Whereas the overall routing process has time complexity of $O(m + \log \frac{n}{m})$, where the value of m is the guarantee for efficient routing.

In addition, hubs act as virtual authorities and supervises activities in and between clusters. Their high popularity makes a hub highly trusted by its connected normal users, and gives a hub particular rights for supervising and auditing. By analyzing telephone call graph, each user will be given an initial credibility based on the node degree d and the call frequency f. Each user can then choose its hub from the connected friends and confer full trust based on the provided credibility. Users may prefer to trust its most intimate friends rather than the publicly known virtual authorities, and their communication path will also vary correspondingly. Our ongoing work incorporate adaptive trust computing into the routing algorithm.

```
Algorithm 1: User Lookup
 1  for (i = 1; i ≤ Number_of_Friends; i++) do
 2      if (SID_b = SID_i) then
 3          Return u_i;
 4      else
 5          if (CID_b = CID_a) then
 6              FIntra = 1;
 7          else
 8              FIntra = 0;
 9  if (FIntra = 1) then
10      for (i = 1; i ≤ Number_of_Friends; i++) do
11          Dist_ib = |SID_i − SID_b|;
12          Find a hub user h_i who has the minimum Dist_ib;
13          Return h_i as the next hop;
14  else
15      if (u_a.hub = 1) then
16          for (i = 1; i ≤ Number_of_Friends; i++) do
17              Dist_ib = |SID_i − SID_b|;
18              Find a hub user h_i who has the minimum Dist_ib and CID_i = CID_b;
19              Return h_i as the next hop;
20      else
21          Send UserLookup(Key_b, SID_b, CID_b) to u_a's hub;
```

5.2 Join and Leave

SW-VoIP nodes participate in a P2P VoIP network using normal P2P protocols. However, the construction and maintenance of a small world P2P VoIP system is a user-interactive process which requires timely role reset and user awareness. The process of join and leave in SW-VoIP has its specific operations, which we illustrate in Algorithm 2.

To assure identities and prevent malicious attacks such as Spam, DoS, etc, a user u_a is not allowed to join the small world network until it has communicated with and been recommended by a given number n_r of SW-VoIP members. The criteria that a SW-VoIP member used to judge external users are based on the number of calls made between them ec, and the average duration time those calls a_ed. If $ec \geq 3$ and $a\_ed \geq 10m$, the external user can get that member's recommendation, and if there are recommendations from at least 3 members, the external user can be recommended to join SW-VoIP network.

Receiving the signal of SW-VoIP members' recommendations, a PCrawler acquires a new user u_a's information and call logs, and reports them to the Dispatcher for issuing a SID. The SID can be used by u_a to infer the cluster it belongs to and set its own *cluster Id (CID)* to the appropriate cluster number. u_a then retrieves the corresponding SID and CID of its connected friends, and calculates their popularities, based on which it elects a hub user for future routing. The user thus successfully becomes a member of small world network by setting *Status Flag (SF)* to 1, where $u_a \in Cluster_a$ and $Cluster_a \subseteq SW-VoIP$, and can launch calls using mechanisms provided by SW-VoIP.

When a user leaves the P2P VoIP network, it checks if it is a member of small world network and if so it is a hub user in the SW-VoIP. A normal SW-VoIP user leaves the small world network by simply informing its connected friends and passing the stored keys on to its closest friend under the same hub. When a hub user leaves, all the connected normal users are informed to recalculate the popularity p of their friends and elect the next best alternatives in the same cluster as their hub. In many cases, hub users leave temporarily and come back to the network in a short period of time t where $t \leq T_{temp}$. Informed with the quick return of the former hub user, normal users will resume the previous connections stored in the cache and push back the current ones. If a hub user returns after a long time t where $t > T_{temp}$, it will be regarded as a new user but with a high initial credibility, and its information stored in the cache will be removed. Meanwhile, the PCrawler responsible for that area will report the new hub to the Dispatcher.

5.3 Load Balancing

Hub users with high popularities may receive a large amount of simultaneous messages due to their pivotal roles in the SW-VoIP. Bottleneck caused by multiple TCP connection requests will result in a big delay for users lookup, which becomes problematic due to the timeliness issue of connection establishment. To optimize lookup performance, SW-VoIP is designed to provide a self-stabilizing and priority-based load balancing mechanism at the application layer.

Before joining the SW-VoIP, a user calculates the throughput Th of its host node based on a well-known TCP throughput model proposed by Padhye [13], where MSS is the maximum segment size, rtt is the average round trip time measured by TCP, and plr is the package loss rate. We assume that MSS and plr are constant across the overall P2P VoIP network, and the value of rtt is based on the average waiting time in the tasks queue of connection request and the propagation delay is determined by the network complexity. The throughput Th should be no larger than $(MSS/rtt) * f(plr)$ to ensure successful message delivery.

Suppose that a nodes had queued n_c many tasks when it receives a user lookup message m_c such as $REQUEST$, $RESPONSE$, ACK, etc, and generates n_c TCP connections to its next hops. We calculate n_c as $n_c = \min_{m_c, NC_{max}}$, where NC_{max} is the maximum number of connections that can be made by each node. NC_{max} is determined by the maximum throughput Th_{max} allowed, user's popularity p, ratio r_p of tasks with high priorities, and two self-controlled parameters α and β $[\alpha, \beta \in (0, 1)]$ using the following formulae.

$$NC_{max} = \alpha * Th_{max} + \beta * p * r_p \qquad (3)$$

```
Algorithm 2: Join and Leave
 1  OnJoin()
 2  Join exiting P2P VoIP network;
 3  if (Num_of_Recommenders ≥ n_r) then
 4  |   Acquire small world identity SID and CID;
 5  |   for (i = 1; i ≤ Number_of_Friends; i++) do
 6  |   |   Retrieve SID_i and CID_i from u_i;
 7  |   |   Calculate popularity p_i of u_i;
 8  |   |_  Elect a hub user h with highest p;
 9  |   SFlag = 1;
10  else
11  |_  SFlag = 0;

12  OnLeave()
13  if (SFlag = 1) then
14  |   if (u_a.hub = 1) then
15  |   |   for (i = 1; i ≤ Number_of_Friends; i++) do
16  |   |   |   Inform u_i its leave;
17  |   |   |   Ask u_i to cache its information and elect a new hub with the next highest p;
18  |   |   |   Set Timer(t_i) on u_i;
19  |   |   |_  Close connection;
20  |   else
21  |   |_  Inform all the connected friends its leave and close connection;
22  else
23  |_  Leave existing P2P VoIP network;
```

If a user is popular or a node has a queue of tasks with high priorities, the maximum number of TCP connections will be increased by increasing the parameter α and decreasing β. Otherwise, decrease the maximum number of TCP connection to reduce expense. The maximum number of connections NC_{max} acts as a signal to inform users about its popularity and tasks status.

As stated, a user in SW-VoIP maintains a list of friends from whom it can elect a hub user with high popularity, and decide the next routing hop based on the closeness of its friends. We optimize SW-VoIP by giving message senders or forwarders the ability to avoid congestion, which we call the *next best choice*. If the tasks waiting in the queue reaches the maximum number of connections NC_{max}, a signaling flag $FConn$ is used to inform others that this node is congested. If $FConn$ is set to 1 in node a, and any user b who plans to connect to a will give up and choose a next best alternative c from the friends list as its hub user and connect to it. The user b will continue checking $FConn$ of its next hubs until it finds a task queue without congestion. The process of load balancing and choosing the new hub is described in Algorithm 3.

Sometimes users may receive high priority calls such as Reverse 911 [21]. For the time critical lookup messages, SW-VoIP gives the emergency calls the highest priority without waiting in the queue or a relatively high priority to be queued and processed quickly. Requests originating from friends can also be given a priority to be queued and processed, and the priority of a task can be increased or decreased according to the credibility of the message source.

Algorithm 3: Load balancing

1 **OnReceive()**
2 **if** $(n_c < NC_{max})$ **then**
3 $\quad FConn = 0$;
4 $\quad$ Enqueue the task of TCP connection;
5 $\quad n_c$++;
6 **else**
7 $\quad FConn = 1$;

8 **OnSend()**
9 **if** $(FConn = 0)$ **then**
10 $\quad$ Generate TCP connection request;
11 **else**
12 $\quad$ **for** $(i = 1; i \leq Numb\_of\_Friends; i++)$ **do**
13 $\quad\quad$ Find the next hub h with highest popularity p;
14 $\quad$ Repeat $OnSend$ to h;

---

6 Experimental Evaluation

In this section, we describe the experiemnt we did in order to compare the object lookup performance of our small world P2P VoIP system with other P2P VoIP systems using normal P2P algorithms. We also conduct repeated trial to find the best performance through adjusting corresponding metrics. The results show that our small world P2P VoIP system can use proposed mechanisms with optimized performance.

6.1 Simulation

Our simulations are conducted using real-life call logs of 97 individual mobile phone users over the course of nine months which were collected by the Reality Mining project group at the Massachusetts Institute of Technology (MIT) [14]. Call logs were collected using Nokia 6600 smart phones loaded with software written to record phone information including incoming/outgoing calls, users in proximity, locations, etc.

We deploy a P2P VoIP system testbed of 1000+ nodes on 300+ PlanetLab [2] machines using OpenVoIP [1], which is an open source P2P VoIP system developed by Columbia University supporting well-known DHTs and unstructured P2P protocols. Based on the assumption of steady user call behavior, we generate P2P VoIP traffic in our system by using these real-life call logs. We set up the six time periods as parameters, which is weekday, weekend, school open, winter break, daytime, and nighttime, and generate traffic with reference to these scenarios.

We report the behavior of node degree distribution in Figure 4, which resembles a power-law distribution with a small portion of high-degree users. Because we have the call log collection of only 97 users, the node degree distribution (e.g. maximum 139 friends) shows that we do not cover all the friends of a user in our small world overlay. We also present the node degree distribution of the SW-VoIP overlay user in Figure 4. We simulated traffic using these two call graph and evaluated our algorithms.

We randomly pick 97 nodes on the testbed and launched about 10,000 real-time P2P VoIP user lookup messages among those nodes, according to who is making or receiving calls at one time, how many calls are made during certain period of time, and how much duration time was used for each call. The small world overlay is constructed based on the information extracted from the call logs on these nodes through distributed PCrawler system, and in turn used to facilitate communications in the P2P overlay.

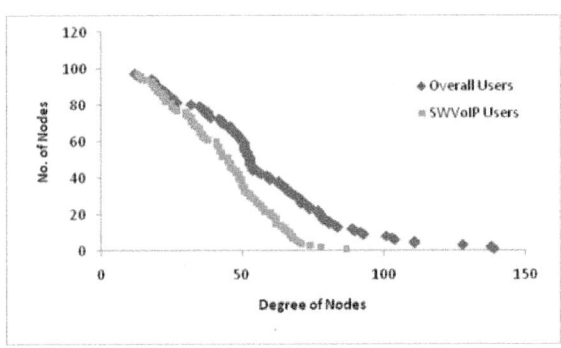

Fig. 4. SW-VoIP Users Call Degree Distribution

Each user in the small world overlay maintains a list of friends, who have called or have been called. If no such friend exists, a normal P2P overlay is used for VoIP communication and the subsequent discovery of new friends. On receiving a user lookup request and confirming itself, the recipient user updates its friends list and stores the call initiator's information carried by the user lookup object. We illustrate this meta-data information in user lookup object in Figure 5, which is a prerequisite for successful routing in small world overlay.

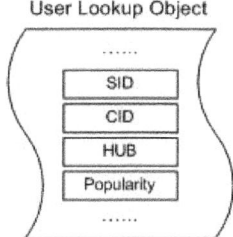

Fig. 5. Small World Meta-data User Info

6.2 Algorithm Evaluation

We evaluated our algorithms by comparing user lookup performance on SW-VoIP to OpenVoIP using Kademlia [11]. Based on the call log data set, we explored the performance test on 97 nodes of the SW-VoIP and the OpenVoIP system. We measured the average hop count and system performance by varying the number of clusters m = 1, 2, 4, and 8 respectively, as the number of hub users varies in the order of 3, 20, 30, 38, correspondingly. This evaluation helps us estimate the optimal number of clusters and reduce user lookup time. The analysis of MIT data set shows that on average 10 percent of the users are on the phone at any time. Therefore we randomly choose 10 nodes and performed a list of users lookups on them at the same time. We illustrate the probability density function of hops for the users lookup under SW-VoIP and OpenVoIP systems in Figure 6, and summarize the average lookup delay in Table 1, which show that our SW-VoIP has a lower average number of hops and lower average delay in user lookup than the OpenVoIP system.

It is important to observe that SW-VoIP has a higher percentage of 1 or 2 hops but a lower percentage of multi-hops $h(h > 2)$ than OpenVoIP. It was also observed that SW-VoIP of 2 or 4 clusters has a better system performance than others, which shows that the efficiency of user lookup is tightly related with the partition of users and the closeness of overall hubs.

Table 1. Average User Lookup Performance

	OpenVoIP	SWVoIP 1	SWVoIP 2	SWVoIP 3	SWVoIP 4
AVE HOP	2.275	1.75	1.72	1.73	1.85
AVE Delay (ms)	470.6575	454.419	324.907	204.83	491.465

We observed that there is a larger number of intra-cluster calls than inter-cluster calls. We generate these two kinds of user lookup traffic on 4 clusters of the SW-VoIP system respectively, and illustrate the system performance in Figure 7 and Figure 8. For intra-cluster calls, most of the routings are processed in $1 \sim 3$ hops with the delay of $0 \sim 400$ ms, whereas inter-cluster routings have the scattered test results ranging from $1 \sim 6$ hops with $0 \sim 1000$ ms delays.

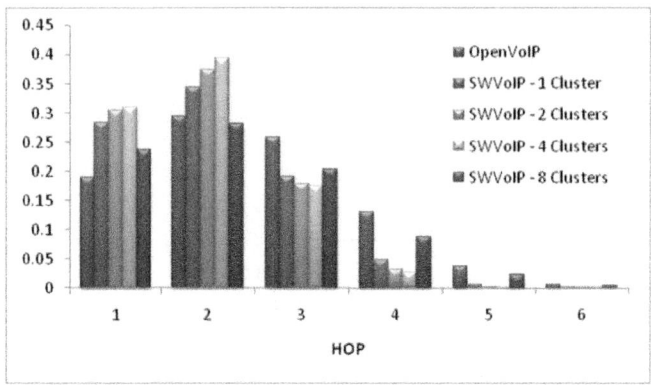

Fig. 6. Probability Density Function of User Lookup Hops

The test results conform that the utility of our SW-VoIP system that message routing are conducted using users' intimated social relationship, and the closer the users are connected, the faster the messages are routed.

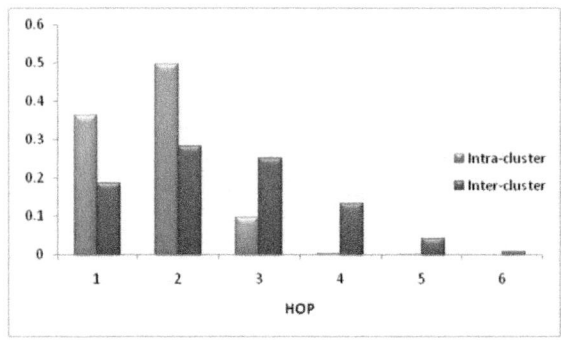

Fig. 7. Intra- and Inter-cluster Hops Comparison

We further explored user lookup performance under heavy network traffic. We attempted to build as many connections as possible by simultaneously launching user lookup traffic on 97 nodes in OpenVoIP and SW-VoIP respectively based on real-life call logs. The traffic generated include messages issued not only from those 97 users but also from additional 52 users in P2P VoIP network, as shown in user call degree distribution of Figure 4. We assume that these 97 users in SW-VoIP are divided into 4 clusters as in earlier described experiments, and load balancing mechanism has little effect on the average link traversal of our SW-VoIP communications. We compared two systems by evaluating user lookup delay under up to 10,000 simultaneous messages. It is observed in Figure 9 and Figure 10 that not only SW-VoIP has less average delay than OpenVoIP, but also

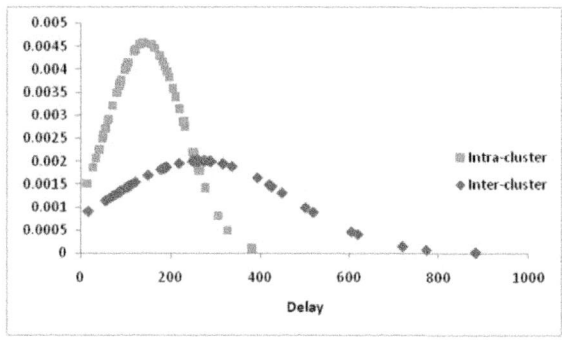

Fig. 8. Intra- and Inter-cluster Delay Comparison

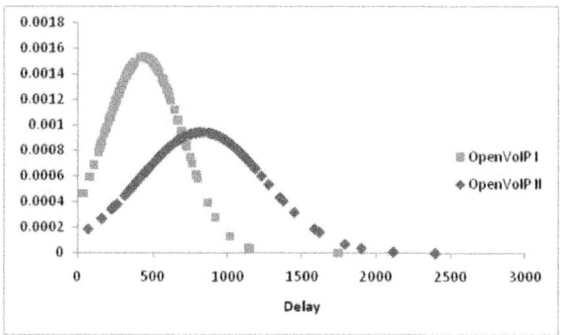

Fig. 9. OpenVoIP Delay under heavy traffic

delays in SW-VoIP increases much less than those in OpenVoIP, which shows that SW-VoIP has a better load balancing capability than OpenVoIP.

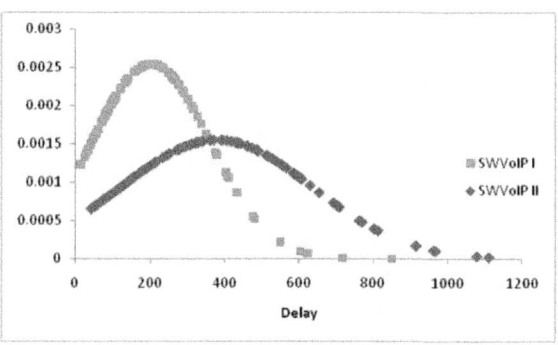

Fig. 10. SWVoIP Delay under heavy traffic

We noticed that system performance may be affected to some extent by parameters setting of PCrawlers such as the frequency of information updating, the rate of simultaneous peer crawling, the number of PCrawers in SW-VoIP, etc. For example, user lookup delay will be increased by up to 10,000 ms if we deploy a PCrawler for each user and concurrently update address information in every 10 ms. Our ongoing experiments are addressing these quantitative issues.

7 Conclusion

We have devised a novel model to leverage the small world properties for Peer-to-Peer VoIP communications based on the telephone users specific social behavior. In order to experimentally validate the utility of our model, we constructed a user oriented small world overlay on top of a P2P VoIP system, and use self-optimization methods to collect data, analyze user behavior, and utilize small world structural properties to facilitate telecommunication. Moreover, the design of SW-VoIP system aims to provide distributed control, epidemic updating, and adaptive trust computing to P2P VoIP users, which are important for the deployment of telecommunication services. Experiments were carried out to optimize clustering users in small world overlay, and compare the performance of SW-VoIP with other P2P VoIP system using normal P2P routing mechanisms such as Kademlia. Initial experiential results reported in this paper show that our model improves system performance by reducing average routing hop and user lookup delays.

Acknowledgments. This work is partially supported by the National Science Foundation under grants CNS-0751205 and CNS-0821736.

References

1. Openvoip: an open peer-to-peer voip and im system, http://www1.cs.columbia.edu/~salman/peer/
2. Planetlab: an open platform for developing, deploying, and accessing planetary-scale services, http://www.planet-lab.org/
3. Skype, http://www.skype.com/
4. Auber, D.: Tulip: A huge graph visualisation framework. In: Mathematics and Visualization (2003)
5. Bryan, D.A., Lowekamp, B.B., Jennings, C.: Sosimple: A serverless, standards-based, p2p sip communication system. In: IEEE, AAA-IDEA (2005)
6. Hui, K.Y.K., Lui, J.C.S.: Small-world overlay p2p networks: Construction and handling dynamic flash crowd. Computer Networks: The International Journal of Computer and Telecommunications Networking (2006)
7. Iamnitchi, A., Foster, I.: Interest-aware information dissemination in small-world communities. In: HPDC (July 2005)
8. Jennings, C., Lowekamp, B., Rescorla, E., Baset, S., Schulzrinne, H.: Resource location and discovery (reload) base protocol. draft-ietf-P2PSIP-base-10 (August 2010)

9. Khambatti, M., Ryu, K.D., Dasgupta, P.: Structuring Peer-to-Peer Networks Using Interest-Based Communities. In: Aberer, K., Koubarakis, M., Kalogeraki, V. (eds.) VLDB 2003 Ws DBISP2P. LNCS, vol. 2944, pp. 48–63. Springer, Heidelberg (2004)
10. Marti, S., Ganesan, P., Garcia-Molina, H.: SPROUT: P2P Routing with Social Networks. In: Lindner, W., Fischer, F., Türker, C., Tzitzikas, Y., Vakali, A.I. (eds.) EDBT 2004. LNCS, vol. 3268, pp. 425–435. Springer, Heidelberg (2004)
11. Maymounkov, P., Mazières, D.: Kademlia: A Peer-to-Peer Information System Based on the XOR Metric. In: Druschel, P., Kaashoek, M.F., Rowstron, A. (eds.) IPTPS 2002. LNCS, vol. 2429, pp. 53–65. Springer, Heidelberg (2002)
12. Nanavati, A.A., Gurumurthy, S., Das, G., Chakraborty, D., Dasgupta, K., Mukherjea, S., Joshi, A.: On the structural properties of massive telecom call graphs: findings and implications. In: CIKM (2006)
13. Padhye, J., Firoiu, V., Townsley, D., Kurose, J.: Modelling tcp throughput: A simple model and its empirical validation. In: Communications Architectures and Protocols (August 1998)
14. Phithakkitnukoon, S., Dantu, R.: Inferring Social Groups Using Call Logs. In: Meersman, R., Tari, Z., Herrero, P. (eds.) OTM-WS 2008. LNCS, vol. 5333, pp. 200–210. Springer, Heidelberg (2008)
15. Ratnasamy, S., Francis, P., Handley, M., Karp, R., Shenker, S.: A scalable content-addressable network. In: SIGCOMM (2001)
16. Rosen, B., Schulzrinne, H., Polk, J., Newton, A.: Framework for emergency calling using internet multimedia. draft-ietf-ecrit-framework-10 (July 2009)
17. Rosenberg, J., Schulzrinne, H., Camarillo, G., Johnston, A.R., Peterson, J., Sparks, R., Handley, M., Schooler, E.: Sip: session initiation protocol. RFC 3261 (June 2002)
18. Seedorf, J.: Lawful interception in p2p-based voip systems. In: Principles, Systems and Applications of IP Telecommunications, IPTComm (July 2008)
19. Singh, K., Schulzrinne, H.: Peer-to-peer internet telephony using sip. In: NOSSDAV (June 2005)
20. Singh, K., Schulzrinne, H.: Using an external dht as a sip location service. Columbia University Technical Report CUCS-007-06 (February 2006)
21. Sorensen, J.H., Sorensen, B.V., Smith, A., Williams, Z.: Results of an investigation of the effectiveness of using reverse telephone emergency waning systems in the october 2007 san diego wildfires. In: Oak Ridge National Laboratory/TM 2009 (June 2009)
22. Stoica, I., Morris, R., Nowell, D.L., Karger, D.R., Kaashoek, M.F., Dabek, F., Balakrishnan, H.: Chord: a scalable peer-to-peer lookup protocol for internet applications. ACM Transactions on Network 11(1) (2003)
23. Stoica, I., Morris, R., Karger, D., Kaashoek, F., Balakrishnan, H.: Chord: A scalable peer-to-peer lookup service for internet applications. In: SIGCOMM (August 2001)
24. Watts, D.J.: Collective dynamics of small-world networks. Nature (1998)

When Will You Be at the Office?
Predicting Future Locations and Times

Ingrid Burbey and Thomas L. Martin

Bradley Department of Electrical and Computer Engineering
Virginia Tech, Blacksburg, VA
{iburbey,tlmartin}@vt.edu

Abstract. The purpose of this paper is to predict people's future locations or when they will be at given locations. These predictions support proactive, context-aware and social applications. Markov models have been shown to be effective predictors of someone's *next* location [1]. This paper incorporates temporal information in order to predict *future* locations or the times when someone will be at a given location. Previous models use sequences of location symbols and apply Markov-based algorithms to predict the next location symbol. In our model, we embed temporal information within the sequence of location symbols. To predict a future location, we use the temporal information as the previous state (or context) in the Markov model to predict the location that is most likely at that given time. To predict *when* someone will be at a location, we use the location as the context and predict the time(s) the person will be at that location. The model produces up to 91% accuracy for predicting locations, and less than 10% accuracy for predicting times. We show that prediction of location and prediction of time are two very different problems, because the number of predictions produced by the Markov model differ greatly between the two variables. A heuristic algorithm is proposed which incorporates additional context to improve predictions of future times to 43%.

Keywords: location-prediction; time-prediction.

1 Introduction

One of the goals of ubiquitous computing is context-awareness, enabling our devices to become our invisible assistants instead of one more device to be managed. Our proactive devices need to "infer the intent of the user" [2]. Knowing the location of a device (and therefore the location of the owner) contributes to knowing the context and the activity of the user. The ability to *predict* the location of the user can support context-aware applications such as assistive devices, reminder systems, social networking applications, and recommender systems.

The ability to predict *when* someone will be at a given location can also support context-awareness. This knowledge can help friends rendezvous or coworkers meet. It can detect anomalies, for example, sending an alert after discovering that someone did not arrive when expected. Collaboration between co-workers can be coordinated when one knows the temporal routines of the other.

This paper examines how to predict future *locations*, such as where someone will be at a given time. In addition, we examine how to predict future *times*, such as when someone will be at a given location. We embed temporal information into location sequences and begin by using a Markov model to predict symbols in the sequence. We then improve upon the Markov temporal predictions with a heuristic model which uses more context than is available in the Markov model.

Our experiments were run on data collected from PDAs which were toted by freshman on the UCSD campus in 2002 [3]. We parsed the data using two levels of temporal granularity: 1-minute timestamps and 10-minute timestamps. One-minute timestamps reflect movement, while the 10-minute granularity focuses on our significant locations or destinations. The experimental results show that a first-order Markov model with fallback can predict future locations with an accuracy of up to 91% for the 10-minute timestamps and 85% for the 1-minute timeslots.

Predictions of future times, such as when someone will be at a given location, are not as successful. Experimental results with the same data return a prediction accuracy of less than 10% unless the model tests against all of the predictions returned, both good and bad, which often (85% of the time) covers the entire 24-hour day. A heuristic algorithm is then used which can incorporate further context, such as a time and/or location that was observed earlier in the day. The heuristic algorithm improves the prediction of future times up to 43% accuracy.

The remainder of this paper is organized as follows. Section 2 discusses several applications for location and time predictions. Section 3 reviews related work in predicting next locations and/or future times. Section 4 reviews the Markov Model. Section 5 contains the details of the data, the experiments using both the Markov model and the heuristic model, and the results. Section 6 summarizes and concludes the paper.

2 Applications

Location prediction can inform many types of proactive, context-aware applications. For a single user, location prediction can provide reminders and recommendations. When predictions are shared with others, social networking can be enhanced, assistive support can be given, resources can be reserved and advertising can be targeted. This section discusses each of these applications in further detail.

For a single user, location prediction can be used for reminders and recommendations. A simple example is a reminder to pack up library books in the morning on a day you are expected to go past the library. The Magitti Recommendation System [4] remembers its owner's and other's past activities to recommend activities that the owner may enjoy at her current location. With predictive capability, a recommendation system such as Magitti could additionally recommend activities for future locations and remind one what to bring.

Predictive, proactive devices can provide 'cognitive assistance' [5] to those who may need help due to mental disability, aging or simple distraction. Predictions can be used to detect anomalies, such as when a user deviates from the expected path or routine [6].

The "Opportunity Knocks" system [5] tracks a user as he rides the bus home to his destination. If he accidentally boards the wrong bus, or gets off at the wrong stop, the system notices the discrepancy, politely 'knocks' to alert the user, and directs him on where to go to catch the correct bus. The current implementation of the system asks the user to identify his destination before he boards the bus. With predictive ability, the application could either predict the user's final destination or suggest a reduced set of likely destinations for the user to select.

Sharing predictions with a caregiver can promote independence. In [7], Chang, Liu and Wang developed a social networking system to assist people with mental illness. The system includes alarms for situations where the user does not arrive at her expected destination at the expected time or if she deviates from her expected path home. Predictions of future destinations could inform assistive technology applications such as these.

Social networking can also be enhanced using predictions. When predictions are shared, a friend can determine when to 'run-into' or serendipitously rendezvous with someone (or avoid them). Common destinations could be used for social match-making [8]. Merchants could use predictions to target advertising, perhaps offering a coupon to lure you to a different restaurant for lunch than your usual stop.

Privacy can be supported by location-prediction if it replaces tracking. For example, a corporation that tracks employees in order to find the closest technician in case of a problem can conceivably use location-prediction instead to produce a list of technicians who are likely to be nearby.

Ashbrook and Starner, in [9], suggest several multiple-user applications of location information in addition to those already mentioned. One application is the exchange of favors, such as errands that someone else could possibly do in your place. Another application could support coordinating a meeting of several people.

Computer-supported-collaborative-work (CSCW) is an important area of research, especially with the trend toward remote work. When co-workers are co-located, they learn each other's routines and availability. This ambient information gets lost when workers are in different locations. Begole, Tang and Hill [10] developed algorithms to predict when office workers would be online, indicating that they were in the office and available for communication. This is an application of predicting *time*, such as when someone arrives in the office in the morning or returns from lunch.

3 Related Work

Prediction of the next cell in a sequence of locations has been investigated in the networking and communications arenas in order to improve resource reservation and quality of service [11-17]. These efforts focus on predicting the next cell and do not concern locations that will occur further out into the future. Smart Homes predict in order to reduce paging messages and allocate resources, such as light or heat in the next room the occupant will enter [18-21]. These experiments use a limited location space and are domain-dependent in that they can use knowledge of the geometry of the network or the home in order to constraint and improve the algorithm. In addition, these works can assume continuous location updates. The inhabitants of a Smart Home do not leave the kitchen and appear in the bedroom without walking down the hallway.

In [22], Zeibart et al. successfully predict driving destinations given a partially traveled route. Our goal of pedestrian destination prediction is similar. However, we cannot assume that location updates are continuous in time. Locative devices, such as our cell phones, will occasionally be turned off, get left behind or enter areas where location systems do not have coverage, such as GPS 'urban canyons.' Applications such as Facebook [23] and Twitter [24] rely on user updates, which mean that there will be significant time lapses between location updates. In addition, to support privacy concerns, any location-logging application must allow the user to delete records or stop recording if the user desires [25-27].

The authors of [28] use a combination of a Markov model and a heuristic model to predict the number of users at each base station in a cellular network. They refer to the single best prediction returned as the 'hard decision' and the set of all possible predicted locations as the 'soft decision.' In our work on predicting future times, we also consider hard versus soft decisions.

4 Markov Models for Prediction

Popular data-compression algorithms, such as Lempel-Ziv (used in PKZIP and gzip), use probabilistic models based on Markov theory. These algorithms use historical data to predict the next symbol in a sequence. When compressing, these algorithms use the least number of bits to encode the most predicted symbols or sequences of symbols in order to reduce the size of the output. These algorithms can also be applied to prediction. Begleiter et al. applied several variable-order Markov models to prediction tasks [29]. One of their test applications was the prediction of music, which was of interest to us because the representation of music includes notes, their starting times and durations, which is similar to the temporal information we embed in our location sequences. They achieved good results with the Prediction-by-Partial-Match (PPM) algorithm, which is the model we chose to apply to our problem. One advantage of the PPM algorithm is that it falls back to lower-orders automatically, which incorporates the findings of Song et al., that a low-order Markov model with fallback was the best predictor for location sequences [1]. Cleary et al. provide an excellent tutorial on using the PPM model for prediction in [30].

5 Method

This section of the paper describes the raw data used, the pre-processing steps, modifications made to the Markov model, and the implementation of the heuristic model.

5.1 Location Using 802.11 Access Points

IEEE 802.11 access points can be used as location beacons [31-33]. They are ubiquitous in urban and suburban areas and many mobile devices include built-in

IEEE 802.11 wireless capability, making IEEE 802.11 an inexpensive location-determining technology. Currently, our requirement is room-level or building-level location granularity, which is well supported by available IEEE 802.11-based location systems. In this project, we use the Access Point ID as the location label. We do not take the significant step of converting the access point ID into a useful place name, which in itself is a difficult problem [34, 35].

As part of the Wireless Topology Discovery (WTD) project at UCSD, researchers issued PDAs to 300 freshmen and collected WiFi access traces [3]. The data were collected during an 11 week trace period during the fall of 2002.

While a student's device was powered on, WTD polled the IEEE 802.11 wireless status every 20 seconds. It recorded a timestamp, access point (AP) ID, signal strength, whether or not the device was using AC power and whether or not the device was associated with the access point. Access points that were sensed, but not associated, were also recorded. Later, we used the sensed access points to create a list of access points that were neighbors of the associated access points and were likely to be in close proximity.

5.2 Preprocessing

The raw data from the WTD experiment combines all logs from all users into one file. Our first step was to divide the raw data into individual files for each user. Records that had the same time and date stamp were compared and the record with the highest signal strength was retained and any other records (usually of sensed, non-associated, access points) were discarded. Contiguous records were combined, where contiguous is defined as records with the same user number, the same access point and where the starting time was within one minute of the previous record's starting time. (Recall that polls are approximately 20 seconds apart.) The duration of each session was calculated and included in the output. These combined sessions were not allowed to run over a one day boundary. (In reality, there were no sessions over 20 hours long.)

We wanted two different views of the data. The first is movement data, which represents the information that would be recorded by an always-on mobile device. The second is destination data, which represents the locations that a user might disclose using a social networking application. These locations are the *significant locations* in someone's day. In this work, significant locations are determined solely by a length of stay of at least 10 minutes [36]. In the future, this definition may expand to include recurrence [35]. Throughout the rest of this paper, we label the movement data as "MoveLoc" (for "Movement Locations") and the destination data is labeled "SigLoc" (for "Significant Locations").

The MoveLoc Dataset. The starting time timestamps in the MoveLoc dataset are rounded to the closest preceding minute. We wanted the MoveLoc dataset to include as much movement data as possible, so all sessions, even those with durations of less than 20 seconds, were included. If more than one session occurred in the same one-minute window, then the session with the longest duration was used for that one-minute timeslot and the others were discarded.

Fig. 1 shows the MoveLoc data for User 3. The color bar shows that this user visited 30 locations over the 10 week recording period. Time-of-day is along the x-axis. One can observe that this user had some regular locations between noon and 2:00pm, and some of the movement between locations is indicated by the change in colors/shades.

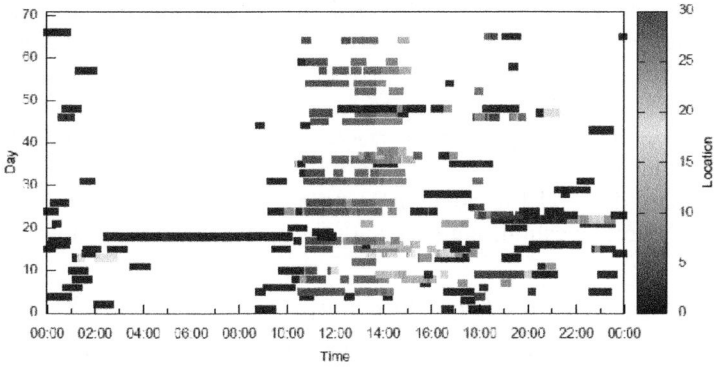

Fig. 1. User 3's MoveLoc (1-minute or less) data

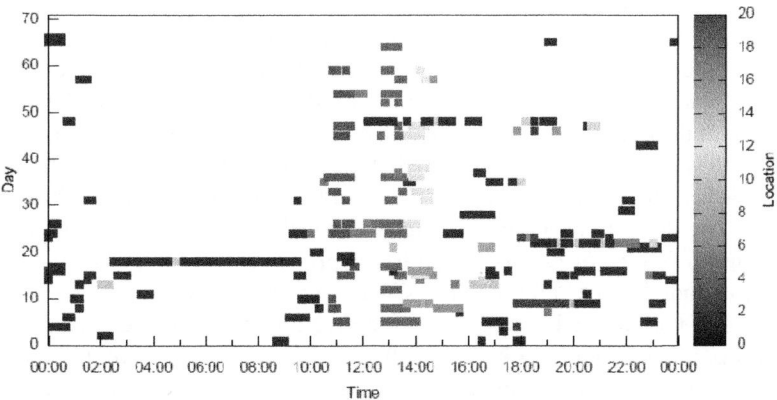

Fig. 2. User 3's SigLoc (10-minute) data

The SigLoc Dataset. The SigLoc dataset is comprised of places where the user spent at least 10 minutes. The first step in creating this view was to remove all sessions with durations of less than 10 minutes. The starting times of the remaining sessions were rounded down to the closest 10 minutes. The SigLoc data for User 3 is shown in Fig. 2. One can observe that the number of locations dropped from 30 to 20 and the transitions between locations have been removed. (Please note that the colors or shades assigned to various locations are not the same for the two figures.)

Finally, the MoveLoc and SigLoc data were then stored in two formats: 16-bit symbols for the Markov model and a human-readable form for the SEQ model.

5.3 PPM for Predicting Location

A PPM (Prediction-by-Partial-Match) implementation of a variable-order Markov Model was used as the basis for the prediction model. Source code from [37] was used as the basis of our implementation. Due to the large number of possible timeslots, the data were pre-processed to encode times and locations as 16-bit symbols, with one range for times and another for locations. It is not necessary to restrict the symbols to a given range; however it is useful for validating the results returned by the model.

When using the model to prediction future locations, the data are represented as a sequence of (time, location) pairs. Each user's data are in a separate file, which has the format:

$$date; \{time_0, location_0, time_1, location_1, \ldots time_n, location_n\}$$

Ten weeks of data are partitioned into test sets, or runs, where each run consists of five weeks of training data and one week of testing data, based on earlier tests [38]. For example, one run could consist of User 3's data from weeks 2 – 6 for training and User 3's data from week 7 for testing.

The model is trained for the first order. (A similar experiment using third order is reported in [38]). Once the model is trained on five weeks of data, the following week is used to test the model. The test data are parsed into individual (time, location) pairs. For each pair, the time is fed into the trained model as the input (context), and the model returns a list of predicted locations with their probabilities, sorted with the highest probability predictions at the top of the list. The model then checks each of the returned predictions against the test location, which is the correct answer. The prediction(s) with the highest probability are checked first (and labeled "Best" in the figures of results.) The lower probability predictions are checked next (and are included in the "all predictions" category in Fig. 3).

If none of the returned predictions are correct, the neighboring access points of the incorrect predictions are checked. Neighboring access points were determined during preprocessing when a list is created of all of the access points which were sensed at the same time as each associated access point. These access points are likely to be in the same building. For example, if, in our test, the correct answer was access point #72, and the Markov model returned a prediction of access point #63, the model would check to see if access point #63 was a neighbor of #72 and likely to be in the same proximity.

If the input time was not found at the first order, the PPM algorithm falls back to the 0th order and returns a list of the most likely locations, regardless of the input time. This means that the model has no information about the time that was entered as the context, so it simply returns the most popular locations.

The results of using the Markov model to predict future locations are shown in Fig. 3. The lowest results are for the MoveLoc dataset which uses 1-minute timeslots. Using 1-minute timeslots increases the number of states in the Markov model up to tenfold compared to 10-minute timeslots. This state explosion increases the number of possible predictions and decreases the probability for each, leading to worse results. For example, if the user arrives at location x at 8:00am and the next day arrives at 8:01am, these times are recorded as separate states in the Markov model, which reduces their individual probabilities. In the SigLoc dataset, which uses wider 10-minute windows, these two records fall into the same 8:00am state, increasing its count, and therefore its probability. The increased size of the timeslot compensates for variability in arrival times by using a longer window of time that is considered to be a correct prediction.

The best results, 91% for the SigLoc data, are achieved when all of the returned predictions are used, regardless of their probabilities. The median number of predictions returned is 2, with a maximum of 13 for the MoveLoc data and 9 for the SigLoc data.

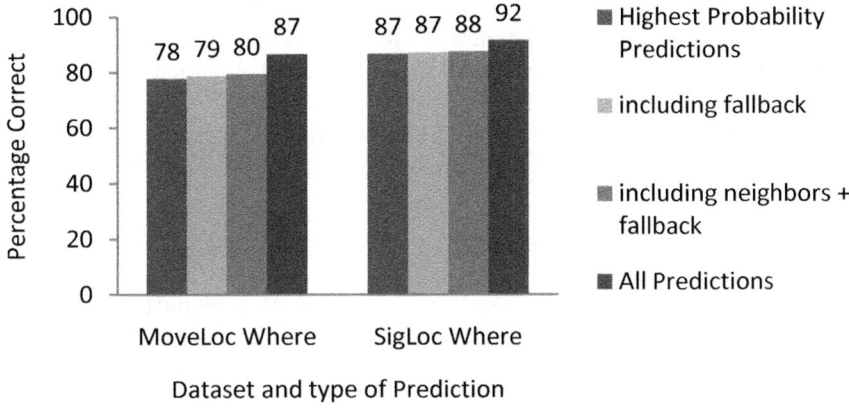

Fig. 3. Predicting Future Locations using the Markov Model

5.4 PPM for Predicting Time

We now turn to the converse problem. Instead of asking questions about someone's future location, such as "Where will Bob be at 10:00am?" we ask questions about *when* someone will be at a given location, such as, "When will Bob be in his office?"

We began with the identical Markov Model. Instead of training on data that was formatted as a sequence of times and their corresponding locations,

$$\{time_0, location_0, time_1, location_1, \ldots time_n, location_n\},$$

the sequences are rearranged to form (location, time) pairs:

$$\{location_0, time_0, location_1, time_1, ... location_n, time_n,\}$$

where $location_t$ is the context, or the corresponding location, for $time_t$.

The model was trained and tested with the same runs of five-weeks of training data with a corresponding week of testing data.

Fallback was not used in this model because it does not make logical sense. In the location application of the model, which is answering the 'where' question, it may make sense for an application to fall back to the most common location. For the 'when' question, such as "When will Bob be in Hawaii," it does not make sense for the model to respond, "I have no data for 'Hawaii,' but the most probable time is 10:00am, so I will predict that Bob will be in Hawaii at 10:00am. In such cases, the model should refuse to predict.

If none of the time predictions returned were correct, they were then checked to see if they were within 10 or 20 minutes of the correct time. This corresponds to the checking of neighboring access points that was done for location prediction.

The results for using the Markov model to predict the time someone will be at a given location are shown in Fig. 4. Initial results are dismal: less than 10% of the best predictions are correct for both datasets. Results improve slightly by allowing predictions to be within 10 or 20 minutes of the correct time.

The high value achieved by using all of the predictions returned comes at a price – the returned predictions cover all times of the day! In other words, the model is asked, "When will Bob be at location x?" and the model responds, "Sometime today between midnight last night and midnight tonight." This result is useless for an application.

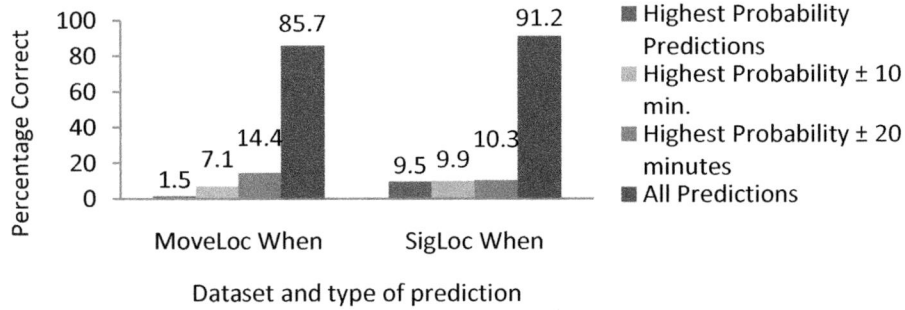

Fig. 4. Predicting Times at Future Locations using the Markov Model

The reason for the differences in our location and time prediction results are apparent when we look at the quantity of predictions returned. The upper bound on the number of predictions returned is the maximum number of states. For time predictions, the maximum number of states is 144 for the 10-minute, SigLoc data, and

1,440 for the 1-minute, MoveLoc data. The maximum number of locations varies by user. Table 1 shows the median number of unique locations and times in the training files. The number of location states is much less than the number of temporal states; hence, the model is more likely to get the location correct. The numbers of predictions returned listed in Table 2 also reflect this discrepancy. The number of predictions returned for location prediction is much less than the number of predictions returned for time prediction.

Table 1. Average Number of Unique Locations and Times for each Training File

Dataset	Ave. # Locations / Training File	Ave. # Times / Training File
MoveLoc	18	646
SigLoc	5	68

Table 2. Number of Predictions Returned for each Test

DataSet	Question	Median # Predictions returned	Max. # Predictions returned
MoveLoc	Where?	2	13
SigLoc	Where?	2	9
MoveLoc	When?	1439	1440
SigLoc	When?	144	144

Recall that the model returns *all* of the predictions found for a given context. The results for "all predictions" in Fig. 4 use all of the predictions returned, which is most cases covers the entire day. Consider Fig. 5, which graphically shows the set of predictions returned for a hypothetical query asking when Alice will be at the office. The graph shows that Alice usually arrives around 9:00am, goes to lunch around noon, and leaves for the day at 5:00pm. The highest probability prediction is the single prediction at 9:00am, which corresponds to the lowest results in Fig. 4. This type of result is useful for applications which need to know the singular times of the day when someone is going to be in a given location. Using all of the reported predictions corresponds to the best results from Fig. 4. These results might be useful to an application that needs to know all of the possible times that Alice might be in her office. However, if Alice left her mobile device in the office overnight one time, these results would always cover the entire day because there would be at least one count for that location at every possible timestamp.

Thresholds. Fig. 5 shows some thresholds that might be used to determine a subset of predictions to be considered. For example, the threshold at 90% would return the prediction that Alice would be likely to be in the office from 8:30am until noon and from 1:10pm until 4:40pm. This broader prediction could be more useful to an application than the single prediction of 9:00am that is returned with a tighter threshold of 0% (which returns the highest probability predictions only).

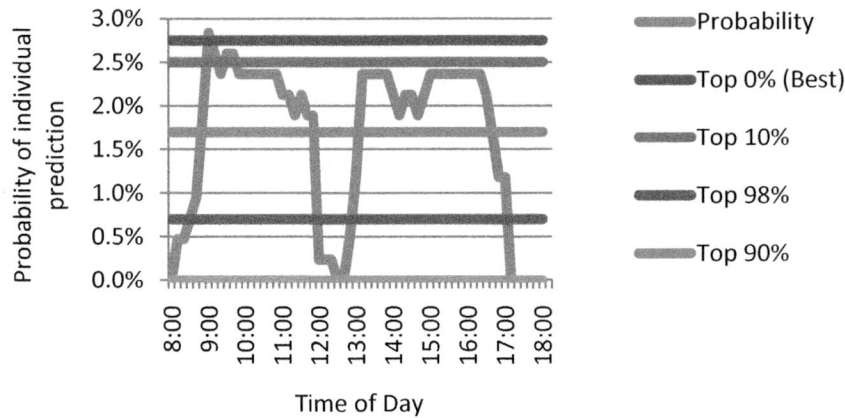

Fig. 5. Thresholds for a Hypothetical set of Predictions

Thresholding was then implemented in the model. Recall that the model returns a list of predictions, each labeled with its probability. The predictions are sorted in order of probability, from the highest probability to the lowest. To invoke a threshold, the top predictions are checked for correctness until their cumulative probability surpasses the probability threshold. For example, a threshold of 10% uses the top predictions until their cumulative probability is greater than 10%. A threshold of 0% uses only the most likely prediction, which is the same as the 'best' predictions in the Fig. 4. A threshold of 100% uses all of the predictions returned, which is the equivalent of the using all of the probabilities, again as shown in Fig. 4. Fig. 6 and Fig. 7 illustrate the results of using a probability threshold for the MoveLoc and SigLoc datasets.

Observing the graphs in Fig. 6 and Fig. 7, one can see there is no apparent optimal threshold value. The value of the threshold is dependent upon the goals of the application. One application may want only predictions with a high probability while another application may want to present all of the possibilities to the user. Imprecise knowledge of user intent may still be useful information [2].

Currently, we do not consider the predictions themselves, only their probabilities. In the future, we may consider combining contiguous predictions into a window of time. For example, if the returned predictions include {8:00am, 8:10am, 8:20am,...,10:00am}, they could be combined into one prediction of 8:00am – 10:00am. Consideration needs to be made about how to compute the resulting probability when contiguous predictions are combined. If Bob only occasionally arrives at the office at 8:00am but he is almost always there at 9:00am, how is that information presented to the user? Should the less-likely 8:00am prediction be included in the results presented to the user? A proper visualization of the results can illustrate these probabilities, either with a temperature graph which uses different colors for different probabilities or a graph similar to Fig. 5.

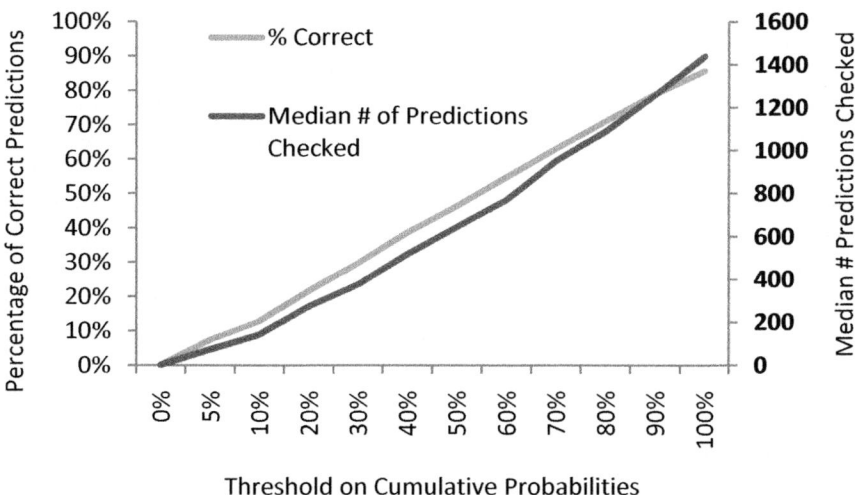

Fig. 6. Applying a Probability Threshold to Time Predictions for the MoveLoc (1-minute) data

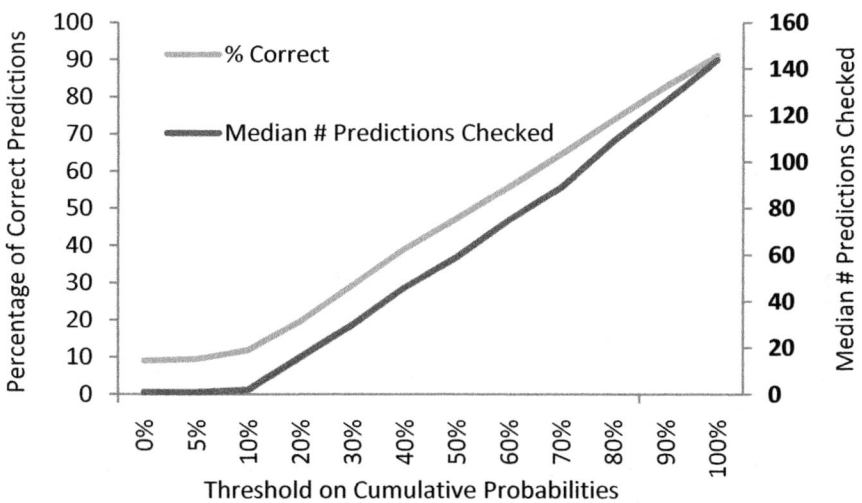

Fig. 7. Applying a Probability Threshold to Time Predictions for the SigLoc (10-minute) data

In addition, when we consider only the probabilities, we are losing some information about the predictions. The PPM model stores counts of the number of times it encounters each pattern in the sequence. Currently, we are testing only against the resulting probabilities and not the counts themselves. The counts could be used to develop a confidence measure about the predictions. The model can return the same probability for two different states, even if it has only one occurrence of the first state and several hundred for the second state. Consider the hypothetical situation where the data has recorded Bob at the mall once at 10:00am, but over the course of 10 weeks, has recorded Bob at the office at all different times between 8:00am and 5:00pm. If the Markov model is queried as to when Bob will be at the mall, it will return a prediction of 10:00am with a probability of 100%. If the same model is queried as to when Bob will be in the office, it will return possibly hundreds of predicted times, and each could have a probability of less than 1%. At first glance, it could appear that the Markov model is more confident that Bob will be at the mall at 10:00am, and that conclusion is erroneous. Displaying confidence increases the user's trust in an application [39]. The counts calculated by the model can be used to determine confidence.

5.5 The Heuristic Model: Traversing the Sequence

Even though the Markov model can be used to predict future times, it is not a suitable solution. Too many predictions are returned and there is a loss of contextual information in statistical models such as the Markov model that are based on counting events. We therefore developed an algorithm which retains more contextual information. We call this model 'Traversing the Sequence' or the SEQ model, because it keeps the entire sequence of times and locations intact and, given contextual information such as a previous time or location, it traverse the sequences to predict a future time at a given location.

The SEQ model represents the user's day as a sequence of (arrival time, location) pairs. Note that this is a different representation than the Markov model. In the Markov model, for example, if a user spends 30 minutes in one location, the result is three records in the SigLoc dataset, one for each 10-minute timeslot. In the SEQ model, this location is represented as one pair with the arrival time and the location and no duration information.

The sequence model is an *unsupervised learning* model, in that the model is not trained on historical data that are labeled with the correct answers. The SEQ model stores the historical data and uses it without training to determine future predictions.

Let us illustrate with a simple example. Our user, Bob, usually goes to a coffee shop at 8:00am, arrives at the office at 9:00am and leaves to go home at 5:00pm. Twice during the last ten days, he skipped the coffee shop and arrived at the office at 8:00am in order to prepare for a 9:00am meeting at another location.

The SEQ model stores these daily sequences without modification. When the model is queried as to when Bob will arrive at the office, it searches each of the sequences to find when Bob arrived at the office and finds that on 8 days he arrived at 9:00am and twice, he arrived at 8:00am. The model would return two predictions: 9:00am with an 80% probability and 8:00am with a 20% probability.

Advantages of the SEQ Model. Because the SEQ model retains historical information, queries with additional context can be asked, such as, "When will Bob be in the office if I just saw him at the coffee shop?" In this case, the model would not search any daily sequences that did not include the coffee shop, and would return a prediction that Bob would arrive at the office at 9:00am with 100% probability. The model can be further expanded to include an earlier time as context (e.g. "When will Bob be at the office if I saw him on the road at 8:15?") or both an earlier time and location.

The model can also be expanded to use other forms of context. If the sequences include a day-of-the-week label, the model can search only the sequences that fall on a particular day of the week or a subset of days (e.g. "When will Bob be in the office on the weekend?")

The SEQ model stays current without retraining. The model can be configured to use only the latest data and ignore outdated sequences. The Markov model used previously requires periodic retraining to incorporate recent data.

The model will refuse to predict if it is given context that it cannot find in a sequence. For example, it will refuse to predict if it is queried about a location it has never seen (e.g. "When will Bob be in Hawaii?") or a previous context that has not been encountered (e.g. "It's 7:00pm, when will Bob be in the office today?")

The SEQ model Experiment. The same raw user data that were used for the Markov model were used for the SEQ model. Instead of using 16-bit symbols to encode time and location as was done for the Markov model, the data for the SEQ model are stored in human-readable, comma-delimited files. The SEQ model was written in Python because of its rapid-development support and ease of handling text strings.

The model was tested with different types of previous context, which we refer to as 'order'. In the 0th order test, the only input is the location. The model is queried as to when the user will be at the test location. This location is referred to as the *quest location*. A previous time can be given as additional context; this is noted as '1-time' order. A previous location can also be used as additional context; this is noted as '1-loc' in the model. And finally, both a previous time and location can be used as additional context, creating the second order.

The additional historical context, such as a previous time, place or both, is used to truncate the daily sequences searched. If a previous time and/or location is given, the index of that time and/or location is used as the left edge of the sequence to be searched for the location in question. The pseudo-code for our SEQ model if given additional context of time (context_time) is listed in Fig. 8.

SEQ Model for Predicting WHEN Someone Will Arrive at the Quest_Location
("When will the user be at the quest_location is she was observed at context_time earlier that day?")

```
Reset counters.
For each daily sequence:
Is the quest_location in the sequence?
If not, increment the count for "Never" and continue onto
the next sequence.
Search the sequence to find the time closest to (equal or
earlier than) the context_time.
If the context time is not found,  increment the count
for "Never" and continue onto the next sequence.
The location of the context_time becomes the left border
of the sub-sequence to be searched.
Search the sub-sequence to find the quest_location.
If not found, increment the count for "Never" and
continue onto the next sequence.
The location of the quest_location becomes the right
border of the sub-sequence to be searched.
Search the sub-sequence to find the context_location.
If not found, increment the count for "Never" and
continue onto the next sequence.
Use the time corresponding to the quest_location as the
predicted time and add it to the counters.
Return the counters.
```

Fig. 8. Pseudo-code for the SEQ Model

Results for the MoveLoc (1-minute) and SigLoc (10-minute) datasets are shown in Fig. 9 and Fig. 10, respectively. The prediction accuracy has improved over the Markov model. Allowing for predictions within 10 or 20 minutes of the correct testing time yields an even higher predictive accuracy.

There are a few causes for the improvement in prediction. First, the SEQ model predicts only arrival times. The Markov model indirectly encodes duration, in that records are repeated in the sequence until the user changes location. The Markov data, therefore, has a chain of multiple records if a user is at a particular location for a length of time. Since the length of the user's stay may vary, the tests on the records at the end of the chain may produce wrong predictions, reducing the average predictive accuracy. (For example, predictions about 5:00pm, when the last arrival time recorded was at 2:00pm, would be incorrect in the current SEQ model.)

The predictions worsen with additional context (the 1loc, 1time and 2 areas of the graphs). There are fewer sequences which match the given context. In some cases, there are no historical sequences that match the additional context and no predictions are made.

The SEQ model uses a smaller dataset of arrival times and locations instead of polled timeslots and locations and returns fewer predictions. This reduction in the number of states improves its performance over the Markov model.

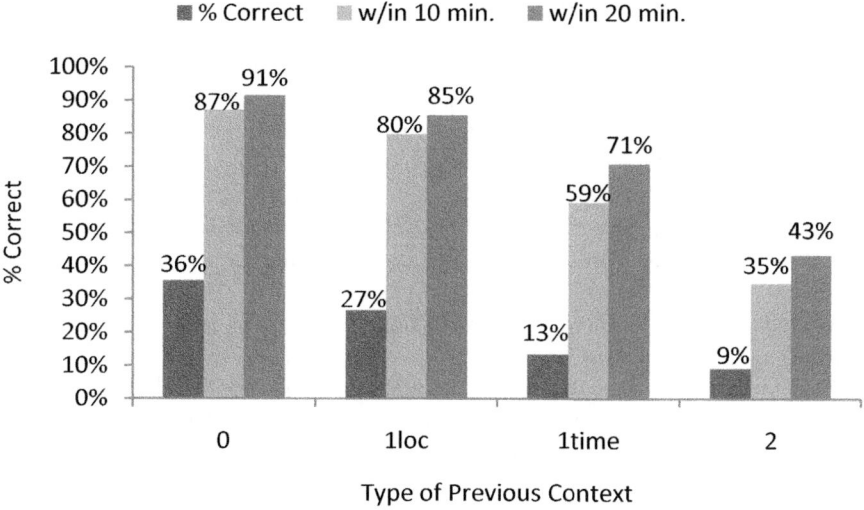

Fig. 9. Results of the SEQ Model on the MoveLoc (1-minute) data

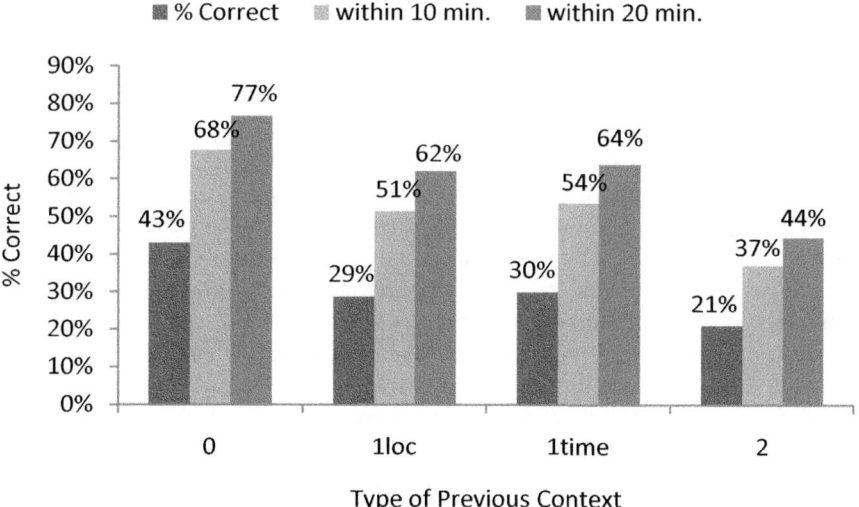

Fig. 10. Results of the SEQ Model on the SigLoc (10-minute) data

6 Summary and Conclusion

Our goal is to develop models for predicting people's future locations and times. This problem is different than previous works because we are not limiting the predictions to the prediction of the *next* location or time, but instead predicting locations and times out into the *future*.

We used data collected from PDAs, and used IEEE 802.11 access point IDs for determining location. We represented the data in two formats: one which represented moving data and used 1-minute timeslots, and one which represented destinations or significant locations and used 10-minute timeslots.

We found that a Markov model with fallback, such as PPM, can predict future locations with accuracy up to 91%. However, the same implementation failed to predict future times. A heuristic model, called the SEQ model, which traversed daily sequences of time and location data, improved prediction of time up to 77%. The heuristic model also supports additional context, such as previous times and/or locations. It can easily be expanded to support other contextual information such as day-of-the-week, or any other information that can be appended to the daily sequences.

We have found that prediction of time is a more difficult problem than prediction of location due to the large number of possible states. Sequence predictors that are effective for predicting next or future locations are not suited for predicting future times.

6.1 Future Work

There are many areas for future improvements in this work. The Hidden Markov Model (HMM) may improve the results of location prediction, as the hidden states could compensate for the unobserved locations between destinations or for situations where two access point labels identify the same location. Visualization of the returned information should be done to impart not just the predicted time or location to the user, but also the relative probabilities. Confidence measures should also be included in the visualization. Visualization becomes more complex when predictions from multiple users are combined.

The SEQ model currently models only arrival time at each location. It can be expanded to include duration information. Duration information would support more useful applications. Prediction of arrival times tells *when* someone will arrive, but it may be more useful to know how long someone will be at a given location. Previous work in learning relative time between events was done in [40].

6.2 Conclusion

This paper is about predicting the future, specifically when someone will be at a given location, or where they will be at a given time. Prediction is a sub-case of context-awareness. In this case, we are not attempting to determine someone's current context or activities, but instead, we take a first step towards predicting future context by predicting future locations or future times at a given location.

We see many applications for such predictions, especially when they are shared among friends, co-workers or care-givers. These predictions can replace the ambient knowledge of our friends and co-workers routines that we pick up automatically if we live or work within close proximity of them. This information can help care-givers remotely and unobtrusively monitor their clients, help friends meet, assist co-workers with contacting others. Hopefully, this work is one small contribution towards the invisible, proactive, assistive devices that are part of the vision of pervasive and ubiquitous computing.

Acknowledgements. The authors would like to thank the anonymous reviewers for their insightful and helpful suggestions.

References

1. Song, L., Kotz, D., Jain, R., He, X.: Evaluating location predictors with extensive Wi-Fi mobility data. In: Proc. INFOCOMM, pp. 1414–1424. IEEE, Hong Kong (2004)
2. Satyanarayanan, M.: Pervasive computing: vision and challenges. IEEE Personal Communications 8, 10–17 (2001)
3. UCSD Wireless Topology Discovery Trace, http://sysnet.ucsd.edu/wtd/
4. Bellotti, V., Begole, B., Chi, E.H., Ducheneaut, N., Fang, J., Isaacs, E., King, T., Newman, M.W., Partridge, K., Price, B., Rasmussen, P., Roberts, M., Schiano, D.J., Walendowski, A.: Activity-based serendipitous recommendations with the Magitti mobile leisure guide. In: Proc. CHI. ACM, Florence (2008)
5. Patterson, D.J., Liao, L., Gajos, K., Collier, M., Livic, N., Olson, K., Wang, S., Fox, D., Kautz, H.: Opportunity Knocks: A System to Provide Cognitive Assistance with Transportation Services. In: Davies, N., Mynatt, E.D., Siio, I. (eds.) UbiComp 2004. LNCS, vol. 3205, pp. 433–450. Springer, Heidelberg (2004)
6. Estrin, D., Chandy, K.M., Young, R.M., Smarr, L., Odlyzko, A., Clark, D., Reding, V., Ishida, T., Sharma, S., Cerf, V.G., Lzle, U., Barroso, L.A., Mulligan, G., Hooke, A., Elliott, C.: Internet Predictions. IEEE Internet Computing 14, 12–42 (2010)
7. Chang, Y.-J., Liu, H.-H., Wang, T.-Y.: Mobile social networks as quality of life technology for people with severe mental illness. IEEE Wireless Communications 16, 34–40 (2009)
8. Axup, J., Viller, S., MacColl, I., Cooper, R.: Lo-Fi Matchmaking: A Study of Social Pairing for Backpackers. In: Dourish, P., Friday, A. (eds.) UbiComp 2006. LNCS, vol. 4206, pp. 351–368. Springer, Heidelberg (2006)
9. Ashbrook, D., Starner, T.: Using GPS to learn significant locations and predict movement across multiple users. Personal and Ubiquitous Computing 7, 275–286 (2003)
10. Begole, J.B., Tang, J.C., Hill, R.: Rhythm modeling, visualizations and applications. In: Proc. UIST, pp. 11–20. ACM Press, Vancouver (2003)
11. Chellapa, R., Jennings, A., Shenoy, N.: A comparative study of mobility prediction in fixed wireless networks and mobile ad hoc networks. In: Proc. ICC, pp. 891–895 (2003)
12. Cheng, C., Jain, R., van den Berg, E.: Location prediction algorithms for mobile wireless systems. CRC Press, Inc. (2003)

13. De Rosa, F., Malizia, A., Mecella, M.: Disconnection prediction in mobile ad hoc networks for supporting cooperative work. IEEE Pervasive Computing 4, 62–70 (2005)
14. Erbas, F., Kyamakya, K., Steuer, J., Jobmann, K.: On the user profiles and the prediction of user movements in wireless networks. In: Proc. PIMRC, pp. 2282–2286. IEEE, Lisbon (2002)
15. Hadjiefthymiades, S., Papayiannis, S., Merakos, L.: Using path prediction to improve TCP performance in wireless/mobile communications. IEEE Communications Magazine 40, 54–61 (2002)
16. Pack, S., Choi, Y.: Fast handoff scheme based on mobility prediction in public wireless LAN systems. IEE Proceedings on Communications 151, 489–495 (2004)
17. Wu, S.-Y., Fan, H.-H.: Activity-Based Proactive Data Management in Mobile Environments. IEEE Transactions on Mobile Computing 9, 390–404 (2010)
18. Chan, J., Seneviratne, A.: A Practical User Mobility Prediction Algorithm for Supporting Adaptive QoS in Wireless Networks. In: Proc. ICON, pp. 104–111. IEEE Computer Society (1999)
19. Das, S.K., Cook, D.J., Battacharya, A., Heierman III, E.O.: The role of prediction algorithms in the MavHome smart home architecture. IEEE Wireless Communications 9, 77–84 (2002)
20. Petzold, J., Bagci, F., Trumler, W., Ungerer, T.: Next Location Prediction Within a Smart Office Building. In: Proc. Pervasive 2005, Munich, Germany (2005)
21. Roy, A., Das, S.K., Basu, K.: A Predictive Framework for Location-Aware Resource Management in Smart Homes. IEEE Transactions on Mobile Computing 6, 1270–1283 (2007)
22. Ziebart, B.D., Maas, A.L., Dey, A.K., Bagnell, J.A.: Navigate like a cabbie: probabilistic reasoning from observed context-aware behavior. In: Proc. UbiComp, pp. 322–331. ACM, Seoul (2008)
23. Facebook Really Wants to Know Where You Are, Considers Buying Loopt, http://www.fastcompany.com/1562563/loopt-facebook-acquisition-rumor-location-based-advertising-social-networking
24. Location, Location, Location, http://blog.twitter.com/2009/08/location-location-location.html
25. Barkuus, L., Dey, A.: Location-Based Services for Mobile Telephony: a Study of Users' Privacy Concerns. In: Proc. INTERACT, pp. 709–712. IOS Press, Zurich (2003)
26. Iachello, G., Smith, I., Consolvo, S., Abowd, G.D., Hughes, J., Howard, J., Potter, F., Scott, J., Sohn, T., Hightower, J., LaMarca, A.: Control, Deception, and Communication: Evaluating the Deployment of a Location-Enhanced Messaging Service. In: Beigl, M., Intille, S.S., Rekimoto, J., Tokuda, H. (eds.) UbiComp 2005. LNCS, vol. 3660, pp. 213–231. Springer, Heidelberg (2005)
27. Voong, M., Beale, R.: Representing location in location-based social awareness systems. In: Proc. BCS-HCI, pp. 139–142. British Computer Society, Liverpool (2008)
28. Dufková, K., Boudec, J.-Y.L., Kencl, L., Bjelica, M.: Predicting User-Cell Association in Cellular Networks from Tracked Data. In: Fuller, R., Koutsoukos, X.D. (eds.) MELT 2009. LNCS, vol. 5801, pp. 19–33. Springer, Heidelberg (2009)
29. Begleiter, R., El-Yaniv, R., Yona, G.: On Prediction Using Variable Order Markov Models. Journal of Artificial Intelligence Research 22, 385–421 (2004)
30. Cleary, J.G., Teahan, W.J., Witten, I.H.: Unbounded length contexts for PPM. In: Proc. Data Compression Conference, Snowbird, UT, pp. 52–61 (1995)

31. LaMarca, A., Chawathe, Y., Consolvo, S., Hightower, J., Smith, I., Scott, J., Sohn, T., Howard, J., Hughes, J., Potter, F., Tabert, J., Powledge, P., Borriello, G., Schilit, B.: Place Lab: Device Positioning Using Radio Beacons in the Wild. In: Gellersen, H.-W., Want, R., Schmidt, A. (eds.) PERVASIVE 2005. LNCS, vol. 3468, pp. 116–133. Springer, Heidelberg (2005)
32. Skyhook Wireless, http://www.skyhookwireless.com
33. Krumm, J., Horvitz, E.: LOCADIO: inferring motion and location from Wi-Fi signal strengths. In: Proc. MOBIQUITOUS, pp. 4–13 (2004)
34. Bolliger, P.: Redpin - adaptive, zero-configuration indoor localization through user collaboration. In: Proc. MELT, pp. 55–60. ACM, San Francisco (2008)
35. Yang, G.: Discovering Significant Places from Mobile Phones – A Mass Market Solution. In: Fuller, R., Koutsoukos, X.D. (eds.) MELT 2009. LNCS, vol. 5801, pp. 34–49. Springer, Heidelberg (2009)
36. Ashbrook, D., Starner, T.: Learning significant locations and predicting user movement with GPS. In: Proc. International Symposium on Wearable Computers, pp. 101–108 (2002)
37. Arithmetic Coding + Statistical Modeling = Data Compression, http://www.dogma.net/markn/articles/arith/part2.html
38. Burbey, I., Martin, T.L.: Predicting future locations using prediction-by-partial-match. In: Proc. MELT, pp. 1–6. ACM Press, San Francisco (2008)
39. Antifakos, S., Kern, N., Schiele, B., Schwaninger, A.: Towards improving trust in context-aware systems by displaying system confidence. In: Proc. MobileHCI, pp. 9–14. ACM Press, Salzburg (2005)
40. Gopalratnam, K., Cook, D.J.: Online Sequential Prediction via Incremental Parsing: The Active LeZi Algorithm. Intelligent Systems 22, 52–58 (2007)

PosQ: Unsupervised Fingerprinting and Visualization of GPS Positioning Quality

Mikkel Baun Kjærgaard[1] and Kay Weckemann[2]

[1] Department of Computer Science
Aarhus University, Denmark
mikkelbk@cs.au.dk
[2] Institute for Informatics
Ludwig-Maximilian University Munich, Germany
kay.weckemann@ifi.lmu.de

Abstract. GPS positioning does not provide pervasive coverage and the accuracy depends on the local environment. When deploying and managing position-based applications it is important to know when to depend on GPS and when to deploy supplementary means of positioning, such as local or inertial positioning. This paper proposes PosQ, a system for unsupervised fingerprinting and visualization of GPS positioning quality. PosQ provides quality maps to position-based applications and visual overlays to users and managers to reveal the positioning quality in a local environment. The system reveals the quality both as it changes over time, in 2D and 3D, and for each type of GPS receiver. Our evaluation provides evidence that the collected quality maps are accurate, that they remain informative over time, that they capture the differences among GPS receivers, and that they can be efficiently collected by participating devices.

Keywords: global positioning system, positioning quality, fingerprinting, efficiency.

1 Introduction

Position-based applications use context information to provide position-dependent functionality. They are developed for a variety of domains e.g. location-based games [4,7], indoor and outdoor services for museums and amusement parks and professional applications for different working domains [10]. The positioning technologies used to enable such applications have to balance coverage, positioning accuracy as well as deployment and operation costs.

GPS positioning is today using state-of-the-art receivers available in most outdoor areas with high accuracy and in some indoor areas with lower accuracy as reported by Kjærgaard et al. [8]. This study also concludes that indoor GPS positioning coverage and accuracy mainly depend on the local environment, e.g. the building materials and number of walls, with some fluctuations due to the movement of people, satellites and atmospheric variations. However, in practice, it is difficult for people to judge for their local environment where and when they can rely on GPS, e.g. indoor availability can depend on room sizes, number of windows, types of walls and roofs and the existence of inner court yards or atriums. In the negative case they have to augment the GPS positioning with, e.g., WiFi [3,11], Powerline [15], GSM [18] or inertial [5] positioning. Secondly, it is difficult to judge what happens if the local environment changes later on, e.g. changes to the roof or new windows, movement of walls or changes in the number of people present. As manual evaluation of GPS availability is both time consuming and has to be continuously repeated it is relevant to automate this process. Thirdly, people using an application and entering an area are interested in judging the positioning quality and maybe decide to switch to inertial positioning [5]. Fourthly, position-based applications can use the information to adjust their functionality to the accuracy in an area [16] and the differences among GPS receivers.

This paper proposes PosQ, a system for unsupervised fingerprinting and visualization of GPS positioning quality. The system provides quality maps to position-based applications and visual overlays to users and managers to reveal the positioning quality. PosQ reveals the quality in three dimensions: As it changes over time, in 2D and 3D, and for each type of GPS receiver. The system enables unsupervised fingerprinting by running in the background of a GPS enabled device without any need of attention by the user. It enables people and applications to judge the GPS positioning quality in a local environment with low effort, e.g., in a mixed indoor-outdoor environment in the scope of a city. The PosQ system consists of four key elements:

- GPS receivers that provide estimates of the positioning quality
- PosQ collectors that fingerprint the positioning quality unsupervised as estimated on participating GPS-devices and efficiently send quality reports to the system's quality database
- Quality maps built from the database
- Visual overlays generated from the maps that can be viewed in popular earth or map viewers

We make the following contributions in this work: First of all, we propose a system for unsupervised fingerprinting of GPS positioning quality that can visualize positioning quality both over time, in 2D and 3D, and for different types of GPS receivers. Secondly, we evaluate the system to provide evidence that the collected quality maps are accurate, that they remain informative over time and that they capture the differences among GPS receivers. The evaluation takes into account different local environments and their positioning quality conditions (outdoor, indoor, and deep indoor). Therefore, several experimental evaluation were executed in a shopping mall, a cultural history museum and a botanic museum with different motion patterns. Thirdly, we demonstrate how the system enables the design of positioning solutions, mapping between different GPS receivers and visualization of positioning quality over time and in 3D. Fourthly, we propose and evaluate methods for efficient collection of quality reports and show that the methods can significantly decrease the number of quality reports needed when updating quality maps in real-time.

The remainder of this paper is structured as follows: In Section 2, we discuss related work. Subsequently, we introduce the PosQ system in Section 3. Afterwards, we present the results of evaluating fingerprinting of positioning quality in Section 4 and discuss the usage of the system in Section 5. Methods and evaluation results for efficient collection of fingerprints are then presented in Section 6. Finally, in Section 7 we provide summarizing conclusions and discuss directions for future work.

2 Related Work

Related work presented in the following is divided into *estimating positioning quality* and *using information about positioning quality*.

2.1 Estimating Positioning Quality

A GPS receiver will, in addition to position and time, estimate the accuracy of a position fix. The *accuracy* of a position fix intuitively denotes how close the fix is to the correct, but unknown, position [10]. Basic algorithms for estimating the accuracy are described in textbooks such as Misra et al. [13], but receivers generally use proprietary algorithms that take into account information such as the channel-to-noise ratios, signal strengths, number of satellites, *dilution of precision (DOP)* (a measure for the satellites' geometric strength for positioning), range residuals, variations in low level measurements and multipath indicators. PosQ collects such estimates to fingerprint the positioning quality. We use the accuracy estimates produced by the receivers proprietary algorithms because only a subset of the above information can be accessed using the location API's on mobile phones (e.g., Nokia and HTC). Also experimental observations presented by Kjærgaard et al. [8] for indoor environments indicate that the estimation can not be based on a single value such as the *dilution of precision* factor because in their experiments low accuracy was better explained by multi-path and weak-signal effects than the geometric strength of the current satellite constellation.

To predict GPS coverage in cities, Steed [17] proposes a system that can predict the binary information of whether GPS is likely to be available or not. The system uses a city model and satellite orbits to calculate how many satellites are likely to be available in the different parts of a city. The calculations only take line-of-sight conditions into account. Therefore, the system cannot provide the fine-grained positioning quality information in different conditions as pursued in this paper.

To estimate the accuracy of outdoor GSM positioning, Dearman et al. [6] propose two methods: A machine-learning method that uses a linear regression technique on basic signal strength features and a ground-truth method that requires the supervised collection of test data. For indoor WiFi-based location fingerprinting, Lemelson et al. [12] propose several methods to estimate the positioning accuracy based on features of location fingerprints and the online signal strength samples. This paper focuses on GPS positioning, but the system can easily be used to collect information from other positioning technologies.

2.2 Using Information about Positioning Quality

The imperfection of positioning technology and its impact on position-based applications have been studied from several perspectives.

Benford et al. [4] study the problem in the context of a location-based game "Can you see me now?" where runners in the real world chase online players in a fixed game zone. The runners are equipped with GPS-enabled devices for positioning. Based on this experience, they identify five strategies for how to deal with imperfections or "seams" in their vocabulary. The five strategies are to remove, hide, manage, reveal or exploit imperfections. These strategies have also been studied in the context of the location-based game Hitchers based on GSM cell positioning [7]. Benford et al. [4] give evidence that the runners observed imperfections themselves with ongoing playtime and then integrated this knowledge into their gameplay strategies, for instance, by "hiding" in areas of low accuracy, where the runners' positions shown to the online players would "jump" or frequently disappear. To reveal imperfections in their game, Benford et al. propose that the system should distribute knowledge of positioning quality to runners and the technical crew. They propose to provide maps that show areas of good or bad positioning quality. As an example, a visualization for GPS with color maps is given, with data derived from log files.

Dearman et al. [6] have studied the impact of revealing positioning imperfections to quantify if people can take advantage of them. The authors based their conclusions on experiments with visualizing positioning accuracy for GSM positioning as circles. They give evidence that enhancing or replacing the sole target position with a circle, intuitively representing the size of the search area, efficiently supports people in their search which resulted in decreased time until the task was finished.

Oppermann et al. [14] motivate that authoring tools for position-based applications must visualize the imperfections of infrastructures. They propose that authoring tools should provide three layers of information: The physical world layer (e.g. maps and GIS), the infrastructure layer (e.g. visualizations of the infrastructure) and the content layer (e.g. regions, events and assets). They describe an authoring tool built on this model and use it to author a location-aware game named Tycoon based on GSM positioning. To provide the information for visualizing the infrastructure, they drive around the deployment area and map GSM cells. They point out that this method does not scale and argue that a system is needed which collects this information unsupervised during the game and can handle changes over time.

In comparison, this paper proposes PosQ, a system for unsupervised fingerprinting of GPS positioning quality that, both prior to deployment and during use, can provide information about positioning quality. The system represents changes over time, in 2D and 3D, and shows differences among GPS receivers. The system can provide visual overlays that can either be shown to users and administrators or imported into authoring tools to visualize the infrastructure.

3 PosQ - Fingerprinting Positioning Quality

The PosQ system consists of four key elements: GPS receivers that estimate positioning quality, PosQ collectors that efficiently report quality information to a server-side quality database, quality maps built from the database, and visual overlays generated from the maps (see Figure 1). Here, a visual overlay means a transparent information visualization layer enhancing the underlying map. In the following subsections, we describe each of these elements and how they are designed to help meeting PosQ's goals of unsupervised fingerprinting and visualization of GPS positioning quality.

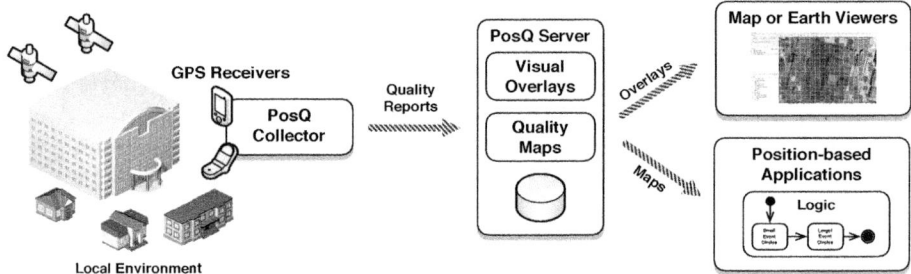

Fig. 1. The four key elements of the PosQ system

3.1 GPS Receivers

As described under related work, most GPS receivers will deliver – in addition to position and time estimates – the accuracy of the position fix. Such estimates are provided both by the built-in GPS receivers in HTC and Nokia mobile phones (available in their standard location APIs) and dedicated GPS receiver from manufactures such as U-Blox (available in PUBX NMEA messages) and Sirf (available using the Sirf binary protocol). PosQ collects such estimates for both the horizontal and vertical accuracy to fingerprint the positioning quality.

3.2 PosQ Collectors

The PosQ collector is a software component deployed on participating devices that collects quality tuples (time, position, horizontal accuracy, vertical accuracy) from either external or internal GPS receivers. The fingerprinting process is therefore fully unsupervised. Depending on the scenario, the information might be stored at the devices until collection is finished or be continuously reported to the server-side quality database. For the latter case, PosQ collectors apply aggregation as well as threshold and area-based updating strategies to limit the number of quality reports. These strategies are described later on.

The collectors can be deployed on the devices as part of a position-based application or as a stand-alone client. In the former case, the collector will enable the user or administrator to judge the positioning quality from visual overlays or for application developers to adapt the logic of their application to the positioning quality in a certain area. It might also be deployed as part of a positioning client, e.g., the client provided by Skyhook Wireless [1]. In the latter case, deploying the collector as a stand-alone client will support administrators in fingerprinting the coverage to judge the quality and know where to augment the positioning with other positioning means to reach a certain coverage and accuracy level.

3.3 Quality Maps

The quality map component provides an expressive and compact interface to the information stored in the quality database. The component aggregates the information stored in the database with regard to location, time stamp and GPS receiver type. Location means that position fixes which are geographically close are combined, because they are likely to have similar error estimates whereas far away positions are independent of each other. Time stamp, because GPS errors are similar within short periods of time. GPS receiver type, because GPS receivers differ in positioning coverage and accuracy. The component also provides meta information about the up-to-dateness of the information.

The component can provide both 2D and 3D quality maps of selected areas. The maps are represented as multidimensional histograms with either two or three dimensions. The histograms are built by aggregating the quality information placed within a histogram bin using the GPS position. The size of the bins is a configurable parameter of the system, but the values should be chosen small enough for the bins to only cover areas with similar positioning qualities. Therefore, for the two horizontal dimensions the size should be chosen to match the size of buildings or building parts whereas for the vertical dimension it should be chosen to match, e.g., the separation of floors.

The quality tuples aggregated by the component will, in high accuracy areas, contain accurate GPS positions and therefore be assigned to the correct bins. However, in low accuracy areas quality tuples might not be positioned within the correct bin and therefore be assigned to a neighboring bin, an effect that we call *pollution* of quality tuples. A *Polluted area* means an area that a calculated position is wrongly mapped to. This pollution of neighboring bins will depend on the accuracy: lower accuracy will increase the polluted area around the true position, but on the other side, an increased area will also lower the absolute impact on each neighboring bin. PosQ counters this problem by design by aggregating quality information over space and time. This means that in high accuracy areas visited by a collecting device, bins will contain many correct tuples. There may also be some polluted tuples, but because the correct tuples outnumber the polluted ones, the bins will be correctly mapped as having high accuracy. To further reduce pollution we include a weighting step in the aggregation that takes into account how jumpy the GPS positions are calculated from the recent history of GPS positions. To aggregate information over time, a weighted average is used that values recent information higher than older information.

To further reduce the pollution problem we have implemented a predictive component following the ideas of Steed [17] that based on information about satellite orbits and a city model estimates for each bin if more than three satellites can be received by line of sight. The information produced by this component is used to mark high accuracy areas prior to the addition of the fingerprinting data thus removing the pollution problem in these areas. We will in the rest of this paper not focus on this component as it is not able to provide detailed quality maps for urban canyons and indoor areas because it does not model signal reflection and attenuation.

In the case that a city model is not available the system might not have any data for a high accuracy area, in this case it could happen that only pollution from a nearby low accuracy area is present. However, firstly, this information will quickly be corrected when a collecting device enters this area and secondly, marking a good area as bad is less a problem than if the system was marking bad areas as good. Within low accuracy areas, the bins might pollute each other, which can slightly skew the values, but never change a low accuracy bin to be shown as a high accuracy one. In other words: Pollution is conservative of accuracy, meaning that it will not falsely upgrade a low-accuracy bin, even if it might downgrade one.

3.4 Visual Overlays

To inspect the positioning quality in an area, PosQ provides visual overlays that color an area depending on either the horizontal or the vertical accuracy of the associated accuracy bin. The overlays are produced in the widely used KML [19] format supported by many map and earth viewers. The map coloring is scaled from pure green for high accuracy (zero meters) over red (above ten meters) to black for low accuracy (forty meters or more) to easily communicate the properties of an area. The visual overlays can both be provided in 2D and 3D and be animated to show timely changes.

3.5 Implementation

The PosQ Collectors have been implemented for several operating systems and run on different types of devices, as listed in Table 1. The PosQ server-side

Table 1. PosQ Collectors implemented for different operating systems, programming languages, devices and GPS receivers

Operating Systems	Languages	Device	GPS		
			Chip	External	Built-in
Windows XP	Java	Asus Eee	U-Blox EVK-5H	•	
Symbian OS	Python	Nokia N95	TI NaviLink NL5350		•
Android OS	Python	Dev Phone 1	Qualcomm MSM7201A		•

quality map and visual overlay components have been written in Java and run in a standard OSGi [2] platform. The communication between the collectors and the server uses TCP connections. To provide easy access for position-based applications, the server-side components are accessible as webservices.

4 Evaluation of Quality Fingerprinting

PosQ's goal is to accurately fingerprint GPS positioning quality. To provide evidence for this ability, we will in this section present results from deploying PosQ in a shopping mall, a cultural history museum and a botanic museum. We will, firstly, argue that the quality maps provided by PosQ are accurate. Secondly, that the quality maps stay valid as the GPS satellite constellation changes over time. Thirdly, that the system can fingerprint the difference in accuracy among GPS receivers. For this presentation, we will focus on 2D quality maps and horizontal accuracy as this is the most common usage of GPS positions.

4.1 Quality Maps are Accurate

To argue that the quality maps provided by PosQ are accurate, we will discuss the results from using the Windows XP PosQ collector with an external U-Blox GPS to fingerprint three areas of a shopping mall: An outdoor area with generally high accuracy (less than three meters), an indoor area with medium accuracy (less than ten meters), and an indoor area with low accuracy (more than ten meters). The visual overlay produced by PosQ from the collected data is shown in Figure 2, together with the collection routes walked in each area. Ideally, in the high accuracy area the bins should be green and sit over the route, in the medium accuracy area the bins should be darker green or light red (worse than ten meters) and be slightly off the route and in the low accuracy area the bins should be mainly red and black (worse than forty meters) and be some bins off the route. As one can see from the figure, these differences indeed show up on the figure as, e.g., in the low accuracy area the bins are less often placed directly over the route as the positioning errors misplace fingerprints in neighboring bins. This even partly pollutes the bins around the medium accuracy area, but none of the medium-accuracy bins is polluted, because the many correctly placed tuples within these bins will outnumber the few polluted ones.

To study the absolute accuracy of the quality maps in greater detail we compare the PosQ quality map values with Root Mean Square (RMS) deviations from ground truth for the bins visited by the collector. However, the used U-blox receiver's documentation does not state which error measure or quantile the error estimates are supposed to match. We have chosen to compare to RMS because it is a standard error measure and is the measure reported by the JSR-179 location API. The RMS errors are calculated from the differences between the GPS positions stored in the collected tuples and manually recorded ground truth positions by the collector. Figure 3 plots in a scatter plot for each visited bin the RMS deviations from ground truth and the corresponding quality map value.

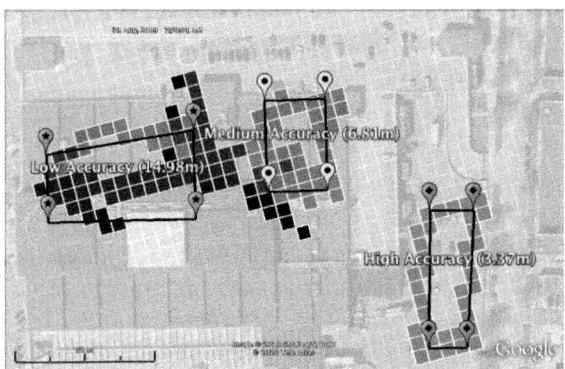

Fig. 2. Visual overlay for a shopping mall as a colored map with scale from bright green (high accuracy) over red to black (low accuracy)

The comparison shows that the map values are for most bins very predictive for the real error. In the low- and medium-accuracy areas, the bin values have a tendency to overestimate the real errors by a couple of meters. The increased deviation for low accuracy area can both be attributed to increased pollution but also that in weak signal conditions receivers will have poorer measurement (e.g., fewer satellites and more fluctuations) to use for the error estimation.

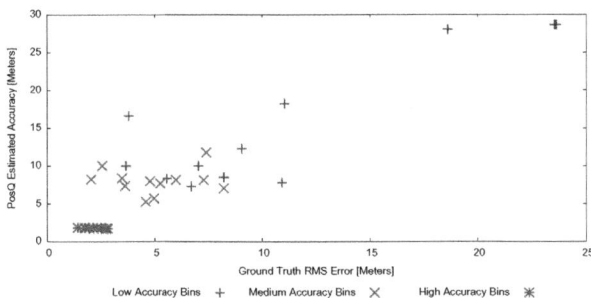

Fig. 3. Comparison of ground truth accuracy with PosQ estimated accuracy for individual bins

4.2 Quality Maps Remain Informative When Satellite Constellation Changes

GPS satellites are moving in their orbits around the earth and therefore the constellation of satellites visible over an area is changing all the time. The orbits of the satellites are designed in such a manner that the same constellation repeats nearly twice a day. For quality maps to be a useful tool, they have to remain informative when the satellite constellation changes. A main argument why they might remain informative is that the impact of the local environment (e.g., indoors because of attenuation) has a high influence on the accuracy. The experimental observations presented by Kjærgaard et al. [8] for indoor environments support this claim, as their results indicate that indoors low accuracy is better explained by multi-path and weak-signal effects than the geometric strength of the current satellite constellation measured as the *dilution of precision* factor.

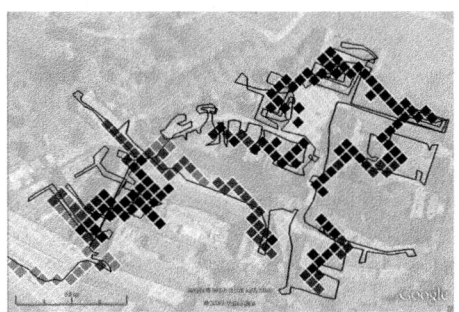
(a) First Satellite Constellation

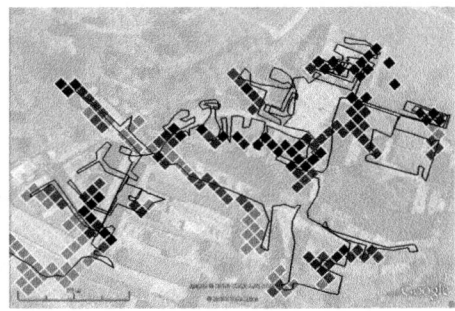
(b) Second Satellite Constellation

Fig. 4. Visual overlay collected with a G1 device for two different satellite constellations

To support the argument that quality maps remain informative, we have collected fingerprints in a cultural history museum using the Android PosQ collector on a G1 device for two different satellite constellations. We collected the fingerprints on two different days at a different time of day to make sure that the constellations were different. The visual overlays produced by PosQ for each of the two quality maps are shown in Figure 4(a) and 4(b), respectively. Each day we walked the route shown in black on the overlays. The route covers local environment with different conditions such as: outdoors, outdoors in building shadows, indoors in buildings of light building materials (e.g. wood) and buildings of heavy building materials (e.g. bricks and stones).

In both figures one can observe that the G1 receiver is not able to provide complete coverage within this area. If one compares the areas in terms of coverage, the high accuracy areas in the lower left corners and in the middle left part of the overlays correspond. For the low accuracy areas in the upper middle and the upper right corner parts, both overlays mark this area as bad. The placement of the colored bins is a bit offset in the low accuracy areas which is due to the fact that the higher errors will offset the GPS positions in different directions. Therefore, the general trends remain the same, but there are some variations in other areas, where generally the accuracy on the overlay for the second constellation is slightly better than for the first constellation. These variations can both be attributed to changes in the number of people around and the changed constellation. PosQ handles these changes by updating its information as devices reenters an area.

4.3 Quality Maps Capture the Differences among GPS Receivers

The coverage and accuracy of a GPS receiver depend, among other things, on the receiver design, the receiver shielding which protects it from nearby electronic components, and the antenna [13]. Quality maps can capture such differences in GPS receiver performance, for instance, to determine that the same GPS receiver chip mounted in two different phone models might not result in the same accuracy because of differences in antennas or interference from nearby electronic components. To evaluate if the quality maps can capture the differences among GPS receivers, we collected fingerprints using the Windows XP PosQ collector with an external U-blox GPS chip at the same time as with the Android PosQ collector on a G1 device. The devices was carried several centimeters apart to avoid any interference. A limitation of the comparison is that the Android location API does not specify an exact error measure for the accuracy estimates.

The visual overlay produced by PosQ for the quality map is shown in Figure 5. The U-blox receiver has a good coverage in the area and nearly all parts of the walked route are colored, except for some bins in the area with several floored brick houses around the upper middle part. The accuracy is generally high, especially when compared to the accuracy of the G1 for the same constellation in Figure 4(b). If one compares the U-blox and the G1 results, it becomes apparent that even though the levels of accuracy and coverage are different, the same areas have been identified as difficult, for instance, the upper middle part.

5 Applications of PosQ

PosQ can collect quality maps that are accurate and remain valid over time, and it can capture differences in GPS receivers. This opens up possibilities for a number of applications. Firstly, PosQ can be used to design a positioning solution. Secondly, PosQ can help developers adapt their position-based applications to different GPS receivers. Thirdly, PosQ can visualize positioning quality over time and in 3D. We detail these in the remaining of this section.

5.1 Designing a Positioning Solution

In order to deploy a successful position-based application for the cultural history museum, an adequate positioning solution is required. To determine if GPS could be used, we collected fingerprints with PosQ by walking a tour through the museum for 35 minutes as described in the previous section. The result (as one can judge from the overlays in Figure 4 and 5) is that the dedicated U-blox receiver provided positioning in nearly all areas whereas the G1 phone only provided positioning in roughly half of the area. Therefore, a second positioning solution other than GPS is needed to provide coverage in the case of the G1 mobile phone. Because many smart phones today implement GSM and WiFi positioning, these are relevant second options. However, the accuracy of GSM positioning on phones is generally poor [11] and therefore the best secondary

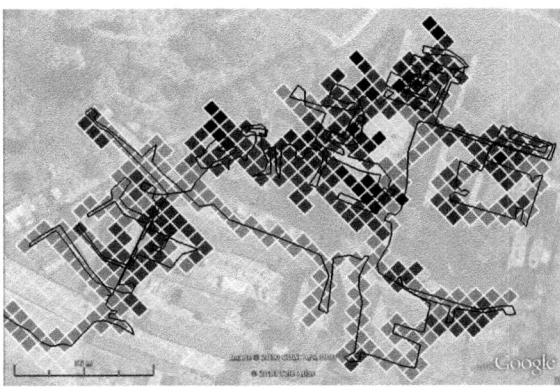

Fig. 5. Visual overlay collected with a U-blox receiver for the second satellite constellation

option is WiFi positioning. As companies providing WiFi positioning (e.g. Skyhook Wireless [1]) allow users to report the location of their own access point, a solution would be to deploy access points in the area and then enter them into such databases.

The cultural museum site is today not covered by WiFi access points and therefore one has to select a placement of access points for positioning. Because the cultural museum shows houses built between 1700 and 1920, the access points must not be mounted in visible locations. Therefore a goal is to add WiFi Positioning where it is needed, using a minimal number of access points. The visual overlays produced by PosQ can thus be used to select an optimal placement of access points. Other positioning options such as bluetooth or IR beacons could be considered using the same approach.

5.2 Mapping between Different GPS Receivers

Many position-based applications define trigger zones which are physical areas where some application logic should be executed as the user enters the area. A developer of such an application would often only design and test on a restricted number of mobile phones. The developer might use the results of Randell et al. [16], who advise that the size of trigger zones should be at least two times the accuracy of the positioning system. To customize the game play, the developer might further increase the size of trigger zones. In this process the developer can use the visual overlays produced by PosQ to decide on the zone sizes in specific areas as also suggested by Oppermann et al. [14]. The problem is: What happens when a user has a phone with another GPS receiver that might either be better or worse than the one used in the designing phase? A solution based on PosQ is to run PosQ on the mobile device and let PosQ compare the fingerprints for the designer's phone with the fingerprints collected by the user's phone to estimate differences in accuracy. We have implemented such a quality mapper extension to PosQ that enables the system to find such mappings.

To test the quality mapper, we used our fingerprints from the deployment of PosQ in a botanic museum and tested with a mapping from the G1 (designer's phone) to a Nokia N95 (user's phone). The case of N95 actually requires the system to find two different mappings as the GPS performance of the N95 depends on whether the keyboard is slid out or not. In the botanic museum case, the keyboard was not slid out, which contributed to the poor coverage of the N95 compared to the G1. The quality mapper compares the values of bins with fingerprinting data from both phones and applies linear regression to find the parameters for a linear model $y = ax$ where y is the values on the user's phone and x the values on the developer's phone. For the given scenario, the mapping parameter a was estimated to be 15.32 which means that the developer's phone is estimated to be fifteen times more accurate that the user's phone. In comparison, when the keyboard is slid out the factor is only 3.71. In an application you would use the factor to scale the size of the trigger zones to adjust to the new phone.

The reason why developers may want to adapt application logic using PosQ instead of the raw estimates from the GPS is that they can not judge the quality of a specific GPS from a single value. By comparing fingerprints between the developer's test phone and the user's phone, you can adapt to the differences between the two phones and not just the current GPS error.

5.3 Visualizing Positioning Quality over Time and in 3D

Visualization of positioning quality is relevant both for end users (as shown by Dearman et al. [6]) and for designers and developers (as argued for by Oppermann et al. [14]). Previous work has considered 2D visualizations for the horizontal accuracy, but, e.g., Oppermann et al. [14] pointed out that also changes over time are relevant. PosQ supports views over time by providing animated visualizations which can show how positioning quality changes as reported by collectors for different points in time.

Because PosQ is based on a 3D quality model and collects information about both the horizontal and the vertical accuracy, the system can visualize the vertical accuracy in 3D. For the cultural museum, a visualization of the 3D model

Fig. 6. Visual overlay in 3D for vertical accuracy

is shown with vertical accuracy information in Figure 6. The 3D visualization both provides hints about the curvature of the terrain but also changes for the buildings where the collecting person visited several floors. The places where the vertical accuracy is above ten meters (red to black) is actually co-located in most cases with the places shown with a black circle where the person entered a building and visited several floors within it. The floor changes makes the GPS more uncertain about the height compared to buildings where no floor changes occur.

6 Efficient Collection of Fingerprints

There are several reasons why PosQ should support the efficient updating of quality information in real-time. Firstly, it can be used as motivated by Benford et al. [4] in the context of location-based games to visualize the positioning quality for players and system administrators during play. Secondly, when using positioning for critical decision making, e.g. when the police or fire fighters enter an unknown building to build up a quality map over where positioning is possible and with what quality. In this section, we will focus on 2D quality maps and horizontal accuracy.

It would be problematic to support this by just letting PosQ collectors continuously report all collected information to the quality database. Firstly, the quantity of collected data adds up to 140 MB per hour, which would use a significant amount of bandwidth and result in scalability problems for the quality database as the number of participating devices increases. Sending this amount of data from the mobile devices would also be a problem because of the power consumption of the radio, e.g. the 3G radio in a N95 phone uses around 1.1 watt which if used continuously would decrease the battery life with a factor of twenty [9].

6.1 Updating Strategies

To address this, PosQ collectors implement several smart updating strategies for positioning quality. Firstly, the collectors apply aggregation to tuples to limit the amount of data per update. Secondly, they apply threshold-based updating to only send updates if the quality deviates from former values above a configurable threshold. Thirdly, they apply area-based updating to send an update when the device has moved through a configurable number of bins.

Figure 7 illustrates the steps involved. The PosQ collector demands GPS positioning and uses this initial position to request data for the bins (1). The quality database returns three parameters for each bin to the client: the aggregated accuracy, a time quality and a measurement quality (2). The collector processes a GPS position and the logic determines that a positioning quality update is not needed (3). Later, a GPS position is processed where the client-side logic determines that the used threshold has been exceeded and initiates a positioning quality update (4). The server processes the update and sends back updated bin data (5). Even later, a GPS position is processed where the collector logic determines that the device has moved more that the configured number of bins and initiates a positioning quality update (6). The server processes the update and sends back updated bin data (7).

6.2 Evaluation Setup

We have evaluated these strategies in our shopping mall deployment using the three different areas, as shown in Figure 2: an outdoor area with high accuracy, an indoor area with medium accuracy and medium changes in accuracy, and an indoor area with low accuracy and large changes in accuracy. For each area a path with four corner points was measured. The path in the high accuracy area has a length of 131 meters and passes through 30 bins, in the medium accuracy area the path has a length of 110 meters and passes through 30 bins and in the low accuracy area the path has a length of 168 meters and passes through 36 bins. To know the real position of the GPS receiver during evaluation, the position of the receiver was recorded manually.

Several runs of evaluations were performed along each path to match three motion patterns: (a) walking with stops at all four corner points of a path for one minute each and walking the path twice; (b) continuously walking the path three times with a stop of one minute in the beginning and in the end; (c) standing three minutes at a corner point and then walking to a second corner point to wait three more minutes.

For each evaluation, the PosQ collector and a reference client were run in parallel. The reference collector continuously sends reports to the quality database which means that it provides minimal communication efficiency but maximal information quality. The threshold-based strategy was configured with a deviation threshold factor of 1.5 and the area-based strategy with an area update

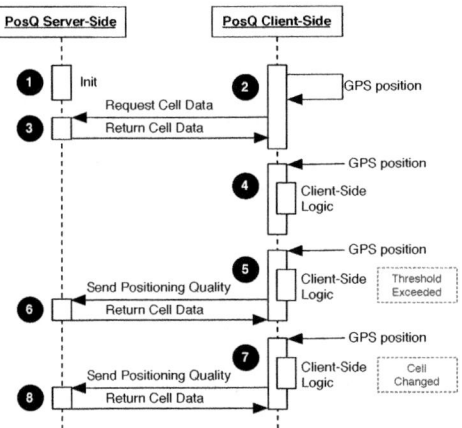

Fig. 7. The steps in the communication between the PosQ collector and the quality database

distance of two bins. These parameters were chosen to test the strategies in a difficult case with tight tracking of the quality in the area due to the small deviation threshold and the short update distance. The goal of PosQ is to use fewer messages than the reference client, while only causing a small or negligible loss in information quality.

6.3 PosQ Decreases the Number of Quality Reports

PosQ is able, when compared to the reference client, to decrease the number of needed quality messages as indicated by the results shown in Table 2 for *motion pattern a*. The number of messages sent by the reference client is linear with time whereas the PosQ client only sends updates when determined by the collector logic. However, because each PosQ update also includes a message with new parameters back to the collector each update is counted as two messages. The PosQ client is in the high accuracy area able to save a factor 9.3 compared to the reference client where as in the low accuracy area it is able to save a factor of 2.9. The difference is both due to that in the high accuracy area there were 47 threshold updates compared to two in the low accuracy area and that in the high accuracy area there were 69 area updates, while only 46 in the low accuracy area. The increased number of updates is due to the more varying accuracy throughout the area as can be seen from Figure 8(b). The results for the medium accuracy area (a factor of 4.8) is in between the two others.

Figure 8 shows the results of the evaluation for *motion pattern a* in both the low and high accuracy areas. Our results indicate that PosQ only impacts the information quality with a small loss compared to the reference client. In both areas, the updates delivered by PosQ tightly match the information reported by the reference client. This is also true for the medium accuracy area (not shown). Figure 8(a) and 8(b) also show when updates are due to bin changes as small arrows. For the low-accuracy area, there were 69 area updates and 46 area updates in the high-accuracy area, which is more than the 37 and 31 area updates necessary, respectively. This could of course be lowered by using a larger area update distance.

The motion patterns have an impact on PosQ performance, because most updates are produced in motion and fewer updates during standing periods. The reason is that the estimated accuracy is more stable when the receiver is

Table 2. Reference compared to PosQ for the number of messages for *motion pattern a* on the routes shown in figure 2

	Accuracy of Area		
	Low	Medium	High
Reference Messages	677	902	893
PosQ Messages	232	188	96
Improvement Factor	2.9	4.8	9.3

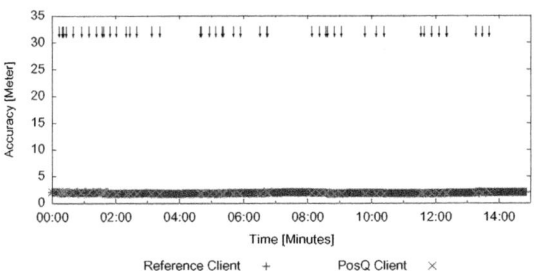

(a) Estimated accuracy in high accuracy area (arrows indicate area updates)

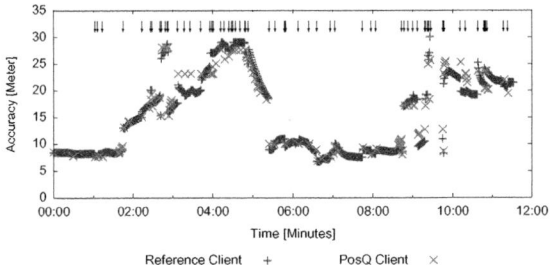

(b) Estimated accuracy in low accuracy area (arrows indicate area updates)

Fig. 8. Results for Positioning Quality

kept motionless and only few erroneous area updates are produced. This can also be noticed from our evaluations for *motion pattern c*, where most updates are produced for the motion-full *pattern b* compared to the nearly motionless *pattern c*. In summary, PosQ is able to decrease the number of needed quality messages with only a small loss in information quality.

7 Conclusion

This paper proposed PosQ, a system for unsupervised fingerprinting and visualization of GPS positioning quality. The system provides quality maps to position-based applications and visual overlays to users and managers to reveal the positioning quality. PosQ reveals the quality in three dimensions: As it changes over time, in 2D and 3D, and for each type of GPS receiver. The system enables unsupervised fingerprinting by running in the background of a GPS enabled device without any need of attention by the device user. It enables people and applications to judge the GPS positioning quality in a local environment with low effort, e.g., in a mixed indoor-outdoor environment in the scope of a city.

Our evaluation of the system provided evidence that the collected quality maps are accurate, that they remain informative over time and that they capture the differences among GPS receivers. The evaluation took into account different local environments and conditions. We demonstrated how the system enables the design of positioning solutions, mapping between different GPS receivers and visualization of positioning quality over time and in 3D. Furthermore, we proposed and evaluated methods for efficient collection of quality reports and showed that the methods can significantly decrease the number of quality reports needed when updating quality maps in real-time. As the PosQ collectors can be deployed on phones as part of applications the only disadvantage for users is that they have to accept the few extra resources that the system uses to realize the benefits of the system.

In our ongoing work we are going to address mainly four issues. Firstly, to integrate the system with other forms of positioning such as WiFi and GSM positioning to fingerprint their positioning quality. Secondly, integrate PosQ with an authoring tool for position-based applications and extend the tool with methods for making use of PosQ's capabilities to provide visual overlays over time and in 3D. Thirdly, study in more detail the problem of pollution of neighboring bins. Fourthly, consider how to integrate privacy protection measures for data contributors.

Acknowledgements. The authors acknowledge the financial support granted by the *Danish National Advanced Technology Foundation* under J.nr. 009-2007-2.

References

1. Skyhook wireless, http://www.skyhookwireless.com/
2. Alliance, O.: Open Services Gateway Initiative (2009), Specification download http://www.osgi.org/Download/Release4V42 (Online, cited February 18, 2010)
3. Bahl, P., Padmanabhan, V.N.: RADAR: An In-Building RF-based User Location and Tracking System. In: Proceedings of the 19th Annual Joint Conf. of the IEEE Computer and Communications Societies (2000)
4. Benford, S., Crabtree, A., Flintham, M., Drozd, A., Anastasi, R., Paxton, M., Tandavanitj, N., Adams, M., Row-Farr, J.: Can you see me now? ACM Trans. Comput.-Hum. Interact. 13(1), 100–133 (2006)
5. Constandache, I., Choudhury, R.R., Rhee, I.: Towards Mobile Phone Localization without War-Driving. In: Proceedings of the Annual Joint Conf. of the IEEE Computer and Communications Societies (2010)
6. Dearman, D., Varshavsky, A., de Lara, E., Truong, K.N.: An Exploration of Location Error Estimation. In: Krumm, J., Abowd, G.D., Seneviratne, A., Strang, T. (eds.) UbiComp 2007. LNCS, vol. 4717, pp. 181–198. Springer, Heidelberg (2007)
7. Drozd, A., Benford, S., Tandavanitj, N., Wright, M., Chamberlain, A.: Hitchers: Designing for Cellular Positioning. In: Dourish, P., Friday, A. (eds.) UbiComp 2006. LNCS, vol. 4206, pp. 279–296. Springer, Heidelberg (2006)
8. Kjærgaard, M.B., Blunck, H., Godsk, T., Toftkjær, T., Christensen, D.L., Grønbæk, K.: Indoor Positioning Using GPS Revisited. In: Floréen, P., Krüger, A., Spasojevic, M. (eds.) Pervasive 2010. LNCS, vol. 6030, pp. 38–56. Springer, Heidelberg (2010)

9. Kjærgaard, M.B., Langdal, J., Godsk, T., Toftkjær, T.: EnTracked: Energy-Efficient Robust Position Tracking for Mobile Devices. In: Proceedings of the 7th Int. Conf. on Mobile Systems, Applications, and Services, pp. 221–234 (2009)
10. Küpper, A.: Location-Based Services: Fundamentals and Operation. Wiley (October 2005)
11. LaMarca, A., Chawathe, Y., Consolvo, S., Hightower, J., Smith, I., Scott, J., Sohn, T., Howard, J., Hughes, J., Potter, F., Tabert, J., Powledge, P., Borriello, G., Schilit, B.: Place Lab: Device Positioning Using Radio Beacons in the Wild. In: Gellersen, H.-W., Want, R., Schmidt, A. (eds.) PERVASIVE 2005. LNCS, vol. 3468, pp. 116–133. Springer, Heidelberg (2005)
12. Lemelson, H., Kjærgaard, M.B., Hansen, R., King, T.: Error Estimation for Indoor 802.11 Location Fingerprinting. In: Choudhury, T., Quigley, A., Strang, T., Suginuma, K. (eds.) LoCA 2009. LNCS, vol. 5561, pp. 138–155. Springer, Heidelberg (2009)
13. Misra, P., Enge, P.: Global Positioning System: Signals, Measurements, and Performance, 2nd edn. Navtech (2006)
14. Oppermann, L., Broll, G., Capra, M., Benford, S.: Extending Authoring Tools for Location-Aware Applications with an Infrastructure Visualization Layer. In: Dourish, P., Friday, A. (eds.) UbiComp 2006. LNCS, vol. 4206, pp. 52–68. Springer, Heidelberg (2006)
15. Patel, S.N., Truong, K.N., Abowd, G.D.: PowerLine Positioning: A Practical Sub-Room-Level Indoor Location System for Domestic Use. In: Dourish, P., Friday, A. (eds.) UbiComp 2006. LNCS, vol. 4206, pp. 441–458. Springer, Heidelberg (2006)
16. Randell, C., Geelhoed, E., Dix, A.J., Muller, H.L.: Exploring the Effects of Target Location Size and Position System Accuracy on Location Based Applications. In: Fishkin, K.P., Schiele, B., Nixon, P., Quigley, A. (eds.) PERVASIVE 2006. LNCS, vol. 3968, pp. 305–320. Springer, Heidelberg (2006)
17. Steed, A.: Supporting Mobile Applications with Real-Time Visualisation of GPS Availability. In: Brewster, S., Dunlop, M.D. (eds.) Mobile HCI 2004. LNCS, vol. 3160, pp. 373–377. Springer, Heidelberg (2004)
18. Varshavsky, A., de Lara, E., Hightower, J., LaMarca, A., Otsason, V.: Gsm indoor localization. Pervasive and Mobile Computing 3(6), 698–720 (2007)
19. Wernecke, J.: The KML Handbook: Geographic Visualization for the Web, 1st edn. Addison-Wesley Professional (November 2008)

SensOrchestra: Collaborative Sensing for Symbolic Location Recognition

Heng-Tze Cheng, Feng-Tso Sun, Senaka Buthpitiya, and Martin Griss

Department of Electrical and Computer Engineering
Carnegie Mellon University
{hengtze.cheng,lucas.sun,senaka.buthpitiya,martin.griss}@sv.cmu.edu

Abstract. Symbolic location of a user, like a store name in a mall, is essential for context-based mobile advertising. Existing fingerprint-based localization using only a single phone is susceptible to noise, and has a major limitation in that the phone has to be held in the hand at all times. In this paper, we present SensOrchestra, a collaborative sensing framework for symbolic location recognition that groups nearby phones to recognize ambient sounds and images of a location collaboratively. We investigated audio and image features, and designed a classifier fusion model to integrate estimates from different phones. We also evaluated the energy consumption, bandwidth, and response time of the system. Experimental results show that SensOrchestra achieved 87.7% recognition accuracy, which reduces the error rate of single-phone approach by 2X, and eliminates the limitations on how users carry their phones. We believe general location or activity recognition systems can all benefit from this collaborative framework.

Keywords: Collaborative sensing, mobile phone sensing, localization, context-awareness, context-based advertising.

1 Introduction

Context-based mobile advertising matches advertisement and e-coupons with potential customers according to their locations, activities, or interests [20, 21, 25]. Symbolic location of a user, such as a store name, is important since a user can get exclusive coupons based on their frequent visits, or receive product recommendations from similar shops. In this work, we focus on the problem of how to recognize the store (e.g. a café, an electronics store, or a clothing store) that a user is in when the user is shopping at a mall with a variety of stores, using only microphones and cameras on mobile phones.

GPS does not work in this scenario since most stores in malls are indoors. Although extensive work has been done in indoor localization [2, 15, 16, 23], most systems are still limited by infrastructure or specific hardware requirements, thus they are unlikely to be widely deployed in every store. Even with an indoor

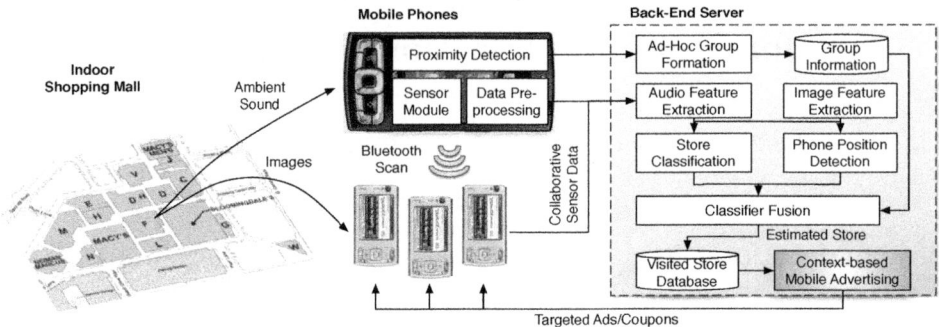

Fig. 1. The system architecture of SensOrchestra

localization system with room-level accuracy, slight errors in coordinates often place a phone at the wrong side of a wall dividing two adjacent stores [1].

Recent research, SurroundSense [1], achieved promising results on symbolic location inference using mobile phones to fingerprint ambient sound, light, color, and motion signatures. However, in order to capture images and audio, a major limitation faced by SurroundSense is that the phone has to be held in the hand at all times, which may not be the way general users usually carry their phones. Simply putting a phone in one's pocket renders the microphone muffled and the camera useless [17]. Since only one single phone is used to detect a user's surroundings, if the phone is facing a source of noise, it is difficult to eliminate the influence of noise and recover from an incorrect inference.

To address the challenges mentioned above, we have developed SensOrchestra, a multi-phone collaborative sensing framework for symbolic location recognition. Using Bluetooth-based proximity detection, SensOrchestra groups nearby phones to sense the ambience together, and then combines the correlated sensor data using a classifier fusion model for location inference. Using multiple phones implies concurrent sensing of the same environment from different positions, thus increasing the chance of getting more useful raw data for context inference and becoming less susceptible to ambient changes. This eliminates the major limitation that the phone has to be held in the hand at all times to recognize a location. Furthermore, SensOrchestra achieves promising results using only microphones and cameras, which are the basic sensors on most of today's mobile phones, without the need of any custom-made hardware or specific infrastructure.

A natural concern is that as the number of stores increases, the fingerprint-based approach seems impractical because it is unlikely that we can find unique audiovisual signatures for every store in a city. However, it should be noted that our approach does not need to be able to differentiate all the stores in distant locations. Although existing GPS or GSM-based positioning system was shown to be unable to identify different neighboring stores [1], it is accurate enough for determining locations in a macro-scale, such as the name or a specific part of a mall. Once the macro-location is known, the candidate symbolic locations can be confined to a limited set. The importance of fingerprint-based indoor localization system is thus to discriminate the ambient signatures of the fine-grained symbolic locations in the same macro-location, such as several neighboring shops in the same part of a mall.

The main contributions of this paper are:

- The design, implementation, and evaluation of SensOrchestra, a collaborative symbolic location recognition system that combines sensor data from multiple nearby phones.
- The design of a classifier fusion model to integrate estimates from multiple phones, and compensates the incorrect estimates caused by ambient noise.
- The investigation of multiple audio and image features, and a complete experiment with realistic setting to compare the effect of different phone positions and multi-phone sensor data fusion.

The paper is organized as follows. In Section 2, we discuss and compare related work. The system design and the method for collaborative location recognition are elaborated in Section 3, and the implementation details in Section 4. In Section 5 and 6 we discuss the dataset used for evaluation and the experimental results. Discussion and conclusion are elaborated in Section 7 and 8, respectively.

2 Related Work

In the field of physical indoor localization [16], the Cricket system [23] achieved centimeter-scale localization using RF and ultrasound beacons installed in the surroundings. While effective for high-budget applications, this kind of system is unlikely to be widely installed in every store in a city. The RADAR [2] system achieved 5-meter accuracy using Wi-Fi fingerprinting, with a tradeoff that careful calibration of Wi-Fi signal strengths are needed at many physical positions in the building, which may not scale over wide areas. In SensOrchestra, we use only mobile phones that people already have. Therefore, our system is low-cost and can be applied to most stores, without the need for additional infrastructure or custom-made hardware.

There has been an emerging interest in user context inference using mobile phones in recent years. While CenceMe [18] infers user activities and social context using microphone and accelerometer on mobile phones, SoundSense [17] achieves general sound classification (ambient sound, music, speech) with over 90% accuracy and learns new sound events in different users' daily lives. SurroundSense [1] incorprates more modalities, including microphone, camera, accelerometer, and Wi-Fi, to achieve ambience fingerprinting for symbolic localization. However, one common feature is that no benefit is gained when there is more than one phone nearby. Building on their ideas, we extend their work by introducing a multi-phone collaborative framework that removes the limitation on where users put their phones and lessens the susceptibility of a single phone to ambient noise.

Similar to fingerprint-based localization, the notion of scene classification has been investigated extensively in the field of computer vision and audio recognition. In [5], probabilistic Latent Semantic Analysis (pLSA) is used to model the latent topics in the images, which are then classified into different scene categories like coast, mountain, or city. In [9,11], both time- and frequency-domain features of audio are extracted to recognize high-level context like locations (street, home, office) or sound events (rain, thunder, train). A major difference is that these works focus mainly on novel statistical learning techniques, using high-quality audiovisual data recorded by standalone cameras or microphones. On the contrary, our work addresses real-world challenges when most phones are resource-limited and not in good positions to sense the environment.

There has also been related work exploring the idea of collaborative or participatory sensing [3,7,19,22]. The PIER system [19] uses participatory sensing to collect GPS traces and calculate personal environmental impact and exposure. In VUPoint [3], nearby phones are grouped together to collaboratively record video of social events triggered by ambient changes, but they did not use the recorded video to infer high-level user context. In addition, rather than prompting the user to actively record events or take sensor readings, SensOrchestra senses the ambience in the background without the need of user's attention. This introduces additional challenges because the sensor data can be noisy and less informative. Different from them in terms of primary goal, sensor data type, and statistical learning methods, our work is one of the earliest attempts to use collaborative sensing for symbolic location recognition.

3 System Design and Method

Given a set of audiovisual observations of the ambience sensed by multiple nearby phones, the goal of SensOrchestra is to estimate the symbolic location that a user is most likely at. The system architecture is shown in Figure 1. We describe each part of the system as follows.

3.1 Proximity Detection and Group Formation

The purpose of proximity detection is to assign the phones in the same symbolic location to the same group for collaborative sensing. Among several candidate techniques, such as the acoustic grouping using short high-frequency chirps outside the audible frequency range [3], we adopt Bluetooth for proximity detection because it is simple and available for most devices without the need for additional processing. While Bluetooth signal can sometimes pass through walls, other state-of-the-art proximity detection techniques can be further incorporated to ensure that all the phones in a group are in the same store.

Our approach works as follows. First, a phone performs a background Bluetooth scan and transmits a list of Bluetooth addresses of discovered devices to the server. The Bluetooth address for each device is a unique, 48-bit address (e.g. 00:12:d2:41:35:e4). After receiving the address list, the server clusters two

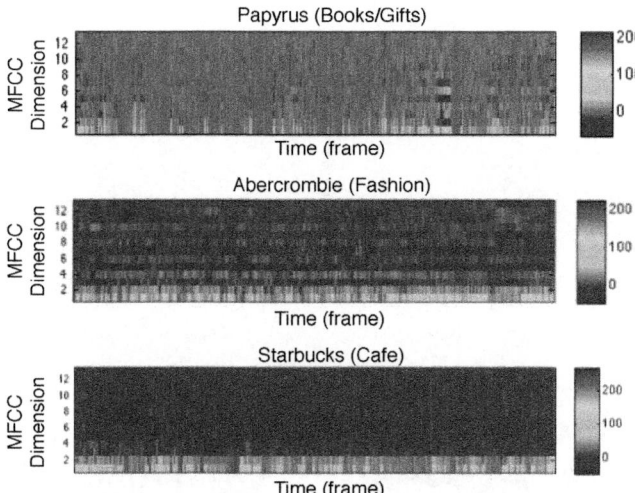

Fig. 2. An example of MFCC features extracted from three stores. The color shows the feature value of each dimension at each time frame.

devices in the same group if both of them appear in each other's list. Since the range of Bluetooth is roughly 10 meters (32.8 feet), it is suitable for forming a group in a typical-size store. The reason why we do not use Bluetooth pairing is the concern of intrusiveness and time. If a mobile phone prompts a user to pair with another device every time a new device is found, it would be highly intrusive, time-consuming and thus undesirable. To preserve the privacy of other mobile phone users, the server discards the Bluetooth address coming from any device that is not running SensOrchestra.

3.2 Mobile Phone Sensing

After an anonymous ad-hoc group is formed, each phone records audio at an 8 kHz sampling rate and take one image every 8 seconds. Each sensing session is 30 seconds. After a sensing session ends, the sensor data and the list of discovered Bluetooth addresses are transmitted to the server through 3G or Wi-Fi connection.

3.3 Feature Extraction

Audio Feature Extraction. For audio features, we adopt Mel-frequency cepstral coefficients (MFCC), one of the most important features for audio signal processing, speech recognition, and auditory scene analysis [9, 12]. MFCC describes the short-term power spectrum over frequency bands from 20 to 16000 Hz, with finer details in the bands to which human ears are sensitive. For each

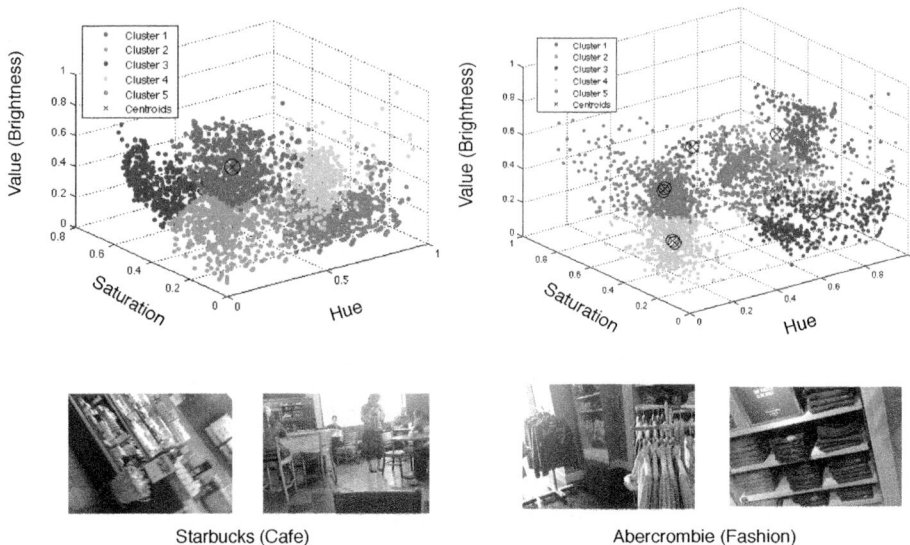

Fig. 3. Example of images taken from two stores and the corresponding dominant color distributions

30-second audio, we extract MFCC using a window size of 256 samples (32 ms when the audio sampling rate is 8kHz). The resulting feature is a 13-dimensional vector for each time frame. We calculate mean and standard deviation of each dimension over the 30-second interval, resulting in a 26-dimensional feature vector (13 for mean, 13 for standard deviation) for each sensing session. Examples of MFCC features are shown in Figure 2, from which we can roughly observe that different stores have different signatures (e.g. high-frequency coffee machine sounds in a cafe) in each dimension of MFCC.

Image Feature Extraction. Since different stores differ in light and dominant colors, for image features we adopt the dominant color distribution [1, 10], a widely used feature for image classification. In each 30-second time frame, we first concatenate the 4 images taken in the session into one combined image, and convert the image from RGB into Hue-Saturation-Value (HSV) color space. The reason that we use HSV color space is because of its similarities to the way humans tend to perceive color, and it is less sensitive to shadow and shading [5]. Since the dimension of the combined image is very high, dimension reduction or a clustering algorithm is needed to extract the compact information for analysis while discarding the redundancy. Therefore, in each image we cluster all pixels by K-means clustering algorithm [13], so that the characteristics of each image can be represented by a small number of clusters. The resulting clusters roughly represents the dominant colors (e.g. red, brown, white, etc.) in a particular store. As shown in Figure 3, the images from the two stores differ in cluster centroid

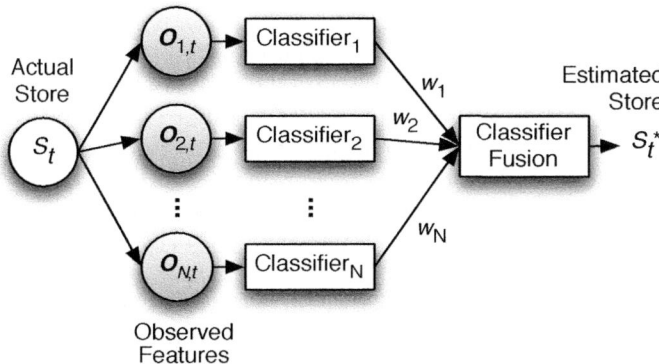

Fig. 4. Classifier fusion model for multi-phone collaborative location recognition

(HSV value of the dominant colors) and cluster size. In our implementation, K is empirically set to 5.

To compare the similarity between the color distribution of two images, we adopt the notion of color similarity measure described in [10] and [1]. The intuition is that two images, I_1 and I_2, are similar if they both have many pixels of similar colors. Let $\{C_{1,1}, C_{1,2}, ..., C_{1,K}\}$ and $\{C_{2,1}, C_{2,2}, ..., C_{2,K}\}$ represent the set of pixel clusters in image I_1 and I_2, respectively. The intuition leads to the following definition of color similarity S:

$$S(I_1, I_2) = \sum_{i,j} \frac{1}{\delta(i,j)} \frac{n(C_{1,i})n(C_{2,j})}{n(I_1)n(I_2)} \quad (1)$$

where $\delta(i,j)$ denotes the Euclidean distance between the centroid of $C_{1,i}$ (the ith cluster of I_1) and $C_{2,j}$ (the jth cluster of I_2). $n(\cdot)$ represents the total number of pixels in a cluster or an image. In other words, if two images are similar, the color cluster centroids are close to each other in the HSV color space, and a large number of pixels belongs to the same dominant color cluster in both images.

3.4 Phone Position Detection

In SensOrchestra, we do not constrain the positions of phones when they are sensing the ambience. Therefore, before using the audio and image features for location recognition, we first detect the information of phone position (inside or outside the pocket) to determine the reliability of sensor data from a certain phone. We use average audio energy and mean of image luminance as features. The intuition is that if the phone is in the pocket (or generally covered by clothing or bags), the images will be dark and the average audio volume recorded will be much lower. A two-class support vector machine (SVM) classifier [6] is trained to classify whether the phone is "in the pocket" or "outside the pocket." The information is used in the next step for a final decision on location recognition.

3.5 Classifier Fusion Model

The classifier fusion model is shown in Figure 4. Suppose there are N users in store S_t at time t, observing features $O_{1,t}, O_{2,t}, ..., O_{N,t}$, respectively. For each individual observation, we use a k-Nearest Neighbor (k-NN) classifier [4] to estimate the most likely store. Specifically, for each sample of testing audio or image features, we calculate the distance from each of the samples in the training set. For MFCC, Euclidean distance between two feature vectors is used. For color distribution features, the pair-wise distance is computed by the similarity measure defined in equation (1).

After each classifier outputs an estimate, we use a weighted majority vote approach for classifier fusion. The weighting is trained by the relative classification accuracy using the data collected from "inside-the-pocket" versus "outside-the-pocket." The intuition is that if the data were sensed from a phone inside the pocket, the estimate based on the data is less reliable because the microphone was muffled. Thus, its vote is multiplied by a lower weight w_{pocket}; otherwise, the vote is multiplied by $(1 - w_{pocket})$. According to the experimental results, we set w_{pocket} to 0.4. Finally, based on the weighted sum of votes from different phones, the store class with the most votes is the final estimate S_t^*.

After a sequence of estimates S_t^* is produced, we apply a temporal-smoothing process to take temporal continuity into account. The intuition is that a user's location trace is continuous in time and unlikely to switch swiftly between several stores. Specifically, using a moving window of the size of 3 consecutive estimates, the final smoothed store decision $S_{smooth,t}^*$ at each time t is defined as:

$$S_{smooth,t}^* = \begin{cases} S_{t-1}^* & \text{if } S_{t-1}^* = S_{t+1}^* \text{ and } S_t^* \neq S_{t-1}^* \\ S_t^* & \text{otherwise} \end{cases} \quad (2)$$

4 Implementation

The client-side program of SensOrchestra is implemented in Python for Symbian S60 [26] platform v1.4.5 on Nokia N95 phones. The background Bluetooth scanning part is implemented using the code from the Personal Distributed Information Store (PDIS) project [24] at Helsinki Institute for Information Technology. The audio is recorded in WAV format with a sampling frequency of 8 kHz, and the images are stored in 640 × 480 JPEG format. The sensor data and the timestamp information are transmitted to an Apache Server, handled by a PHP script, and stored in a MySQL database. The sensor data from the phones within the same group are analyzed by the classifier fusion model implemented in MATLAB, and then the phones retrieve the recognition result from the server. The Support Vector Machine classifier used in phone position detection is implemented using LIBSVM [8].

Table 1. The effect of phone positions (inside pocket vs. outside pocket) on store classification accuracy

Phone Positioin	Inside Pocket	Outside Pocket
Classification Acc.	59.5%	79.8%

Table 2. Comparison of store classification accuracy using single phone vs. multi-phone collaborative sensing

Feature	Single-Phone Approach					Collaborative Sensing	
	Phone1	Phone2	Phone3	Phone4	Avg.	4-Phones	+Temporal
Color Distribution	68.8%	62.5%	59.4%	51.2%	60.3%	70.1%	73.0%
Audio MFCC	75.6%	73.6%	82.6%	70.7%	74.9%	**84.8%**	**87.7%**

5 Dataset

We evaluated SensOrchestra on a dataset containing 536 sensing sessions. The dataset was collected in 10 different stores at the Stanford shopping mall. Six graduate students from Carnegie Mellon University and Stanford University participated in the data collection and annotation tasks at different times (daytime and evenings on both weekdays and weekends). Four different Nokia N95 phones are used, each running the client-side sensing program of SensOrchestra. Each session consists of a 30-second audio and four images. To facilitate future research in the related field, we will make our dataset available on our website.

Throughout the experiment, there were 4 users visiting one store at the same time, each carrying one phone, browsing items in the store like other customers. To simulate a realistic situation, the users were asked to generally keep their phones in their pockets, and occasionally take out their phones for a while. The users holding their phones were asked to mimic general customers' phone usages, e.g. making phone calls or reading messages. After each sensing session, the user annotated the phone position and the store name, which serve as ground-truth labels in experiments.

Due to the realistic settings, most of the data are low quality images caused by motion blur, non-informative views (e.g. other customers or a close-up view of a wall), and unusual camera angles. In addition, most of the audio are recorded when the phones are put in the users' pocket, resulting in muffled sounds and loss of acoustic details of the surroundings. However, the results show that our approach is reasonably robust to overcome these difficulties.

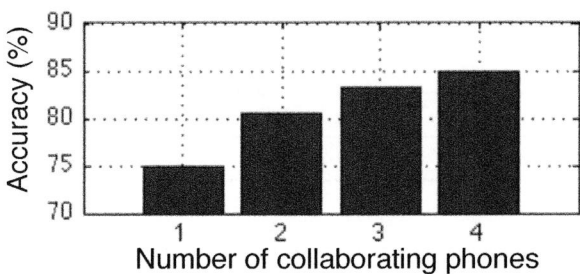

Fig. 5. The effect of the number of collaborators on store classification accuracy using audio features

6 Evaluation

Given each sample (audio and image features extracted from one sensing session), the system classifies the sample as one of the stores in the training set. The results are validated using 10-fold cross-validation. The classification accuracy metric used in this section is calculated by:

$$accuracy = \frac{\#\ correctly\ classified\ samples}{\#\ total\ samples} \qquad (3)$$

6.1 Phone Position Detection Results

We first validate the hypothesis that two classes of phone positions (either "inside-the-pocket" or "outside-the-pocket") can be accurately detected. Experimental results shows that either audio energy or mean image luminance is discriminative enough to detect the phone position with 92% and 99% accuracy, respectively. In the classifier fusion step, we determine the weighting of each phone based on the phone position detected by image luminance.

6.2 Impact of Phone Position on Classification Results

Next, we investigate how the store classification accuracy will be affected by where users put their phones. We divided the dataset into two subsets depending on whether a sample is sensed by a phone inside or outside the pocket, and ran a store classification experiment on each subset. Only the result using audio features is available since the images are all dark when cameras are put in pockets. The result in Table 1 shows that putting a phone in the pocket severely degrades the classification accuracy since the microphone is muffled. The result also supports our claim that using only one phone can result in poor performance if the user keeps the phone in the pocket most of the times, which suggests the solution of collaborative sensing.

Table 3. Confusion matrix of store classification results (shown in percentage) using collaborative sensing and audio features only

Ground Truth	Classification Results (Classified As)							
	Cafe	Body	Electr.	M.Fash.	W.Fash.	Books	Home	Dining
Cafe	88.8	0	0	0	0	0	0	11.2
Body Care	0	88.2	7.4	0	2.9	1.5	0	0
Electronics	0	9.7	61.3	0	6.4	3.2	19.4	0
Men Fashion	4.4	0	0	95.6	0	0	0	0
Women Fashion	4.2	4.2	12.5	0	75.0	4.2	0	0
Books	0	0	0	0	5.3	94.7	1.3	0
Home Decor	0	6.9	5.2	0	0	22.4	63.8	1.7
Dining	2.0	0	0	0.7	0	0	0	97.8

6.3 Collaborative Store Classification Results

In this section, we validate our main hypothesis that the collaborative sensing approach of SensOrchestra can improve store classification accuracy over the single-phone approach. For the single-phone approach, we ran experiments by switching the testing set among data from the four phones carried by four users (Phone1 to Phone4). For the collaborative sensing approach, each sample is connected with other samples that were sensed at the same time but from different nearby phones. The individual estimates are aggregated using the classifier fusion approach, which generates the final classification decision. 10-fold cross-validation was applied to both cases.

As shown in Table 2, image features (dominant color distribution) yield lower accuracy, probably caused by the low quality of images in the dataset because there was no limitation on phone positions or orientations when the pictures were taken. When using the single-phone approach with only audio features (MFCC) for classification, each phone achieved different performance, with an average accuracy of 75%. Using the 4-phone collaborative sensing approach, we improved the average accuracy to 85%. The result supports our hypothesis that sensor data from different phones actually complement each other. Some errors caused by ambient noise can be compensated if the majority of classification estimates are correct. After the temporal-smoothing process, we achieved 87.7% accuracy using only audio features, which is similar to the accuracy reported in [1] using 4 modalities (audio, image, motion, Wi-Fi), but we do not require the phones to be held in the hand at all times. Also, we used fewer modalities, which implies substantial saving on energy consumed by sensors.

We also compare the results using different numbers of collaborators. As shown in Figure 5, while accuracy gradually increases as more phones' estimates are combined, even adding only one collaborator can improve the accuracy by 6%. More insights can be drawn from the confusion matrix shown in Table 3. The system is effective in recognizing café and the dining restaurant, but can still misclassify one as the other sometimes because of the high similarity of these

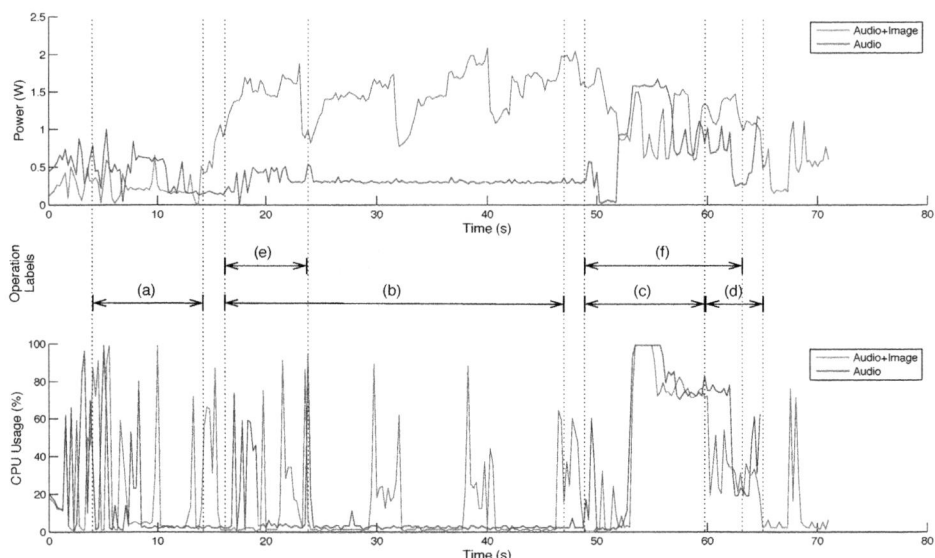

Fig. 6. The power and CPU usage of the Nokia N95 phone when running SensOrchestra phone sensing software. The operations performed on the phone are labeled as follows: (a) Bluetooth scanning; (b) Microphone/Camera sensing; (c) Sensor data transmission (audio only); (d) Location recognition result reception; (e) One shot of the camera; (f) Sensor data transmission (audio and image).

two kinds of store. Another example is that the electronics store (Sony Style) is frequently misclassified as the home decoration store (Brookstone). A possible reason is that Brookstone actually has a section selling home appliances and electronics, probably resulting in images and sound signatures similar to Sony Style.

6.4 Energy and CPU Measurements

In addition to store classification performance, we now evaluate the energy consumption and CPU usage of SensOrchestra when running on the Nokia N95 phone. The measurements are collected using the Nokia Energy Profiler toolkit [14], with a sampling rate of 4 measurements per second.

Figure 6 shows the power and CPU usage during one sensing interval. Two configurations are tested for comparison: 1) Both microphone and camera are turned on to sense and transmit audiovisual data; 2) Only the microphone is turned on and the audio is transmitted. When the program starts, it takes 10 seconds on average to perform the Bluetooth scan for proximity detection, with power consumption of 0.5 W on average. Although slightly increasing the need of power and CPU, Bluetooth scan is essential for ad-hoc group formation for collaborative sensing. Then, the phone starts to capture audio and image for

30 seconds. If using only microphone to record audio, the energy consumption rate is 0.3 W. While audiovisual data provide more information for context inference, camera shots consume more than 0.9 W, which is three times as much as the power consumed by audio recording. Also, audio features are shown to be more robust than image features in Section 6.3. Therefore, to reduce energy consumption, a possible solution would be turning off the camera or adopting state-of-the-art duty cycle management techniques [27].

Sensor data transmission between the phone and the server is also a power-consuming operation, which increases the CPU usage to over 80%, and drains 1.2 W in average. One solution is moving the feature extraction part to the phone to save bandwidth in data transmission, with a tradeoff of increasing the computational burden on the phone.

7 Discussion

7.1 Transmission Bandwidth

We first discuss the bandwidth required for data transmission. For the current implementation, a 30-second WAV file is 472 kB, and a JPEG image is about 60 kB each on average. Each Bluetooth address in the discovered address list is only 48 bit, so the additional overhead is negligible. Therefore, the total amount needed for transfer is approximately 472 kB + 4 × 60 kB = 712 kB, which takes about 10 seconds in average to upload with Wi-Fi or 3G connection. If only audio data are used, the data size for transmission can be reduced by 33%.

7.2 Response Time

Response time is also an important factor for a location recognition system, since context-based advertisement or coupons may need to be delivered to the user in a timely fashion. We define response time as the interval from the time that sensor data are collected, to the time that location recognition result is received. As shown in Figure 6, after the sensor data are collected, the system response time is about 10 to 15 seconds (operation (c) and (d)), depending on the network connection and the transmitted data size. Considering the whole sensing and inference interval, the system can update the user's location once every minute. While this can be improved by reducing the sensing period, we believe it is timely enough for symbolic location recognition since most customers are unlikely to switch from store to store every minute. Extracting features locally on the phone can further improve the response time by reducing the data size for transmission.

7.3 Applications to Mobile Advertising

We now discuss the potential of applying the system to context-based advertising. On the user's side, SensOrchestra is a realistic solution since most of the time there is more than one person shopping in a store and carrying his/her mobile phone. A user can simply opt out if there is any privacy concern. On the advertisement provider's side, SensOrchestra is low-cost and scalable enough to be deployed to a wide variety of stores since it requires only a reasonable amount of audio/image samples (less than 70 sensing sessions for each store in our experiments) provided by each store for training, without the need for additional infrastructure. Furthermore, our multi-phone sensing framework enables the opportunity of mobile group advertising. Since customers in proximity are grouped together, the advertisement provider can send group-targeted coupons (e.g., Three users in the same store may receive a "buy-two-get-one-free" coupon for them even if they do not know each other). Leveraging the strength of such social connections or incorporating other context information can be interesting directions for mobile context-based advertising.

8 Conclusion and Future Work

In this paper, we have presented the design and evaluation of SensOrchestra, a collaborative sensing framework for symbolic location recognition that leverages multiple nearby mobile phones. We showed that using only a single phone is not robust enough to overcome muffling and ambient noise. Experimental results validate our hypothesis that recognition accuracy improves as more phones collaborate. Using the proposed classifier fusion approach to combine correlated estimates from different phones, we are able to achieve 87.7% accuracy with only audio features and without any assumption on how users carry their phones during the sensing process.

Our ongoing work is to balance the tradeoff between energy consumption, transmission bandwidth, and accuracy. Improving the reliability of proximity detection is another important issue. We are also applying the collaborative sensing framework to group activity recognition, which provides richer information for context-based mobile advertising.

Acknowledgments. This research is supported by a grant from Nokia Research Center and by the CyLab Mobility Research Center at Carnegie Mellon under grant DAAD19-02-1-0389 from Army Research Office. The views and conclusions contained here are those of the authors and should not be interpreted as necessarily representing the official policies or endorsements, either express or implied, of Army Research Office, Carnegie Mellon University, or the U.S. Government or any of its agencies. We appreciate the advice of Asim Smailagic, Pei Zhang, Ying Zhang, and Patricia Collins.

References

1. Azizyan, M., Constandache, I., Roy Choudhury, R.: SurroundSense: mobile phone localization via ambience fingerprinting. In: Proceedings of the 15th Annual International Conference on Mobile Computing and Networking, pp. 261–272. ACM (2009)

2. Bahl, P., Padmanabhan, V.N.: RADAR: an in-building RF-based user location and tracking system. In: Proceedings of Nineteenth Annual Joint Conference of the IEEE Computer and Communications Societies, INFOCOM 2000, vol. 2, pp. 775–784. IEEE (2000)
3. Bao, X., Choudhury, R.R.: VUPoints: Collaborative Sensing and Video Recording through Mobile Phones. In: Proceedings of The First ACM SIGCOMM Workshop on Networking, Systems, and Applications on Mobile Handhelds, pp. 7–12 (2009)
4. Bishop, C.M.: Pattern Recognition and Machine Learning (Information Science and Statistics). Springer-Verlag New York, Inc., Secaucus (2006)
5. Bosch, A., Zisserman, A., Munoz, X.: Scene classification using a hybrid generative/discriminative approach. IEEE Transactions on Pattern Analysis and Machine Intelligence 30, 712–727 (2008)
6. Burges, C.J.C.: A tutorial on support vector machines for pattern recognition. Data Mining and Knowledge Discovery 2, 121–167 (1998)
7. Burke, J., Estrin, D., Hansen, M., Parker, A., Ramanathan, N., Reddy, S., Srivastava, M.B.: Participatory sensing. In: Workshop on World-Sensor-Web (WSW 2006): Mobile Device Centric Sensor Networks and Applications, pp. 117–134 (2006)
8. Chang, C.-C., Lin, C.-J.: LIBSVM: a library for support vector machines (2001), Software available at http://www.csie.ntu.edu.tw/~cjlin/libsvm
9. Chu, S., Narayanan, S., Kuo, C.-C.J.: Environmental sound recognition with time-frequency audio features. IEEE Transactions on Audio, Speech and Language Processing 17(6), 1142–1158 (2009)
10. Deng, Y., Manjunath, B.S., Kenney, C., Moore, M.S., Shin, H.: An efficient color representation for image retrieval. IEEE Transactions on Image Processing 10, 140–147 (2001)
11. Eronen, A.J., Peltonen, V.T., Tuomi, J.T., Klapuri, A.P., Fagerlund, S., Sorsa, T., Lorho, G., Huopaniemi, J.: Audio-based context recognition. IEEE Transactions on Audio, Speech, and Language Processing 14(1), 321–329 (2006)
12. Fang, Z., Guoliang, Z., Zhanjiang, S.: Comparison of different implementations of MFCC. Journal of Computer Science and Technology 16(6), 582–589 (2001)
13. Kanungo, T., Mount, D.M., Netanyahu, N.S., Piatko, C.D., Silverman, R., Wu, A.Y.: An efficient k-means clustering algorithm: Analysis and implementation. IEEE Transactions on Pattern Analysis and Machine Intelligence 24, 881–892 (2002)
14. Kuulusa, M., Bosch, G.: Nokia Energy Profiler Version 1.2 (2009), Software available at http://www.forum.nokia.com/Library/Tools_and_downloads/Other/Nokia_Energy_Profiler/
15. Lin, H., Zhang, Y., Griss, M., Landa, I.: WASP: An Enhanced Indoor Locationing Algorithm for a Congested Wi-Fi Environment. In: Fuller, R., Koutsoukos, X.D. (eds.) MELT 2009. LNCS, vol. 5801, pp. 183–196. Springer, Heidelberg (2009)
16. Liu, H., Darabi, H., Banerjee, P., Liu, J.: Survey of wireless indoor positioning techniques and systems. IEEE Transactions on Systems, Man and Cybernetics, Part C (Applications and Reviews) 37(6), 1067–1080 (2007)
17. Lu, H., Pan, W., Lane, N., Choudhury, T., Campbell, A.: SoundSense: scalable sound sensing for people-centric applications on mobile phones. In: Proceedings of the 7th International Conference on Mobile Systems, Applications, and Services, pp. 165–178. ACM, New York (2009)

18. Miluzzo, E., Lane, N.D., Fodor, K., Peterson, R., Lu, H., Musolesi, M., Eisenman, S.B., Zheng, X., Campbell, A.T.: Sensing meets mobile social networks: the design, implementation and evaluation of the cenceme application. In: SenSys 2008: Proceedings of the 6th ACM Conference on Embedded Network Sensor Systems, pp. 337–350. ACM (2008)
19. Mun, M., Boda, P., Reddy, S., Shilton, K., Yau, N., Burke, J., Estrin, D., Hansen, M., Howard, E., West, R.: PEIR, the personal environmental impact report, as a platform for participatory sensing systems research. In: Proceedings of the 7th International Conference on Mobile Systems, Applications, and Services, pp. 55–68 (2009)
20. Narayanaswami, C., Coffman, D., Lee, M.C., Moon, Y.S., Han, J.H., Jang, H.K., McFaddin, S., Paik, Y.S., Kim, J.H., Lee, J.K., Park, J.W., Soroker, D.: Pervasive symbiotic advertising. In: HotMobile 2008: Proceedings of the 9th Workshop on Mobile Computing Systems and Applications, pp. 80–85. ACM (2008)
21. Partridge, K., Begole, B., Alto, P., Road, C.H.: Activity-based Advertising: Techniques and Challenges. In: Proceedings of the 1st Workshop on Pervasive Advertising, pp. 2–5 (2009)
22. Paxton, M., Benford, S.: Experiences of participatory sensing in the wild. In: Ubicomp 2009: Proceedings of the 11th International Conference on Ubiquitous Computing, pp. 265–274. ACM (2009)
23. Priyantha, N.B., Chakraborty, A., Balakrishnan, H.: The cricket location-support system. In: Proceedings of ACM International Conference on Mobile Computing and Networking, pp. 32–43 (2000)
24. Rimey, K.: Personal Distributed Information Store (PDIS) Project (2004), Software available at http://pdis.hiit.fi/pdis/download/
25. Sala, M.C., Partridge, K., Jacobson, L., Begole, J.: An Exploration into Activity-Informed Physical Advertising Using PEST. In: LaMarca, A., Langheinrich, M., Truong, K.N. (eds.) Pervasive 2007. LNCS, vol. 4480, pp. 73–90. Springer, Heidelberg (2007), http://www.springerlink.com/index/U5692H972H232382.pdf
26. Scheible, J., Tuulos, V.: Mobile Python: Rapid prototyping of applications on the mobile platform. Wiley Publishing (2007)
27. Wang, Y., Lin, J., Annavaram, M., Jacobson, Q.A., Hong, J., Krishnamachari, B., Sadeh, N.: A framework of energy efficient mobile sensing for automatic user state recognition. In: MobiSys 2009: Proceedings of the 7th International Conference on Mobile Systems, Applications, and Services, pp. 179–192. ACM (2009)

Activity-Aware Mental Stress Detection Using Physiological Sensors

Feng-Tso Sun[1], Cynthia Kuo[1,2], Heng-Tze Cheng[1], Senaka Buthpitiya[1], Patricia Collins[1], and Martin Griss[1]

[1] Carnegie Mellon University
{lucas.sun,hengtze.cheng,senaka.buthpitiya,
patricia.collins,martin.griss}@sv.cmu.edu
[2] Nokia Research Center
cynthia.kuo@nokia.com

Abstract. Continuous stress monitoring may help users better understand their stress patterns and provide physicians with more reliable data for interventions. Previously, studies on mental stress detection were limited to a laboratory environment where participants generally rested in a sedentary position. However, it is impractical to exclude the effects of physical activity while developing a pervasive stress monitoring application for everyday use. The physiological responses caused by mental stress can be masked by variations due to physical activity.

We present an activity-aware mental stress detection scheme. Electrocardiogram (ECG), galvanic skin response (GSR), and accelerometer data were gathered from 20 participants across three activities: sitting, standing, and walking. For each activity, we gathered baseline physiological measurements and measurements while users were subjected to mental stressors. The activity information derived from the accelerometer enabled us to achieve 92.4% accuracy of mental stress classification for 10-fold cross validation and 80.9% accuracy for between-subjects classification.

Keywords: Mental stress, electrocardiogram, galvanic skin response, physical activity, heart rate variability, decision tress, Bayes net, support vector machine, stress classifier.

1 Introduction

Stress is a physiological response to the mental, emotional, or physical challenges that we encounter. Immediate threats provoke the body's "fight or flight" response, or acute stress response [5]. The body secretes hormones, such as adrenaline, into the bloodstream to intensify concentration. There are also many physical changes, such as increased heart rate and quickened reflexes. Under healthy conditions, the body returns to its normal state after dealing with acute stressors.

Unfortunately, many of the stressors in modern life are ongoing. Chronic stress can be detrimental to both physical and mental health. It is a risk factor for hypertension and coronary artery disease [22,12]. Other physical disorders, including irritable bowel syndrome (IBS), gastroesophageal reflux disease (GERD), and back pain, may be caused or exacerbated by stress [16]. Chronic stress also plays a role in mental illnesses, such as generalized anxiety disorder and depression [11].

Chronic stress is difficult to manage because it cannot be measured in a consistent and timely way. One current method to characterize an individual's stress level is to conduct an interview or to administer a questionnaire during a visit with a physician or psychologist. This method provides only a momentary snapshot of the individual's stress level, as most individuals cannot accurately recall the history of the ebb and flow of their stress symptoms [3].

Continuous monitoring of an individual's stress levels is essential for understanding and managing personal stress. A number of physiological markers are widely used for stress assessment, including: galvanic skin response, several features of heart beat patterns, blood pressure, and respiration activity [31,?]. Fortunately, miniaturized wireless devices are available to monitor these physiological markers. By using these devices, individuals can closely track changes in their vital signs in order to maintain better health.

Measuring physiological signals during everyday activity is more difficult than in a rigorous laboratory environment. First, the physiological responses caused by mental stress can be masked by variations due to physical activity [1]. For example, people may have higher heart rate when standing than when sitting. Heart rate may also increase when people are mentally stressed. Hence, using heart rate alone as an indicator to detect mental stress may lead to misclassification. Second, signal artifacts caused by motion, electrode placement, or respiratory movement affect the accuracy of measured recordings. Third, it is also difficult to determine the ground truth of a user's stress level when labeling training data in mobile environment. These factors increase the difficulty of developing a pervasive mental stress detection application for everyday use.

We introduce an activity-aware, multi-modal system that combines accelerometer, ECG, and GSR information to differentiate between physical activity and mental stress. We conducted a user study with 20 participants across three different physical activities: sitting, standing, and walking. With activity information derived from the accelerometer, we achieved 92.4% accuracy for 10-fold cross validation and 80.9% accuracy for between-subject's classification.

In the next section, we describe how we can measure the body's responses to mental stress. Next, we discuss prior work on stress detection. Section 4 describes our experimental protocol and our physiological feature extraction and classification methods. Experimental results are presented in Section 7.

2 Background

The autonomic nervous system (ANS) regulates the body's major physiological activities, including the heart's electrical activity, gland secretion, blood pressure, and respiration. The ANS has two branches: the sympathetic nervous system (SNS) and the parasympathetic nervous system (PNS). The SNS mobilizes the body's resources for action under stressful conditions. In contrast to the SNS, the PNS relaxes the body and stabilizes the body into steady state.

2.1 Heart Rate Variability (HRV) and Stress

Under acute stress, the SNS increases heart rate, respiration activity, sweat gland activity, etc. After the stress has passed, the PNS reverses the stress response [17]. Since the ANS controls the heart, measuring cardiac activity is an ideal, non-invasive means for evaluating the state of the ANS.

An ECG is a recorded tracing of the electrical activity generated by the heart. Figure 1 shows a P wave, a QRS complex, and a T wave in the ECG. The P wave represents atrial depolarization, the QRS represents ventricular depolarization, and the T wave reflects the rapid repolarization of the ventricles [8]. The R-R interval is the time interval between two R peaks and is used to calculate heart rate.

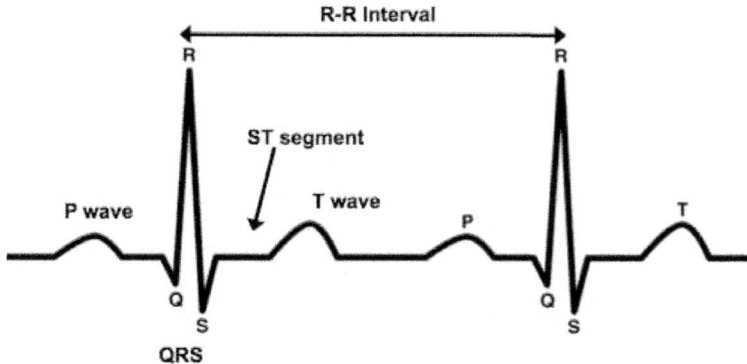

Fig. 1. Electrocardiogram sample

Heart rate variability (HRV) refers to the beat-to-beat variation in the R-R interval. HRV analysis can be categorized into time-domain and spectral-domain analysis. Several time-domain parameters include:

- mean HR: mean heart rate (beats per minute);
- mean RR: mean heartbeat interval (ms);
- SDNN: standard deviation of RR-intervals between normal beats;
- RMSSD: root mean square of the difference between successive RR-intervals; and
- pNN50: the percentage of heartbeat intervals with a difference in successive heartbeat intervals greater than 50 ms.

Three widely used components can be found in HRV power spectrum:

- LF (0.04-0.15 Hz): a low-frequency component that is mediated by both the SNS and PNS;
- HF (0.15-0.4Hz): a high-frequency component mediated by the PNS; and
- LF/HF: LF to HF ratio that is used as an index of autonomic balance.

2.2 Galvanic Skin Response (GSR) and Stress

GSR is a measure of the electrical resistance of the skin. A transient increase in skin conductance is proportional to sweat secretion[6]. When an individual is under mental stress, sweat gland activity is activated and increases skin conductance. Since the sweat glands are also controlled by the SNS, skin conductance acts as an indicator for sympathetic activation due to the stress reaction.

The hands and feet, where the density of sweat glands is highest, are usually used to measure GSR. There are two major components for GSR analysis. Skin conductance level (SCL) is a slowly changing part of the GSR signal, and it can be computed as the mean value of skin conductance over a window of data. A fast changing part of the GSR signal is called skin conductance response (SCR), which occurs in relation to a single stimulus. Widely used parameters for GSR include the amplitude and latency of SCR and average SCL value[2].

3 Related Work

The validity of using ECG and GSR measurements in mental stress monitoring has been demonstrated in both psychophysiology and bio-engineering. HRV analysis based on ECG measurement is commonly used as a quantitative marker describing the activity of the autonomic nervous system during stress. For example, Sloten et al. conclude that the mean RR is significantly lower (i.e., the heart rate is higher) with a mental task than in the control condition while pNN50 is significantly higher in the control condition than with a mental task [26].

Also, conventional short-term HRV features (e.g., a 5-minute sample window) may not capture the onset of acute mental stress for a mobile subject. Salahuddin et al. noted that HR and RR-intervals within 10 sec, RMSSD and pNN50 within 30 sec, high frequency band (HF: 0.15 to 0.4 Hz) within 40 sec, LF/HF, normalized low frequency band (LF: 0.04 to 0.15 Hz), and normalized HF within 50 sec can be reliably used for monitoring mental stress in mobile settings [23]. Hence, mental stress can be recognized with most HRV features calculated within one minute.

Boucsein provided an extensive coverage of early research of GSR related to stress [2]. He showed that slowly changing SCL and SCR aroused by specific stimulus are sensitive and valid indicators for the course of a stress reaction. Setz et al. demonstrated the discriminative power of GSR in distinguishing stress caused by a cognitive load and psychosocial stress by using a wearable GSR device in an office environment [25]. In this study, analysis of the data showed that the distributions of the SCL peak height and the SCR peak rate carry information about the stress level of a person.

Some research has used multiple physiological features to determine the existence of the subject's stress response. Zhai and Barreto applied an interactive "Paced Stroop Test," a psychological test of the subject's mental attention and flexibility, as a stimulus to elicit emotional stress in the subject [33]. The Paced Stroop Test requires the subject to select the font color of a word shown on the screen. The word itself names a potentially different color. The authors proposed to extract features from the subject's physiological response (blood volume pulse, galvanic skin response, skin temperature and pupil diameter) during both the congruent phase (matching color name and font color) and incongruent phase (mismatching color name and font color). An example of the incongruent phase is shown in Figure 3. Three learning algorithms, Naive Bayes, Decision Tree, and Support Vector Machine (SVM), are used to classify relaxed and stressed states. The SVM classifier reached an accuracy of 90.1% with 20-fold cross validation.

Some experiments have been conducted in the real world. For instance, Healey and Picard measured drivers' stress reactions by monitoring multiple physiological signals, such as ECG, GSR, electromyogram (EMG), and respiration in a prescheduled route setting [10]. They used 5-minute intervals of data during the rest, highway, and city driving conditions to distinguish between three levels of driver stress. Heart rate and skin conductance provided the highest overall correlations with drivers' stress level across multiple drivers and driving days, reaching an accuracy of over 97%.

Most previous research considers distinguishing the physiological response to mental stress from subjects at rest. While developing mental stress monitoring algorithms in real-life ambulatory situations, it is crucial to take physical activity (e.g., walking) and posture (e.g., sitting or standing) into account. Cardiovascular variability is highly affected by changes in body posture and physical activity [30]. In Van Steenis et al.'s sample, subjects' mean HR increased significantly from a supine to sitting posture (from 66 to 77 bpm), from a sitting to standing posture (86 bpm), and from a standing posture to dynamic body movements (92 bpm). A major obstacle for ambulatory monitoring is that physiological dysregulation or emotion effects can be confounded by physical activity. Many physiological parameters, including heart rate, respiratory sinus arrhythmia, and skin conductance level, are strongly affected by both anxiety and exercise [32]. The daily routines involve different psychophysiological body activation characteristics. Kusserow et al.'s study showed that the physical-related and mental-related routines that are correlated with heart activity can be characterized and visualized as different activation patterns using RR-intervals and accelerometer data [13]. Schumm et al.'s work validated that it is possible to provoke and measure GSR with the startle event during different walking speeds [24]. However, the faster a person is walking, the more the peak distribution of GSR approaches a uniform distribution. Activity information is also helpful in ECG-based identity authentication area. The perturbation of the ECG signal due to physical activity is a major obstacle in applying the technology in real-world situations. Sriram et al.'s work presented a novel ECG- and accelerometer-based system that can authenticate individuals in an ongoing manner under various activity conditions [27].

Thus, physical activity distorts the result of mental stress detection in a mobile health monitoring scenario. In this paper, we compensate for the effects of physical activities by extracting a set of accelerometer features that characterize different physical activities along with ECG and GSR features. We hypothesize that the accelerometer features provide the necessary auxiliary information for differentiate physical activity and mental.

4 Methodology

In this section, we describe the components of the wireless sensor system we used, the procedure of the experimental environment, and the segmentation of experimental dataset.

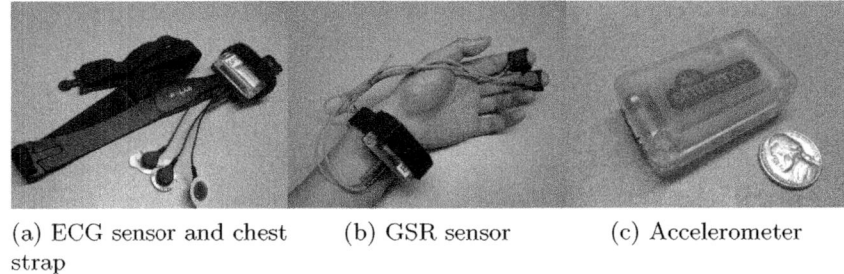

(a) ECG sensor and chest strap   (b) GSR sensor   (c) Accelerometer

Fig. 2. SHIMMER sensors including ECG, GSR, and accelerometer

4.1 Wireless Sensor Network

We used the SHIMMER platform developed by Intel's Digital Health Group. SHIMMER is a small wireless sensor platform with an integrated 3-axis accelerometer designed to support wearable applications. We also used SHIMMER's ECG and GSR daughter boards for data acquisition. The sensor data from the ECG sensor and accelerometer were sampled at 100 Hz, and the data from the GSR sensor were sampled at 32 Hz. Data were transmitted to a PC via Bluetooth connectivity and saved to binary and comma-separated value files. We used three sensor nodes for the wireless sensor network configuration. Photos of the sensors are shown in Figure 2. The ECG sensor node was strapped to an elastic chest belt and three electrodes were placed on the body to form lead II and lead III [1] recording configurations. The GSR sensor was attached on a wrist band. Then, skin conductance was measured at the base of two fingers by measuring the electrical current that flowed as a result of applying a constant voltage. The third sensor node which was placed on the waist belt was used to collect accelerometer data.

---

[1] (Lead II is the voltage between the left leg (LL) electrode and the right arm (RA) electrode), and Lead III is the voltage between the (positive) left leg (LL) electrode and the left arm (LA) electrode.

4.2 Experimental Protocol

20 participants were monitored, 13 men and 7 women. Participants were students, faculty, and staff at our university. A computer application that randomly presented Stroop Color-word interference tests and mental arithmetic problems was provided as stressor. The Stroop test has been widely utilized as a psychological or cognitive stressor to introduce emotional response and autonomic reactivity [28]. Because participants would answer so many questions during the study, we added a variant of the Stroop test to prevent habituation, where participants were asked to select either color-name or font-color. The mental arithmetic is based on the Montreal Imaging Stress Task (MIST), consisting of two levels of difficulty under time pressure [7]. The mental arithmetic will adapt to participants' level or adjust the time limit in order to maintain an appropriate level of stress. The participants completed the mental tasks by interacting with a 19-inche touch-screen. Examples of two mental tasks are shown in Figure 3. When the participant provided an answer before the end of time limit, the feedback "correct" or "wrong" was displayed. The interface also shows the elapsed time and the participant's accuracy rate.

Participants were confronted with mental stress in each of three different conditions: sitting, standing, or walking. Each condition consisted of a baseline measurement with no stressor, measurement during the mental tasks, and a recovery segment:

1. Baseline segment (10 minutes): Listen to meditation music (in seated, standing, or walking position).
2. Mental task segment (10 minutes): Complete Stroop test and mental arithmetic under time pressure while seated, standing, or walking.
3. Recovery segment (10 minutes): Sit in a chair with closed eyes and listen to meditation music

All participants completed three conditions in random order. For the baseline and mental task segments, the participant had to complete the physical activity simultaneously. For example, when the participant was in the mental task segment of walking session, the participant was required to walk on the treadmill at 3 mi/hr and complete the mental task using the touch screen at the same time.

4.3 Data Collection

We collected sensor data from each participant for three physical activity conditions. The data were separated into six datasets. Hence, for each participant, we collected six data sets shown in Figure 4. The dataset for each segment contains 19200 GSR samples and 60000 ECG and accelerometer samples. The dataset for all 20 participants, including the six segments, is around 45 Megabytes.

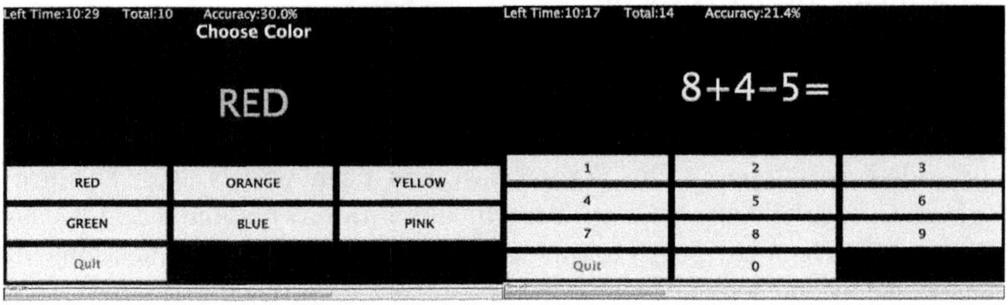

(a) Stroop Test User Interface   (b) Mental Arithmetics User Interface

Fig. 3. Screenshots of the stressor application

Baseline Segment (Sitting Only) 10 min	Stressed Segment (Mental Task + Sitting) 10 min	Recovery Segment (Sitting) 10 min

(a) Sitting condition

Baseline Segment (Standing Only) 10 min	Stressed Segment (Mental Task + Standing) 10 min	Recovery Segment (Sitting) 10 min

(b) Standing condition

Baseline Segment (Walking Only) 10 min	Stressed Segment (Mental Task + Walking) 10 min	Recovery Segment (Sitting) 10 min

(c) Walking condition

Fig. 4. Experimental conditions

5 Data Analysis

5.1 Feature Extraction

For each participant's 60 minutes of data, we segment each channel of data into a 60-second window to obtain the data windows $\omega_1, \omega_2, ..., \omega_{60}$. We denote F_i as the feature vector extracted from the data window ω_i and $F_i(j)$ is the j_{th} feature in the feature vector F_i. We create a set of feature vectors F for each participant's data set. Each segment in the experiment protocol has 10 feature vectors (e.g. $F_1 - F_{10}$ for SitBase, $F_{11} - F_{20}$ for SitStress,..., $F_{51} - F_{60}$ for WalkStress).

We chose a 60-second window for two reasons. First, the HRV features that we used in this study can distinguish between stressed and baseline segments using 60-second windows [23]. Second, the 60-second feature window reduces the impact of misclassified R-peaks by averaging HRV features within the window. All feature extraction algorithms are implemented in MATLAB. To eliminate the artifacts caused by variations in electrode contact and physical motion, we applied both moving average and band-pass filtering techniques. For R-peak detection, we mainly adapted a derivative method [19] with modifications.

HRV analysis: HRV analysis methods can be categorized into time domain and spectral domain analysis. Time domain analysis is calculated directly from RR-intervals over the feature window. Examples of time domain features include mean value of the RR-interval (mean RR), standard deviation of the RR-interval (Std RR), mean value of the HR (mean HR), standard deviation of the HR (Std HR), RMSSD, and pNN50. Moreover, in the spectral domain methods, a power spectrum density (PSD) estimate is calculated for the RR interval series. Frequently used spectral measures are the very low frequency (VLF, 00.04 Hz), low frequency band (LF) and high frequency band (HF), and the ratio LF/HF. These spectral domain features are often interpreted as a measure of sympathovagal balance (autonomic state influence by the sympathetic and parasympathetic nervous system). We first calculated six time-domain features of HRV including mean RR, Std RR, mean HR, Std HR, RMSSD, and pNN50. Then, we applied a Fast Fourier Transform (FFT) to convert the time-domain RR-interval sequence

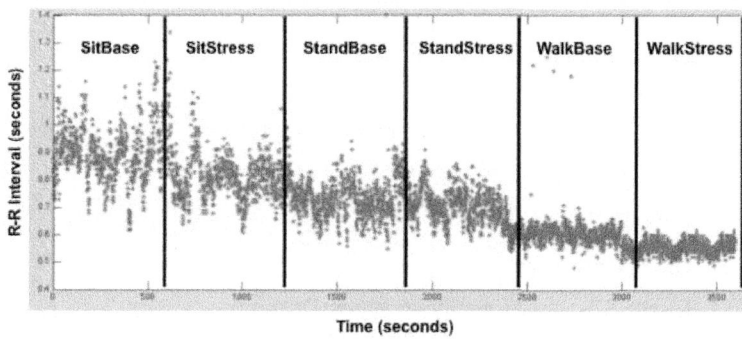

(a) RR interval data of a subject

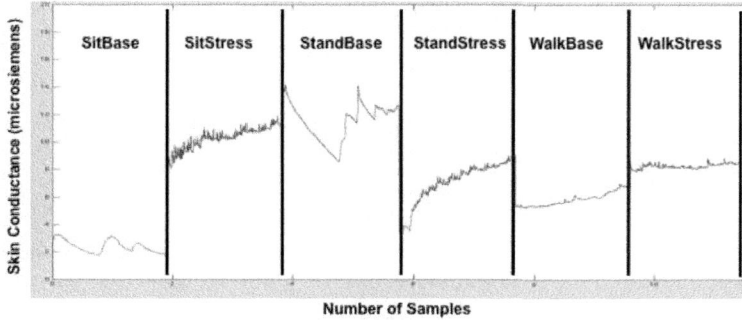

(b) Skin conductance of a subject

Fig. 5. RR interval and GSR data in six experiemental segments

to the power spectrum. The frequency components are used to calculate three spectral-domain features of HRV for each window: LF, HF, and LF/HF ratio.

GSR analysis: Due to the startle response (the physiological response of body to a sudden stimulus), the resistance of the skin can vary. The GSR can measure these subtle differences [29]. All GSR signals were filtered with a 256-point lowpass filter with 3Hz cutoff frequency to reduce noise. We calculated three GSR features: the total number of the startle responses in the segment, the sum of the response magnitude, and the sum of the response duration. These three features characterize the startle response, and Healey and Picard demonstrated their reliability [10]. Two additional features, mean and standard deviation of skin conductance level, are calculated over the feature window. Figure 5 shows the R-R interval and skin conductance recordings of a subject over six experimental segments.

Accelerometer analysis: Olguin and Pentland's work indicated that an accelerometer placed on hip significantly helped classify activities such as sitting, running, crawling, and lying down [18]. Therefore, we placed one accelerometer on the waist belt close to the hip in order to maximize the difference of signal among sitting, standing, and walking activities. For each of the three axial dimensions, we calculated twelve features: mean value, standard deviation, energy, and correlation of each two axes. Table 1 lists the features derived from the ECG, GSR, and accelerometer data.

Table 1. Feature Vectors

	Sensor	Features
$f1$-$f9$	ECG	Mean RR, Std RR, Mean HR, Std HR RMSSD, pNN50, LF, HF, LF/HF ratio
$f10$-$f14$	GSR	Mean SCL, Std SCL, Total magnitude, Duration, and Number of startle responses
$f15$-$f26$	Accel	Mean of X, Y and Z axis Standard deviation of X, Y, and Z axis Energy of X, Y, and Z axis Correlation coefficient of XY, YZ, and ZX

5.2 Feature Normalization

Skin conductivity and heart rate signals are dependent on each individual's initial physiological level. Even when the GSR or HR baseline level is measured from the same individual, these signals are likely day-dependent due to variations in physiology caused by diet or sleep, variations in mental state affected by mood, or variations in the sensor's connectivity with skin [20]. Hence, to eliminate the intra-individual factor, we applied Equation 1 to each feature in feature vector set F. Since we conducted a short recording interval (one hour) for each participant, day-to-day variation caused by mood fluctuations is not considered in this study.

$$F(j)_{norm} = \frac{F(j) - Z_{min}(j)}{Z_{max}(j) - Z_{min}(j)}, \text{where}$$
$$Z_{min}(j) = min\{F_i(j)\}, \forall i \in |F|$$
$$Z_{max}(j) = max\{F_i(j)\}, \forall i \in |F| \qquad (1)$$

Equation 1 describes the normalization process for each feature. The first step is to subtract the minimum value from each feature such that the feature with minimum value becomes 0. Then, the feature values are divided by the overall range in six segments to make all the features lie between 0 to 1. The normalized feature values are fed to the classifiers described in the next section.

6 Stress Classification

We used the WEKA machine learning engine to train classifiers using various learning methods, including the J48 Decision Tree, Bayes Net, and support vector machine (SVM) for stress inference [9]. We divided the training data into two different sets in order to evaluate how activity information may influence the results of stress inference. One set of training data only includes the ECG- and GSR-related features while the second set also includes the accelerometer information. We also evaluated classification performance for between-subjects datasets and within-subject datasets.

6.1 Decision Tree

Decision Tree is a commonly used machine learning technique that uses a divide-and-conquer approach to classify testing data. During the learning stage, the tree structure is constructed. The tree structure has internal nodes and leaves. Internal nodes represent the test conditions while the leaves represent the classification results. We used a J48 Decision Tree for mental stress classification. Again, since we are interested in observing how the accelerometer information affects the accuracy of mental stress classification, we separated the training data into two sets. One dataset includes features extracted from accelerometer, ECG, and GSR recordings. The other dataset only consists of ECG and GSR features. When the Decision Tree is being constructed, the most informative feature (with a higher information gain) will be used near the root. Therefore, we are interested in observing the constructed decision tree structure to see if the accelerometer information is used in the test conditions and provides higher information gain.

6.2 Bayesian Network

We are also interested in using a Bayesian network structure to model the probabilistic relationships among physical and mental stress. Figure 6 shows two

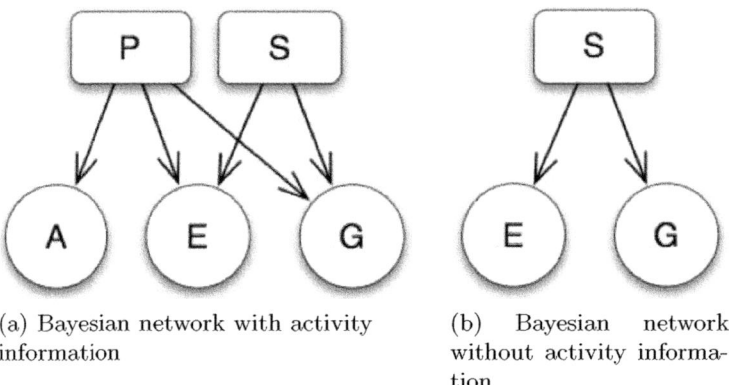

(a) Bayesian network with activity information

(b) Bayesian network without activity information

Fig. 6. Variable S represents a binary stress state and variable P represents the three physical activities. Variables A, E, and G are the accelerometer, ECG, and GSR features, respectively.

Bayesian network structures with and without considering the activity information to predict the existence of mental stress.

By comparing the inference results of the models in Figure 6a and in Figure 6b, we can investigate the effects of activity information in a mobile stress detection scenario.

In Figure 6a, $S \in$ {baseline, stressed} represents a binary stress state, and $P \in$ {sitting, standing, walking} represents the three physical activities. A, E, and G are the subsets of features defined in Table 1. $A = (a_1,...,a_{12})$ is a 12-feature vector corresponding to the accelerometer measure. $E = (e_1,...,e_9)$ is a 9-feature vector related to the HRV parameters. $G = (g_1,...,g_5)$ represents a set of 5 GSR-related features. Equation 2 shows the joint probability distribution encoded by the Bayesian network structure shown in Figure 6a. The Bayesian network structure shown in Figure 6b only uses physiological signals from ECG and GSR to infer the probability of the mental stress state.

$$P(S, P, A, E, G) = P(S) \cdot P(P) \cdot P(A|P) \cdot P(E|P, S) \cdot P(G|P, S) \qquad (2)$$

We ran the K-Means clustering algorithm with the Euclidean distance metric on accelerometer data to automatically label variable P. Because we have three types of activities (sitting, standing, and walking), we set K = 3. Figure 7a shows an example of accelerometer raw data we collected from the experiment. Figure 7b shows the class of activity derived from Figure 7a using the K-Means algorithm.

The probability distribution of S conditioned on the three observed variables (A = a, E = e, and G = g) can be obtained by marginalizing the activity variable P as shown in Equation 3.

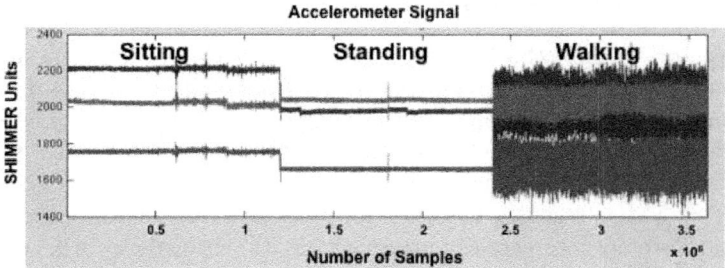

(a) Accelerometer data from one subject (red:x-axis, green:y-axis, and blue:z-axis)

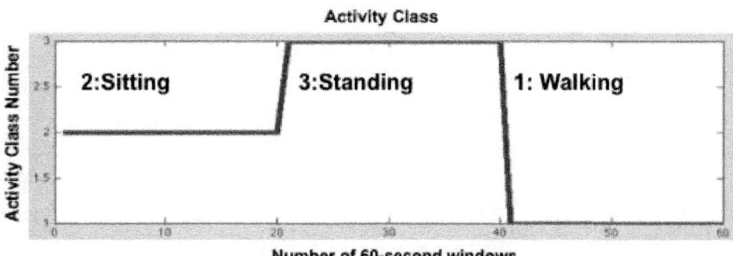

(b) Activity classes derived from accelerometer data using the K-means algorithm

Fig. 7. Accelerometer data of a subject during three activities and the activity class derived by the K-Means algorithm

$$P(S|A=a, E=e, G=g) = \sum_{i=0}^{|p|} P(S, P=p_i|A=a, E=e, G=g) \quad (3)$$

The binary class of the stress state is estimated by maximizing the posterior probability in Equation 4.

$$s_{estimated} = \underset{S}{\operatorname{argmax}} \sum_{i=0}^{|p|} P(S, P=p_i|A=a, E=e, G=g) \quad (4)$$

An evaluation of cross validation for these two Bayesian networks is presented in Section 7.

6.3 Support Vector Machine

Support Vector Machine (SVM) is a classifier that performs classification by constructing a high-dimensional hyper-plane [4]. The constructed high-dimensional hyper-plane is optimized to separate the testing data into two classes. SVM also allows different types of kernel functions to transform testing data points into a higher dimensional space and make the transformed data easier to be classified. Since SVM has recently become a popular machine learning technique for classification, we are interested in investigating its performance with our testing dataset.

7 Experimental Results

In this section, we present the results from two experiments and a comparison of ECG and GSR feature efficacy. First, to investigate how a combination of features affect stress classification accuracy, we design four feature combinations from the measured accelerometer, ECG, and GSR data. Second, we test if the classifiers generalize across subjects by training our classifier on subset of subject data and testing our classifier on the remainder. Finally, we analyze the ECG and GSR features in six conditions across 20 participants.

7.1 Cross Validation with Different Feature Combinations

The first type of feature combination includes data measured from the accelerometer, ECG, and GSR. In each of the other three types of feature combinations, one feature is excluded. For each of the three classifiers described in Section 6, we evaluated its classification accuracy using these four types of feature combinations. Figure 8 plots the results of using 10-fold cross validation. For all of the three types of classifiers, excluding data recorded from the accelerometers degrades in the classification accuracy. The experimental results provide evidence to support our hypothesis that accelerometer data help the classification accuracy in a mobile stress detection scenario; physiological signals are both affected by physical activities and mental stress levels.

Furthermore, since heart rate is highly affected by the intensity of physical activities in our experiment, the classification results are even better for Bayesian network and SVM classifiers without including ECG features compared to the all-feature combination. Unlike ECG, the GSR features are good indicators to identify the presence of mental stress. When GSR features are excluded, the accuracy of each classifier decreases compared to the all-feature combination. We also found that the best classification accuracy (92.4%) is obtained from using the decision tree classifier with the all-feature combination. Moreover, the structure of the decision tree uses the energy of the x-axis from the accelerometer data as the root test condition. Several accelerometer features are also used as test conditions close to the root of the tree. It proves that activity information provides higher information gain in the decision tree learning stage. Table 2 shows more detail of the cross validation results on 1200 samples. The grey cells highlight correctly recognized instances (true-positives).

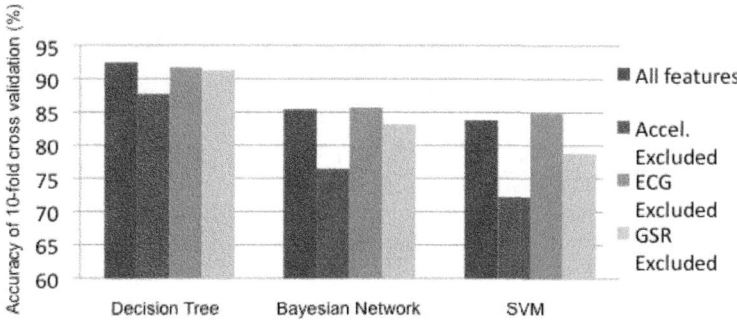

Fig. 8. Accuracy of the three classifiers using different feature combinations

Table 2. Confusion matrix for the combination of three classifiers and different feature combinations

		Decision Tree		Bayesian Network		SVM	
Features		Baseline	Stressed	Baseline	Stressed	Baseline	Stressed
All features	Baseline	553	47	475	125	509	91
	Stressed	44	556	50	550	104	496
Accel. Excluded	Baseline	530	70	412	188	455	145
	Stressed	78	522	95	505	189	411
GSR Excluded	Baseline	539	61	431	169	461	139
	Stressed	44	556	33	567	115	485
ECG Excluded	Baseline	552	48	483	117	510	90
	Stressed	52	543	55	545	92	508

7.2 Between-subjects Experiment

We randomly selected a subset of our twenty subjects and used their data to train our classifier. The, we tested the classifier on the remainder. For a given between-subjects classification setting, we repeated 10 times and calculated the average accuracy. To observe the effect on the size of the subset data, we changed the number of subjects in the training data set from 3, 6, ...to 18. The results shown in Figure 9 demonstrate that the SVM classifier outperforms the other two classifiers for between-subjects classification.

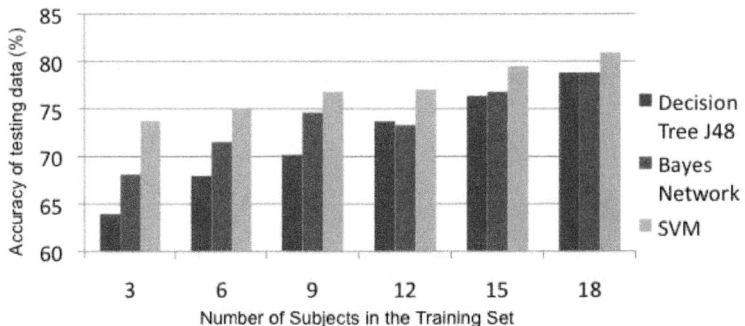

Fig. 9. Classification accuracy between subjects with the three classifiers

7.3 Comparison of HRV and GSR Parameters

Table 3 compares HRV features for the six experimental segments: (SitBase, SitStress, StandBase, StandStress, WalkBase, and WalkStress). Each HRV parameter is calculated by the average value across 20 participants for each segment. The mean RR interval decreases in all of the three mental task segments while the mean HR increases. The trend of mean RR and HR proves their efficacy in distinguishing mental stress across the three physical activities. The pNN50 feature also has a tendency to decrease from the baseline to the mental task segment. In the walking condition, pNN50 significantly decreases compared to the lower-intensity activities (sitting and standing). The standard deviation of RR is relatively high in walking condition and decreases the percentage of heartbeat intervals with difference in successive heartbeat intervals greater than 50 ms.

The last three rows in Table 3 are three spectral-domain parameters. The LF component is an indicator for both sympathetic modulations and cardiac vagal activity. It slightly increases from the baseline during all three activities. The HF component is considered to reflect parasympathetic modulations. There is a large increase in the HF component when the participants are walking. The imbalance of increase on LF and HF causes the LH ratio (LF / HF), a widely adapted index of sympathetic modulations, to decreases from sitting to walking. The HF increases during exercise has also been noted by other researchers [21].

From our analysis of all HRV parameters, we found that mean HR and RR are the most reliable features to recognize mental stress across three physical activities. The standard deviation of RR and HR did not demonstrate a coherent relation to the baseline and stressed segments. Spectral-domain parameters are sensitive to the physical activity conditions. Hence, this explains why excluding HRV features even increases in accuracy compared to the all-feature combination as shown in Figure 8.

Table 4 lists five GSR parameters for each segment. For each startle response, we can indicate its duration and magnitude. The total duration was calculated by accumulating the total elapsed time of the responses in the window. The total magnitude was measured by summing up the difference of the onset and the peak

Table 3. Comparison of HRV parameters in six conditions

HRV Parameters	Sit Base	Sit Stress	Stand Base	Stand Stress	Walk Base	Walk Stress
*Mean RR (ms)	887.59	814	752.07	722.43	586.03	562.94
Std RR (ms)	70.88	85.39	82.44	68.35	92.47	98.94
*Mean HR (bmp)	69.53	75.59	82.84	85.66	107.09	110.79
Std HR (bmp)	5.54	7.56	8.00	9.50	18.98	16.21
*pNN50 (%)	19.54	15.69	12.09	11.38	4.49	4.23
LF (%)	7.04	8.45	7.49	7.77	9.43	9.45
HF (%)	6.25	6.51	6.33	6.73	13.95	15.64
LH Ratio	1.34	1.51	1.45	1.48	0.67	0.71

of each startle response in window. The number of response occurrences over the one minute window was also recorded. Total duration, total magnitude, total occurrence of the responses, and mean GSR level illustrate an obvious increase from baseline to stressed segment. However, the standard deviation does not provide significant change between conditions.

8 Discussion and Conclusion

Previous mental stress studies were conducted in the laboratory with sedentary subjects. However, the controlled setting in a laboratory is not suitable for mobile mental stress monitoring because physical activity affects the measured physiological signals. The main goal of this study was to determine whether activity information can compensate for the interactive effects of mental stress and physical activity, which affect the accuracy of mental stress detection. Therefore, we conducted a user study in which participants completed baseline and mental task segments across three physical activities (sitting, standing, and walking).

This paper presented a multimodal approach to model the mental stress activation affected by physical activities using accelerometers, ECG, and GSR sensors. Our analysis showed that accelerometer data is necessary to improve mental stress detection in a mobile environment. We also noticed that the Decision Tree classifier has the best performance in our experiments using 10-fold cross validation. Decision Tree is recognized as one of the classification methods

Table 4. Comparison of GSR parameters in six conditions

GSR Parameters	Sit Base	Sit Stress	Stand Base	Stand Stress	Walk Base	Walk Stress
*Total duration(second)	3.17	14.30	4.16	13.15	13.72	16.32
*Total magnitude(μSiemens)	0.79	2.04	0.75	3.32	1.69	1.97
*Total occurrence	1.09	6.58	3.13	6.37	5.63	7.47
*Mean GSR(μSiemens)	4.69	4.83	6.19	6.97	6.42	7.22
Std GSR(μSiemens)	0.62	0.53	0.62	0.71	0.63	0.52

with low computational complexity [14]. Therefore, the performance along with the low complexity of the Decision Tree classifier makes it a practical design choice for stress detection on mobile devices.

Furthermore, we also compared how physical activities and mental stress affect HRV and GSR parameters. We found that the GSR features are relatively independent of the three activities, even when participants were walking at a 3mi/hr. The between-subjects experiment demonstrated that we need to use up to 90% of subjects' data to achieve the classification accuracy of around 80%. It indicates that physiological signals tend to be user-dependent; hence, mental stress monitoring applications should also rely on personalized data in the training stage. We plan to further investigate the user-dependent attribute with more participants in the future. This study was limited to three specific activities and a relatively short recording time. The next step is to design a mobile platform enabling participants to wear sensors on a daily basis.

Our activity-aware scheme for mental stress detection can facilitate the development of many affective mobile applications using physiological signals (e.g., stress management, affective tutoring, and emotion-aware human computer interfaces). Including activity recognition techniques to interpret users' emotional states helps produce more feasible wearable sensors in everyday life.

References

1. Bernardi, L., et al.: Physical activity influences heart rate variability and very-low-frequency components in holster electrocardiograms. Cardiovascular Research 32(2), 234 (1996)
2. Boucsein, H.: Electrodermal Activity. Plenum, New York (1992)
3. Breslau, N., Kessler, R., Peterson, E.L.: Post-traumatic stress disorder assessment with a structured interview: reliability and concordance with a standardized clinical interview. International Journal of Methods in Psychiatric Research 7(3), 121–127 (1998)

4. Burges, C.J.: A Tutorial on Support Vector Machines for Pattern Recognition. Data Mining and Knowledge Discovery 2, 121–167 (1998)
5. Cannon, W.: Bodily Change in Pain, Hunger, Fear and Rage: An Account of Recent Research into the Function of Emotional Excitement. Appleton, New York (1915)
6. Darrow, C.: The rationale for treating the change in galvanic skin response as a change in conductance. Psychophysiology 1, 31–38 (1964)
7. Dedovic, K., Renwick, R., Mahani, N.K., Engert, V., Lupien, S.J., Pruessner, J.C.: The Montreal imaging stress task: Using functional imaging to investigate the effects of perceiving and processing psychosocial stress in the human brain. Journal Psychiatry Neuroscience 30, 319–325 (2005)
8. Dubin, D.: Rapid Interpretation of EKG's. Cover Publishing (2000)
9. Hall, M., Frank, E., Holmes, G., Pfahringer, B., Reutemann, P., Witten, I.H.: The WEKA Data Mining Software: An Update. SIGKDD Explorations 11 (2009)
10. Healey, J.A., Picard, R.W.: Detecting stress during real-world driving tasks using physiological sensors. IEEE Transactions on Intelligent Transportation Systems 6(2), 156–166 (2005)
11. Herbert, J.: Fortnightly review: Stress, the brain, and mental illness. British Medical Journal 315, 530–535 (1997)
12. Holmes, S., Krantz, D.S., Rogers, H., Gottdiener, J., Contrad, R.J.: Mental stress and coronary artery disease: A multidisciplinary guide. Progress in Cardiovascular Disease 49, 106–122 (2006)
13. Kusserow, M., Amft, O., Troster, G.: Psychophysiological body activation characteristics in daily routines. In: IEEE International Symposium on Wearable Computers, pp. 155–156 (2009)
14. Lim, T.-S., Loh, W.-Y., Cohen, W.: A comparison of prediction accuracy, complexity, and training time of thirty-three old and new classification algorithms. Machine Learning 39 (2000)
15. Lundberg, U., et al.: Psychophysiological stress and EMG activity of the trapezius muscle. International Journal of Behavioral Medicine 1(4), 354–370 (1994)
16. Monnikes, H., Tebbe, J., Hildebrandt, M., et al.: Role of stress in functional gastrointestinal disorders. Digestive Diseases 19, 201–211 (2001)
17. T.F. of the European Society of Cardiology, The North American Society of Pacing, and Electrophysiology. Heart rate variability: Standards of measurement, physiological interpretation, and clinical use. European Heart Journal 17(2), 1043–1065 (1996)
18. Olguin, D., Pentland, Y.: Human activity recognition: Accuracy across common locations for wearable sensors. In: Proc. 10th International Symposium Wearable Computer, pp. 11–13 (2006)
19. Pan, J., Tompkins, W.J.: A real-time QRS detection algorithm. IEEE Transactions on Biomedical Engineering BME-32(3), 230–236 (1985)
20. Picard, R.W., Vyzas, E., Healey, J.: Toward machine emotional intelligence: Analysis of affective physiological state. IEEE Transactions on Pattern Analysis and Machine Intelligence 23(10), 1175–1191 (2001)
21. Pichon, A., Bisschop, C., Roulaud, M., et al.: Spectral analysis of heart rate variability during exercise in trained subjects. Medicine and Science in Sports and Exercise 36, 1702–1708 (2004)
22. Pickering, T.: Mental stress as a causal factor in the development of hypertension and cardiovascular disease. Current Hypertension Report 3(3), 249–254 (2001)
23. Salahuddin, L., Cho, J., Jeong, M.G., Kim, D.: Ultra short term analysis of heart rate variability for monitoring mental stress in mobile settings. In: 29th Annual

International Conference of the IEEE, Engineering in Medicine and Biology Society, EMBS 2007, pp. 4656–4659 (August 2007)
24. Schumm, J., Bächlin, M., Setz, C., Arnrich, B., Roggen, D., Tröster, G.: Effect of movements on the electrodermal response after a startle event. In: Proceedings of 2nd International Conference on Pervasive Computing Technologies for Healthcare, Pervasive Health (2008)
25. Setz, C., Arnrich, B., Schumm, J., La Marca, R., Troster, G., Ehlert, U.: Discriminating stress from cognitive load using a wearable EDA device. IEEE Transactions on Information Technology in Biomedicine 14(2), 410–417 (2010)
26. Sloten, J.V., Verdonck, P., Nyssen, M., Haueisen, J.: Influence of mental stress on heart rate and heart rate variability. In: International Federation for Medical and Biological Engineering Proceedings, pp. 1366–1369 (2008)
27. Sriram, J.C., Shin, M., Choudhury, T., Kotz, D.: Activity-aware ECG-based patient authentication for remote health monitoring. In: ICMI-MLMI 2009: Proceedings of the 2009 International Conference on Multimodal Interfaces, pp. 297–304. ACM, New York (2009)
28. Stroop, J.: Studies of interference in serial verbal reactions. Journal of Experimental Psychology 18, 643–661 (1935)
29. Tarvainen, M., Koistinen, A., Valkonen-Korhonen, M., Partanen, J., Karjalainen, P.: Analysis of galvanic skin responses with principal components and clustering techniques. IEEE Transactions on Biomedical Engineering 48(10), 1071–1079 (2001)
30. Van Steenis, H., Tulen, J.: The effects of physical activities on cardiovascular variability in ambulatory situations. In: Proceedings of the 19th Annual International Conference of the IEEE, Engineering in Medicine and Biology Society, vol. 1, pp. 105–108 (November 1997)
31. Vrijkotte, T., et al.: Effects of work stress on ambulatory blood pressure, heart rate, and heart rate variability. Hypertension 35(4), 880–886 (2000)
32. Wilhelm, F.H., Pfaltz, M.C., Grossman, P., Roth, W.T.: Distinguishing emotional from physical activation in ambulatory psychophysiological monitoring. Biomedical Sciences Instrumentation 42, 458–463 (2006)
33. Zhai, J., Barreto, A.: Stress detection in computer users based on digital signal processing of noninvasive physiological variables. In: 28th Annual International Conference of the IEEE, Engineering in Medicine and Biology Society, EMBS 2006, pp. 1355–1358 (September 2006)

Dr. Droid: Assisting Stroke Rehabilitation Using Mobile Phones

Andrew Goodney[1], Jinho Jung[2], Scott Needham[1], and Sameera Poduri[1]

[1] University of Southern California, Los Angeles CA 90089, USA
{goodney,sneedham,sameera}@usc.edu,
http://robotics.usc.edu
[2] KJITC, Gyunggi-Do, Korea
visusee@mnd.go.kr

Abstract. In this paper we present our initial work on a mobile phone application for assisting stroke rehabilitation. We believe that using a mobile phone to administer and track stroke rehabilitation is novel. We call our system Dr. Droid and focus on the automated scoring of motions performed by patients being administered the Wolf Motor Function Test (WMFT) by placing a smart phone in a holster at the patients wrist. We have developed a complete software application that administers the test by giving audio and visual instructions. We collect a motion trace by sampling the 3-axis accelerometer available on the phone. We double-integrate the acceleration data and apply a novel reorientation algorithm to correct for mis-alignment of the accelerometer. Using dynamic time warping and hidden Markov models we assign an objective, quantitative score to the patient's exercises. We validate our method by performing experiments designed to simulate the motions of a stroke patient.

Keywords: mobile sensing, gesture recognition, personal health monitoring, telemedicine, stroke rehabilitation.

1 Introduction

Each year in the United States approximately 795,000 people suffer from a stroke. Of those strokes, about 29% are fatal [14]. Many of the remaining 71% of stroke patients will require some form of rehabilitation. Stoke rehabilitation is performed in the hope that the patients will maintain or regain range and detail of motion lost during the stroke.

Damage to the brain caused by a stroke often manifests itself as paralysis or loss of motor control on one side of the body, particularly with one limb. Several tests designed to quantify and monitor the rehabilitation progress of a stroke patient exist. In this work we focus on the Wolf Motor Function Test [11] (WMFT). The WMFT test focuses on range of motion in the upper extremity with a set of 17 exercises such as lifting a soda can to the mouth or turning a key. Table 1 shows a list of these exercises.

Scoring for the WMFT is based on two scores. The first score is the time required to perform the exercise. This time is measured by a test administrator

Table 1. Wolf Motor Function Test task descriptions

1. Forearm to table (side)†(1)
2. Forearm to box (side)
3. Extend elbow (side)†(2)
4. Extend elbow (weight)
5. Hand to table (front)
6. Hand to box (front)
7. Weight to box*
8. Reach and retrieve
9. Lift can†(3)
10. Lift pencil
11. Lift paper clip
12. Stack checkers
13. Flip cards
14. Grip strength*
15. Turn key in lock†(4)
16. Fold towel
17. Lift basket

* Indicates un-timed tasked
† Indicates activities evaluated in this paper

using a stop-watch. The maximum possible score is 120 seconds. The second score is the functional analysis (FA) score. This score ranges from zero (meaning the affected limb can not perform the activity at all), to five (meaning the motion appears normal). The FA score is determined 'off-line' by a specially trained WMFT evaluator using a video of the patient.

A test that requires trained administers and evaluators greatly increases the cost of administering the WMFT and similar tests. Additionally, these tests are usually performed in an office or clinical setting under video observation. This setting maybe uncomfortable for the patient (which may lead to reluctance to attend future testing sessions), while adding additional overhead to the cost.

Researchers have shown that the variations between trained WMFT administrators and evaluators is low and the test reliable [9,4], at the cost of several hours of time for one or two trained practitioners. Additionally, timing errors of up to one second are likely when using a stop-watch to time the exercises [6]. This same work showed that automated timing of the WMFT is possible using wrist mounted accelerometers.

In this work we focus on both automating and scoring the WMFT in such a way as to provide quantitatively meaningful scores, while reducing the costs associated with the test and improving the patients experience.

Dr. Droid is our platform to assist stroke rehabilitation. We have devoped an application that runs on a smart phone worn on the patients wrist during the WMFT. Accelerometers in the device will record the motion of the patient. This acceleration data is then processed to give a quantitive score judging the quality of motion. For the hardware platform we use mobile phones running the Android operating system. Mobile phones are a good choice because they

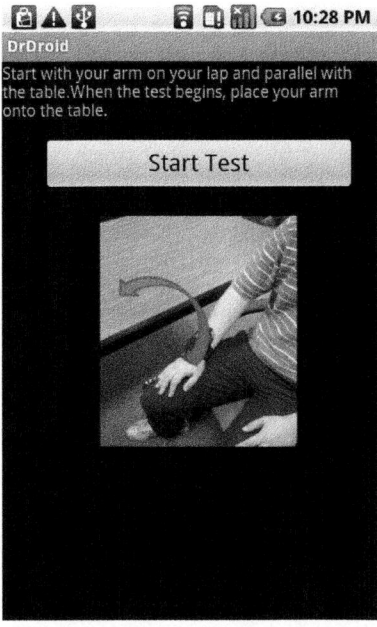

Fig. 1. Screen shot of the Dr. Droid application showing the wrist mounted phone and visual cues

are nearly ubiquitous and provide a rich array of sensor modalities at a cost approaching free. Additionally, the Android operating system provides an open, flexible platform that allows rapid application development. Combined together, these two aspects of the Android mobile phone allow us to provide an inexpensive way to administer the test at a time and location of the patients choosing, with at least as much quantitive information (time and FA score) as is currently available. Additionally, since the application is hosted on a smart phone, we enable a host of other telemedicine and digital health care applications.

There are four main contributions of this work: 1. We use a mobile phone in a novel way to make stroke rehabilitation inexpensive and comfortable for the patient, while also returning standardized, objective scores; 2. We develop two scoring algorithms based on a novel reorientation technique that corrects for errors caused by mis-alignments in the accelerometer sensor data. Thus we allow some flexibility in how the phone is mounted to the patients wrist; 3. We have implemented a complete system running on Android-based smart phones; 4. And finally we provide a quantitive comparison of three scoring algorithms under two testing modalities.

The rest of this paper is organized as follows: section 2 describes the Dr. Droid smart phone application and the scoring algorithms; section 3 describes the experiments conducted to verify our algorithms; section 4 describes related work in the literature; finally, we conclude and discuss the future work we aim to undertake in section 5.

2 Dr. Droid

This section describes the Dr. Droid application and scoring algorithms. First we provide a brief description of the system design and application operation. Second we describe in detail the algorithms we use to arrive at an objective score for a patients motion.

2.1 Assistive Application and Remote Control

The Android operating system provides a Java based programming framework for application development [1]. We used this framework to develop the Dr. Droid application. The Dr. Droid application provides several features. Users first may decide to perform an individual or a series of tests. Dr. Droid then prompts the user with visual, textual and spoken descriptions of the exercise under test. The user then performs the required action. Finally, our scoring algorithms (see section 2.2) compare the performed action to a template and return a set of scores. When performing a series of exercises the user is then presented and prompted with the next test. Finally, a database of past scores is retained, allowing the user to track the progress and report results to a practitioner. Given the average age of stroke patients and considering their ability to control a mobile device we have attempted to develop an as easy to use application as possible.

During development of Dr. Droid, we realized that it can be awkward to push buttons on the phones display while it is holstered at the wrist. As stroke patients in rehabilitation have limited mobility, we also developed a remote-control application using the XMPP protocol. This application also runs on Android phones and allows an assistant (medical practitioner or friend/family) to start and stop the timing of a test or to advance to the next test. XMPP is an open messaging protocol, thus our remote control application is not limited to running on a second phone. The remote control can be run on a desktop, laptop or tablet PC.

2.2 Data Processing and Scoring Algorithms

The core goal of Dr. Droid is to score how naturally a patient moves while performing an excercise. To do so, we perform an analysis that is very similar to gesture recognition. Gesture recognition is used in several related fields, from handwriting recognition on tablet-PCs or PDAs to gaming environments like the Nintendo Wii. However, our problem is slightly different. In gesture recognition the software system is trained with a 'vocabulary' of gestures and an unknown action is evaluated against the vocabulary and the most likely match is chosen as the detected gesture. In this work we know a-priori which gesture (exercise) is being performed, we only need to know how 'well' or 'poorly' it was performed.

Many machine learning algorithms have been used for gesture recognition, see the X-wand [19] project for an example where several were evaluated with the same hardware. In order to provide meaningful scores to the patient, we selected algorithms that provide a numeric metric of the similarity between two gestures.

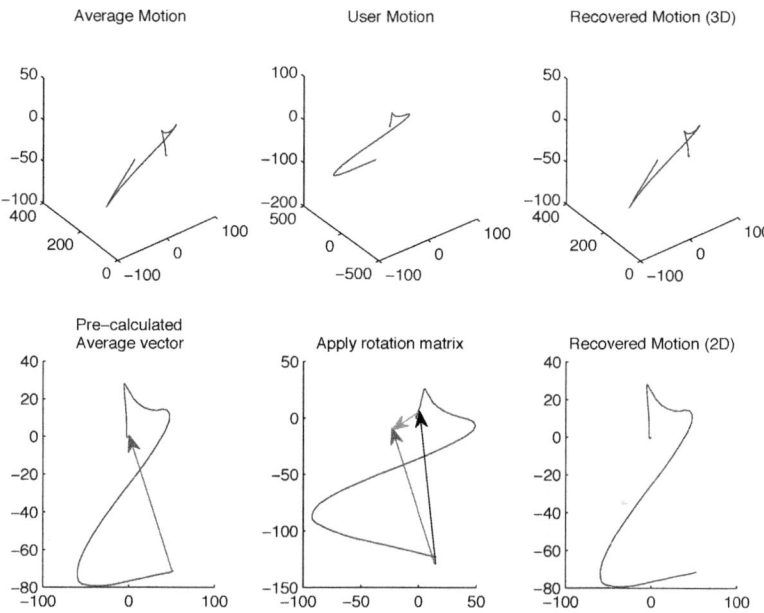

Fig. 2. Graphical illustration of our novel path-based reorientation algorithm

Two algorithms, dynamic time warping [7] and hidden Markov models [5] are a natural fit to this constraint. DTW is based on a dynamic programming technique and finds the minimal cost required to warp one signal onto the other, as long as the difference between two signals can be measured by some distance function. HMMs return the likelihood that a given signal belongs to a given (previously trained) class.

The Dr. Droid application collects acceleration data from the 3-axis accelerometer that is a standard feature of Android smart phones. The data is sampled at 50Hz. The WMFT allows a maximum test duration of 120 seconds, thus our acceleration traces contain no more than 18,000 samples. The phone used for testing, the Google Nexus One, has a 1GHz processor and is capable of scoring the motions using the above algorithms in a fraction of a second.

We include three scoring algorithms primarily because to date we have not had access to stroke patients to collect training data. Therefore, we cannot conclusively calculate which of our scoring methods correlates best with the FA scores given by trained observers. We also hypothesize that our algorithms may pick up on different aspects of a motions natural-ness and thus provide richer feedback about a patient's motion than a single FA score.

Path-Based Orientation Correction. Many difficulties with gesture recognition based on acceleration data are rooted in the orientation problem. If the orientation of the accelerometer is different between the training and test data,

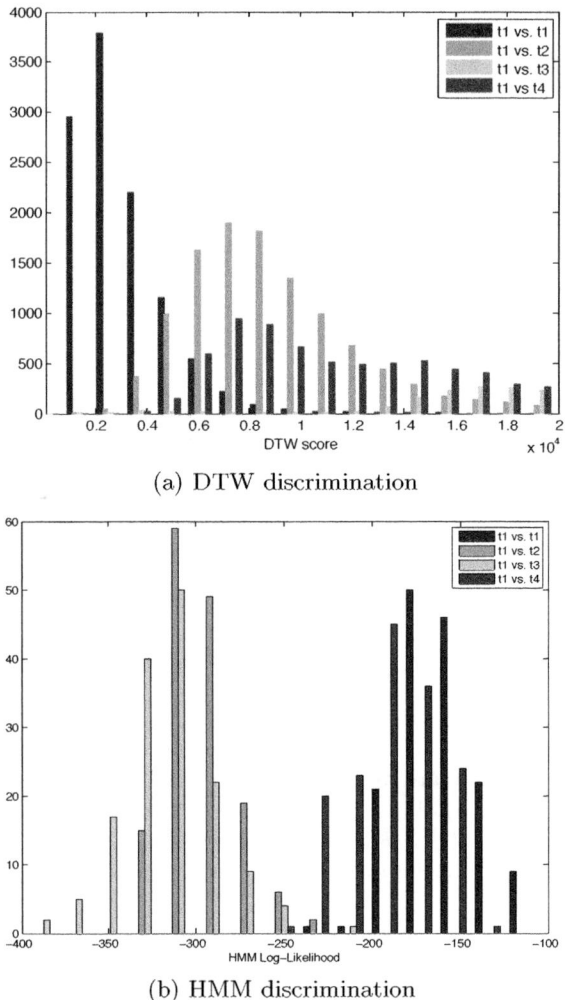

Fig. 3. Histogram showing the ability of the DTW and HMM algorithm to score similar motions with similar scores. Intra-class scores (blue) cluster together, indicating similar scores are given to similar motions.

many gesture recognition algorithms preform poorly. Researchers have developed several techniques to reorient acceleration data [10], but this has proven to be difficult and many systems solve this problem by requiring that for optimal performance, the orientation of the device is constrained. Furthermore, variations in the amplitude of the action require some sort of normalization, which also reduces the accuracy of gesture recognition.

However, the prescribed nature of our testing environment leads to a novel solution for the above limitations. First, we observe that the amplitude of the test is fixed by the definition of the WMFT. Second, we observe that if one considers the starting point for a test as the origin of a 3-D reference frame, then the ending points for two motions performing the same exercise will have approximately the same coordinates. Acceleration data collected with a misoriented sensor will describe the path in a rotated reference frame.

Therefore, in order to reorient the motion traces to best align them with the template action, we need only to calculate the rotation necessary to align the end-point of a motion with the end point of the template. To do so we calculate the cumulative double integration of the acceleration data to obtain the 3-D path traced by the wrist mounted phone. Then we calculate the homogenous rotation matrix that will move the 3-D (x,y,z) coordinate at the end of the trace to the (x,y,z) coordinate of the template motion. The rotation matrix is then applied to every (x,y,z) coordinate in the path to reorient the data.

We found that using this reoriented path data provided the best results with the DTW and HMM algorithms. Please see figure 2 for a graphical illustration of the reorientation algorithm.

Dynamic Time Warping. Given two discrete signals $A[t]$ and $B[t]$ of length N and M respectively, DTW finds the best match between $A[t]$ and $B[t]$ by warping (compressing or stretching) $B[t]$ where necessary to minimize the distance between the signals. DTW is a flexible algorithm in that one may define the distance metric as necessary. We are working with the reoriented 3D path data, so signals $A[t]$ and $B[t]$ comprise a sequence of (x,y,z) points. We thus choose the distance metric to be the Euclidian distance between the points. Then $\forall i \in 1..N,\ \forall j \in 1..M$ we define the N by M *distance matrix*, D as follows:

$$D(i,j) = \sqrt{(A[i]_x - B[j]_x)^2 + (A[i]_y - B[j]_y)^2 + (A[i]_z - B[j]_z)^2}$$

The DTW algorithm then finds the least-cost path through the distance matrix from $D(1,1)$ to $D(N,M)$. Please see [7] for details of the algorithm.

An additional advantage of this algorithm is that it naturally handles sequences of different lengths. We expect the completion time to vary between tests, but do not need to normalize or resample the time series when using DTW. If $A[t]$ and $B[t]$ are the same signal, then the cost is zero. As the signals vary in path and time, the cost increases.

Thus, motions of similar path and time will receive lower scores. Less natural or longer motions will receive larger scores. Therefore, we conclude that the path based DTW score is a similarity metric and we use the cost calculated to traverse the matrix as the DTW component of the Dr. Droid score.

Hidden Markov Models. Hidden Markov models are a well known method for modeling noisy, time-varying processes. They are used in many areas such as speech, gesture and handwriting recognition; natural language processing, and

Table 2. Comparing the average results of our scoring algorithms across two emulation methods

(a) DTW Scoring

Exercise	Avg. Score - Normal	Avg. Score w/ weights	Avg. Score 'poor motion'
1	2079.35	4877.96	19041.85
2	5925.08	3834.49	10499.91
3	20313.27	17464.42	33212.71
4	15911.03	18382.09	160464.70

(b) HMM Scoring

Exercise	Avg. Score - Normal	Avg. Score w/ weights	Avg. Score 'poor motion'
1	-151.14	-184.08	-210.55
2	-300.47	-286.03	-376.95
3	-365.47	-378.73	-498.45
4	-225.32	-304.72	-377.48

(c) Smoothness Scoring

Exercise	Avg. Score - Normal	Avg. Score w/ weights	Avg. Score 'poor motion'
1	411.97	725.14	523.39
2	274.17	327.78	487.05
3	284.16	521.34	673.18
4	2456.56	5393.08	4590.48

bioinformatic applications such as DNA analysis. The process being modeled is assumed to be a Markov process where the state-sequence is unknown, but the emissions from the states are observed. Once trained, a HMM will return the probability that a given sequence of observed emissions was generated by the modeled process. We assume a linear topology for our Markov process.

In order to train the HMM, we need to express the 3D path as a sequence of discreet states. Similar to work done on 2D handwriting recognition [8], we propose the following approach that is novel in the 3D gesture processing literature. From one (x, y, z) point in the path data we discretize the direction of travel necessary to reach the next point. With three axes (x, y, z) and three possible directions $\{-, 0, +\}$ there are 27 possible movements. The series of 3D points is converted into sequence of states labeled with the corresponding state number. This state-sequence representation of the path data is then used in the training and querying of a HMM for each exercise in the WMFT. In order to determine the HMM component of the Dr. Droid score we calculate the log-likelyhood that a given path describes the given exercise.

Smoothness. If one observes a stroke patient performing the WMFT (see [3] for a video), it will be observed that the motion is often jerky or halting. In order to quantify the smoothness of a motion we have developed a third metric based on the average energy of the acceleration signal over time. For this metric we operate directly on the 3-axis accelerometer data. We segment the acceleration signal into frames of 12 samples with an overlap of 6 samples. The energy of the acceleration signal in each frame is calculated and the average of these energy values is returned as the Smoothness component of the Dr. Droid score.

3 Experimental Evaluation

In this section we describe two experiments undertaken to verify the correctness of our scoring algorithms. We first describe the experiments and then provide a quantitative comparison of the results.

The goal of the Dr. Droid scoring metrics can be summarized as follows: *similar motions should receive similar scores.* We first collected training data of natural motion. We chose four exercises from the WMFT (marked with a † in table 1) that are representative of the different types of motion in the WMFT. We called these four exercises 'task 1' through 'task 4.'

Three of the authors then performed the exercises 50 times with the phone holstered at the wrist (see figure 1). These training examples were used to develop the path based DTW algorithm and to train the HMM.

First, we needed to evaluate if the DTW and HMM algorithm would give similar (low) scores to similar motions. To do so we scored all of the training examples against each other, both intra-class and inter-class. See figure 2.2 for a histogram of the resulting scores for the DTW and HMM algorithm with task 1 used intra-class. In these figures we see that indeed the intra-class scores cluster at the lower magnitude end of each graph (log likelihood scores are negative). This implies that our algorithms meet our goal of scoring similar motions with low scores. The results for the other tasks were similar.

Next we performed two experiments to emulate the motion of a stroke patient. In the first experiment we attached 10 pounds of weight to the forearm under test. In the second experiment we executed motions that emulated the difficulty and limited range of motion exhibited by patients as seen in videos of the WMFT. For the DTW algorithm we chose a 'golden example' from among our 150 training examples for each exercise. The golden example was defined as the training example with the minimum average score when compared to all other training examples of a particular task. The golden example is then used as the template signal for our DTW algorithm. For the HMM and Smoothness algorithms we calculated the score as described above.

Table 2 contains a summary of the experiment results. These results overall show that we have met our goal of giving larger scores to unnatural motion. Average scores highlighted in blue indicate scores where the scoring algorithm correctly scored the experimental motions as 'worse' than the training examples. Scores highlighted in red indicate where the scoring algorithm gave a 'better' score to the unnatural motion. These results, while not ideal, are worth discussing.

First we note that all of the red results occur in the weighted experiment. While performing the weighted experiment we noted that the weights made the exercises more difficult, but they did not necessarily make our movement unnatural. Moving the arm when weighted required very deliberate and thoughtful actions, which may have yielded efficient, correct motions.

However, when we look at the results for the 'poor motion' experiment, we see very good results. All of the algorithms score these purposely unnatural motions much higher than the training data. Therefore we can conclude that our scoring algorithms, when used on data featuring a range of motions from very unnatural to natural will give meaningful scores.

4 Related Work

As stroke rehabilitation is an important medical necessity, there is a large body of work in this area. Similar to our research is robotic assisted research. Fasoli et. al and Lo et. al [12],[15] describe robotic therapies that allow the scoring and tracking of patient progress. Closer still to our work is the AutoCITE [16] project. AutoCITE is a computer and specialized workstation deployed at the patients home. AutoCITE allows computerized testing, however it is expensive and once installed, fixed in location.

Specific to the WMFT test, Wade et. al [6] explore using a wrist mounted accelerometer to automate the timing of the WMFT. A recent masters thesis by Avinash Parnandi [2] builds on this work to attempt a similar goal of automating the FA scoring aspect of the WMFT using machine learning techniques, however the system requires a custom sensor board and wearable computer.

Accelerometer based gesture recognition is a well developed field, with Pylvanainen contributing a book chapter [18] to Springer's LNCS. More recently, Liu et al. [13] developed a DTW based gesture recognition for mobile phones, while Prekopcsak [17] used HMM and support vector machines to do gesture recognition on Nokia smart phones. All of these methods operate directly on the acceleration data and thus require orientation constraints.

5 Conclusion and Future Work

In this paper we have presented a application that uses a mobile phone to improve stroke rehabilitation. Using a mobile phone in this manner is novel and provides a less expensive and more flexible rehabilitation process to patients and medical practitioners. We describe Dr. Droid, a complete system that administers the stroke rehabilitation protocol and provides quantitive scores of a patients movement. To generate these scores we present a novel path based reorientation algorithm that improves the performance of our DTW and HMM algorithm. We tested Dr. Droid experimentally to verify the system design and scoring algorithms provide a working user experience and meaningful scores. Finally, we discuss several opportunities for future research.

Dr. Droid to date has served as a proof-of-concept application to test the suitability of the Android mobile phone to this applicaiton.

To prove our scoring algorithms at a minimum provide the same information about a patient as the functional ability score, we need to test our algorithms against data collected from actual stroke patients over the course of their recovery. In order to obtain this data we first hope to expand on the work of a colleague at USC who has kinematic models based on data captured from stroke patients. These models are in the form of joint angle vs. time. The path taken by the wrist and the acceleration data can be recovered from these models. The timing and functional ability scores for these models are also available. If our algorithms prove successful scoring the motions, then we envision a study where practitioners at a stroke rehabilitation clinic solicit volunteers to wear a holstered phone during their WMFT exercises.

We also will work to improve the Dr. Droid application as well. We hope to collect feedback from practitioners and patients regarding the features and usability of the application. We also will improve the remote control application which we see as expanding into an interactive tool to help practitioners, patients and caregivers track the patient rehabilitation progress. We will include automatic timing algorithms to further improve the accuracy of the Dr. Droid scores.

Telemedicine and health monitoring are large and active fields of research. At this point in the development of Dr. Droid we have to date focused on the scoring algorithms. We recognize that health data is considered some of the most sensitive data one might collect. Any number of privacy schemes may be applied to our system and will be added at a later date. Additionally, Internet enabling the application to allow automatic uploading of scores to a patients doctor or therapist is also and obvious extension of our work. We would like to recognize that we have not ignored these aspects in designing Dr. Droid, but thus far they have been beyond the scope of our work.

Acknowledgments. The authors would also like to thank Dr. Eric Wade for his thoughts and guidance.

References

1. http://www.android.com/
2. http://cres.usc.edu/pubdb_html/files_upload/670.pdf
3. Wolf motor function test, youtube video,
   http://www.youtube.com/watch?v=S1Jk88Nd-ZM
4. Assessing wolf motor function test as outcome measure for research in patients after stroke. Stroke 32(7), 1635–1639 (2001)
5. Bishop, C.M.: Pattern Recognition and Machine Learning (Information Science and Statistics). Springer-Verlag New York, Inc., Secaucus (2006)
6. Wade, E., Parnandi, A., Mataric, M.: Automated administration of the automated administration of the wolf motor function test for post-stroke assessment. In: ICST 4th International ICST Conference on Pervasive Computing Technologies for Healthcare 2010 (2010)

7. Ellis, D.: Dynamic time warp in matlab, http://www.ee.columbia.edu/~dpwe/resources/matlab/dtw/
8. Elmezain, M., et al.: A hidden markov model-based continuous gesture recognition system for hand motion trajectory. In: ICPR, pp. 1–4 (2008)
9. Morris, D.M., et al.: The reliability of the wolf motor function test for assessing upper extremity function after stroke. Archives of Physical Medicine and Rehabilitation 82(6), 750–755 (2001)
10. Prashanth, M., et al.: Nericell: rich monitoring of road and traffic conditions using mobile smartphones. In: SenSys 2008: Proceedings of the 6th ACM Conference on Embedded Network Sensor Systems, pp. 323–336. ACM, New York (2008)
11. Wolf, S.L., et al.: Forced use of hemiplegic upper extremities to reverse the effect of learned nonuse among chronic stroke and head-injured patients. Experimental Neurology 104(2), 125–132 (1989)
12. Fasoli, S.E., Krebs, H.I., Stein, J., Frontera, W.R., Hogan, N.: Effects of robotic therapy on motor impairment and recovery in chronic stroke. Archives of Physical Medicine and Rehabilitation 84(4), 477–482 (2003)
13. Liu, J., Wang, Z., Zhong, L., Wickramasuriya, J., Vasudevan, V.: uwave: Accelerometer-based personalized gesture recognition and its applications. In: IEEE International Conference on Pervasive Computing and Communications, pp. 1–9 (2009)
14. Lloyd-Jones, D., Adams, R.J., Brown, T.M., et al.: Heart Disease and Stroke Statistics–2010 Update: A Report From the American Heart Association. Circulation 121(7), 46–215 (2010)
15. Lo, A.C., Guarino, P.D., Richards, L.G., et al.: Robot-Assisted Therapy for Long-Term Upper-Limb Impairment after Stroke. N. Engl. J. Med. 362(19), 1772–1783 (2010)
16. Lum, P.S., Taub, E., Schwandt, D., Postman, M., Hardin, P., Uswatte, G.: Automated constraint-induced therapy extension (autocite) for movement deficits after stroke. J. Rehabil. Res. Dev. 41(3A), 249–258 (2004)
17. Prekopcsak, Z.: Accelerometer based real-time gesture recognition. Poster Preview (2008)
18. Pylvänäinen, T.: Accelerometer Based Gesture Recognition Using Continuous HMMs. In: Marques, J.S., Pérez de la Blanca, N., Pina, P. (eds.) IbPRIA 2005. LNCS, vol. 3522, pp. 639–646. Springer, Heidelberg (2005)
19. Wilson, A., Shafer, S.: XWand: UI for intelligent spaces. In: CHI 2003: Proceedings of the SIGCHI Conference on Human Factors in Computing Systems, pp. 545–552. ACM, New York (2003)

Open Transaction Network: Connecting Communities of Experience through Mobile Transactions

Kwan Hong Lee[1], Dawei Shen[1], Andrew Lippman[1], and Erik Ross[2]

[1] MIT Media Lab, Cambridge MA 02139, USA
{kwan,dawei,lip}@media.mit.edu
[2] Bank of America
Charlotte, NC 28202, USA
erik.s.ross@bankofamerica.com

Abstract. In order to understand the value of social information in the context of mobile commerce, we created the Open Transaction Network (OTN), a collaborative, social transaction system. OTN uses voluntarily contributed transactions to index personal and social experiences in the physical world and to form dynamic communities around purchases. We use mobile phones and Open Spaces as portals to facilitate sharing of transactions. Through real world deployment we investigate the design elements and analyze users' tolerance to sharing such experiences. The sociability threshold, introduced as a measure of user's willingness to share for different categories of products, is found to correlate with price. The system was deployed to over 20 users over a 5 month period to allow the participants to share their in-store purchases. The analysis of empirical data shows that second degree connections are valuable for obtaining recommendations.

Keywords: open information, social networks, transactions, mobile, commerce, advertising.

1 Introduction

As mobile phones become Internet- and transaction-enabled, how we consume (search, find, buy and use) products and services are changing in many dimensions. *Presale* behavior is changing as people begin to research products and services on demand, instead of waiting for an Internet connection. As we traverse through different locations, information on relevant products and services will be broadcast to us. Near *point of sale*, consumers will be able to compare the products in real time and contact their social network for opinions on their choices prior to their purchase. Cash registers may become obsolete as consumers pay immediately with their phones bypassing long lines at the register. After a purchase (*post sale*), people will be able to connect with communities of people who can be real-time resources to provide quality assessments to those considering a similar purchase; to seek out guidance and aid in troubleshooting; and to ask warranty related questions.

We expect the same changes that have occured with the web to occur at a more rapid pace on mobile devices as the existing Internet infrastructure and services extend to the mobile space. The web has gone through three distinct periods of growth; as an information service (initial world wide web), a global transaction platform (electronic commerce), and now in large measure, a social space for sharing dynamic, real-time information and relationships with other people. Such combination of social space and transaction platform has naturally evolved into the mobile environment as smart phones and Internet-enabled devices become the hub of personal communications. Although there are anecdotal evidence accumulated from mobile advertising and SMS flash mobs regarding social-mobile purchase influence, we lack a quantitative system understanding of the effects of social networking in mobile commerce [13].

In this paper we present the Open Transaction Network as a system that provides a better understanding of people's purchase behaviors and the flow of information through mobile transactions. We investigate the design of a mobile, user contributory, social networked financial environment, that allows individuals to share their purchase behaviors with a community of consumers to collaboratively sense the activities and opinions of the market. The system can reveal the potential of passive information diffusion due to its socially networked nature and openness. We discuss its implications on mobile commerce as it may enable more informed financial decisions in geo-local contexts through friends and communities of similar people. We investigate people's willingness to share and the potential benefits and issues related to open transactions. We developed the system and deployed it for almost 5 months to over 20 users (14 female, 8 male) at a corporation where some of the users are known collegues.

The paper is organized as follows. The following section describes related work and the context for understanding financial data in a social milieu. The third section introduces the OTN design principles, system architecture, and user interface considerations. In the fourth section, we present the findings from field deployment and analysis of the collected data. The fifth section discusses the limitations of OTN in the current environment and its potential impacts on mobile advertising. In the final section, we conclude with design guidelines for social mobile commerce and future research directions.

2 Related Work

In the general area of mobile commerce, OTN touches several aspects of related research. OTN principles build upon many previous works on participatory sensing and mobile applications to commerce. Discussions on information revelation are insightful, since the data collected through OTN are through voluntary contributions. Mobile marketing as an application is also considered as well as the impacts of open transactions and social networks.

2.1 Participatory Sensing

Nowadays, data is constantly collected about our lives and activities as we traverse through this world. When we register for a driver's license, open a bank account, eat at a restaurant, fly on a plane and access web sites, we generate data about our activities. [12] describes the information asymmetry problem that creates externalities to data owners, as services that own the data make the data available to third parties while users themselves though possible, do not utilize much of their own data.

Recent literature on participatory sensing and in-situ sensing through mobile devices demonstrate how users can contribute content and data from mobile devices to allow other users and services to understand the environment and the users' behavioral patterns[8]. For example, digital footprinting through mobile phones can be used to track tourist traffic and their activities in different areas in Rome[9]. Mobile digital footprinting allows the city to know which areas are popular on what days of the week and what months of the year. It also allows the city to view where people from different nationalities visit. Such information enables local merchants to tailor their offerings to people of certain nationalities and manage their inventories accordingly.

2.2 Information Revelation

With digital mediation, we can capture the extent and the willingness of people to reveal themselves. Information revelation and self disclosure through computer mediation[15] describe how reciprocity is a key factor in users' revelation. Development of online social networks has led to emergent styles of openly sharing one's personal information for various social motivations: ease of connecting to new communities, exhibitionism through open content, and lightweight maintenance of social relationships. Ease of connecting with people online is shown through Facebook where 30% of people are willing to accept strangers as friends permitting their profiles to become open and visible[10].

People are willing to share their identity and preferences with online merchants when sites are able to provide personal incentives such as discounts for products and services[1]. Especially when there are needs and willingness to help others, people reveal themselves in online communities to provide a network of exchanges, self expression and answers[20] to questions. By diffusing such information through the devices we carry, we can facilitate opportunistic communications in-situ.

[17] indicates that one of the ways to build trust in a mobile transaction environment is to help build a community which allows members to exchange experiences and develop relationships. Contribution from all participants engaged in reciprocal communication forms the foundation of trusting relationship in the new social digital era.

2.3 Communities and Commerce Decisions

The theory of embedded markets[7] describes how purchase behavior is influenced by the social relations that are embedded in those transactions. The total return from the transactions increase or decrease the relationship strength between sellers and buyers depending on the satiation of the transaction. Data on these transactions can provide insights to these social relationships. Unfortunately, usually the seller maintains the history of these transactions with very fragile records (paper receipts or personal memory) maintained by the buyer. Amazon.com and Netflix has been successful in opening up transactions of people with similar purchase behaviors for collaborative filtering and provide users with recommendations during shopping online. More importantly, the lack of information on social relationships and people who made similar purchases make it difficult to understand the possible influences and social causes for the purchase. In the case of OTN like architecture, the open transactions and the social network of consumers can be used to investigate the horizontal embeddedness due to diffusion, gift exchanges or auctions that may bind consumers to each other.

In mobile commerce, numerous work has been done on providing people with just-in-time information for price comparisons, planning people's shopping routes and managing digital payments and coupons. [3] focuses on navigation and availability of products in shopping malls and delivering offers in proximity. Project Aura shows how tagging of the physical world items from the mobile can help acquire just-in-time information in mobile commerce settings[5]. The system had features to publicly share reviews and annotations with the public, but the effectiveness of such social information was not evaluated. [18] is an object annotation system that links physical objects in the retail environment with online content. It discusses the potential utility of extending it to social networks, but has not implemented it to investigate its implications of sharing. We extend these efforts to understand the design elements of incorporating social networks in mobile commerce contexts.

3 Open Transaction Network

Open Transaction Network (OTN) is a system built to investigate the design elements for incorporating social information into mobile commerce. In this section, we delineate the system architecture of OTN, its design principles, and user interface considerations.

3.1 Scenario

One Sunday morning, Jane, a flute instructor, and her friend John, a computer scientist, are driving to Jane's flute class and they notice that Jane's "check engine" light came on. John indicates that he has experienced the same phenomenon in one of his cars many times before and that it usually is not a serious

problem but something that she should get checked sooner than later just to
be safe. He leaves a voice mail for his brother Mike, who used to work for a
car manufacturer, to see if he has any suggestions. In the meantime, they find a
local mechanic open. However, when they speak to the mechanic, he tells Jane
she will need to leave the car overnight to perform the computerized diagnostic
and that it will cost $91. Since Jane is unable to leave her car overnight and
is concerned about the warning indicator, the mechanic suggests going to the
nearby CarZone that can usually run the diagnostic in a few minutes and perform it for free. Jane and John thank the mechanic and head to CarZone. The
CarZone associate performs the computerized diagnostic and the code provides
two possible causes, either the water pump or the engine coolant temperature
sensor. As the CarZone associate is explaining the causes, John gets a call from
Mike. Based on the results of the test, Mike informs John that as long as the
engine temperature is stable, it can wait a few days. With this information Jane
is comforted and decides to take her car back to the mechanic on Monday.

Throughout this interchange, a large amount of distributed embedded knowledge is shared and discovered by the actors, both friends and experts, to objectively assess the problem. OTN is an attempt to alleviate such situations by
allowing users to search their social network for purchase experiences. In this
case, Jane would search "check engine light on" in OTN and get a list of friends
and people that can help her. She would be able to call them directly and talk
to them. In the case when she does not find anybody, she would log her "free"
experience at CarZone in OTN for others to consume. These transactions would
be automatically logged and accessible as transactions become more real time
through digital receipts at the point of sale and card transactions get processed
in real time with services like Yodlee providing standard APIs to retrieve these
card transactions. In addition, as mobile devices become capable of handling
payments, one could pay with the phone and also use it to decide to share it
with friends and public at the point of sale.

3.2 System Architecture

In this section, we present the potential of OTN constructed through people's
participation. This approach can have numerous benefits to the consumer. It
provides easy sensing of the market by utilizing the experiences and knowledge of
other shoppers. When Jane goes shopping for shoes and finds several sales events
at different stores, she publishes this information on OTN. This information is
easily shared with others who might be interested in shopping for shoes and
she becomes a reference for recommendations and information. Such information
reduces the opportunity cost for searching while providing background awareness
to ongoing sales events.

OTN was deployed to a group of real users to understand people's willingness
to contribute and share purchase information with others in their social network.
[18] mentions about the potential for communities forming around products and
we develop the concept further to understand the system design issues to support
such communities.

The system is a mobile application with a web backend. The mobile application provides data entry and easy access to the information collected from the participants. The web backend is linked to various external backends to facilitate recording of transactions (Figure 1).

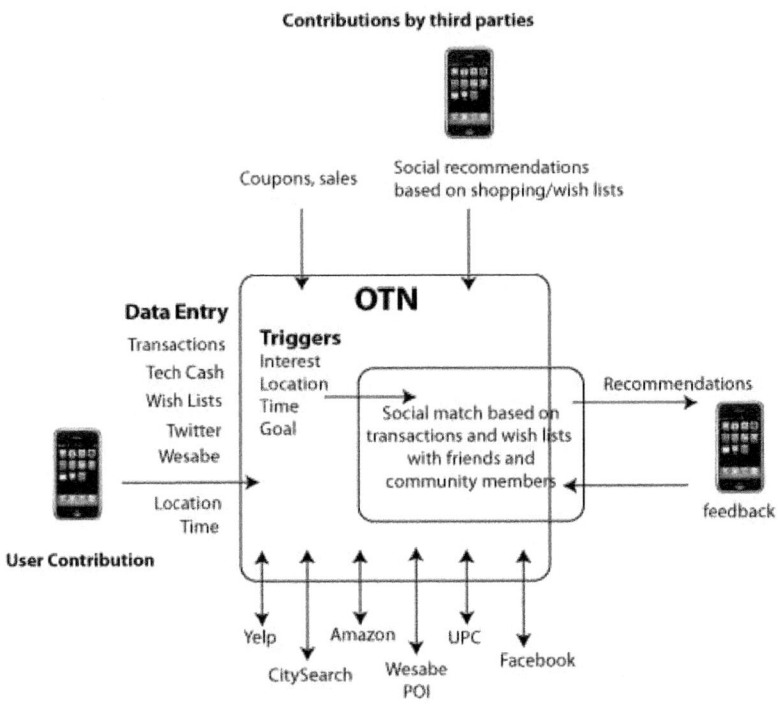

Fig. 1. OTN Architecture

OTN has the potential to convey recommendations in-situ. Moreover, it could track and influence people's decisions in-situ, but at the time of the trial we did not have the mechanisms to track causality of such influences. We will consider this in our future work on digital receipts and digital menus.

3.3 Mobile Application

The mobile application allows one to log one's shopping list and itemized list of purchases to form a social network around these items. Individuals can share this data to inform and guide others in a timely, personal and contextual manner when they are shopping for a product or seeking a service. It can also help people connect opportunistically in a local area to make group purchases, to pick up an item for a friend, and to perform reverse auctions. Beyond consumption, it allows one to be more aware of spending as transactions may be shared with friends or family.

The mobile application has the following core features: a purchase log, a wish list, and friends. The purchase log records any purchases that the participant wishes to capture in OTN. The purchase log records the 1) item name, 2) category, 3) the store name, 4) price, 5) rating and 6) sharing option (friends or OTN). Ideally if there was a digital receipt system or an advanced mobile payment system, the first four items would be entered automatically. However, in the current world, it is rare to find merchants like BestBuy that make itemized level data available through an API. As a result, manual entry was used due to the lack of a standardization for obtaining required information. The wish list is used to record items people are intending to purchase in the future. The friend's view allows one to view friends' open transactions and wishlist items(Figure 2).

The purchase list and the wish list automatically becomes input to the system to filter what users may potentially purchase regularly or what they might want in the future. The data, in return, allows personalizations for fulfilling wish lists that merchants could use for reverse auctions. Not only does it provide control to the consumer, but it also allows vendors to sense the market and control their inventory and prices based on the requirements of users[2].

(a) Search Results   (b) Profile

Fig. 2. OTN iPhone mobile application

3.4 Product Entry and Auto Completion

One of the important features to facilitate user interaction with the system is for the application to recognize what the user is trying to enter, search and auto complete this data based on the location. The key to such an approach is to do a geospatial query for relevance. In order to accomplish this, we used the following architecture.

In order to expedite the product entry, UPC codes are used. One can enter the UPC code in the search bar. The UPC code maps to the name of the item and the manufacturer. This allows a standard way of identifying a product so that all users' information can be correlated. Other information such as price, rating and settings for sharing still must be manually entered.

A shortcoming of UPC codes is that they only exist for manufactured goods. In the case of OTN where many transactions were in categories of food, services, and entertainment, UPC codes are not particularly beneficial. As a result, the utilization of the UPC database was limited for this trial.

As the user types in a query, it is posted to the server. The server looks up the existing cache of search results to see if any similar search has been made. If so, the result candidates are sent as JSON objects back to the client. If there are no cached results, the search query is made against Yelp and City Search (scraped and downloaded on our server). The results returned are parsed to obtain the names and cached in our local database. The results are again returned as JSON objects to the client.

3.5 Extending to Open Spaces

We extended OTN and created a prototype system named "Open Spaces", which allow us to demonstrate the opportunity of using just-in-time information in retail spaces. [14] describes potential uses for having large shared screens in retail shops to track traffic, locate items, and provide feedback to store owners and customers. Open Spaces specifically allows OTN information to be integrated with existing shared spaces that would be available in such pervasive computing environments. Open Spaces is an extension of OTN, where people can browse through and share their shopping lists on a large shared public display.

The large shared public displays present the ability for merchants to engage in sales offers to specific people. The phone broadcasts the wish list to the merchant as the customer opens the OTN mobile app, the store can accordingly provide location information where the items could be found, and also present a just-in-time discount offer to influence the customer's purchase decision.

If another customer who shares interests in the same product approaches, both customers' wished item can be displayed on the screen. The customers can share information with each other and help each other decide on the purchase. The system also allows access to a friend's or a family member's wish list and to potentially purchase the item on his/her behalf since the user is located at the store and has visibility to their friend's needs. The merchant can also provide a group offer to motivate people to coordinate their purchases when two people have same wish item (Figure 3).

Fig. 3. Open Spaces: Wish list viewing and collaborative shopping

3.6 Wish List Sharing Protocol

OTN's wish list sharing protocol allows people to openly share their wish lists with others. It operates by using the Bonjour protocol to share with any entities that are willing to accept the wish list sharing protocol. There are two ways to have it implemented: infrastructure and ad-hoc. The infrastructure mode only publishes a URL, which returns a JSON-formatted wish list. Each wish list item contains information about the item including its name, its brand, the price range, and a description. The ad-hoc mode publishes a JSON-formatted wish list only to neighboring devices that accept the wish list sharing protocol. Applications like Open Spaces could utilize this ad-hoc feature when an Internet connection is not available.

Such wish list functionality enables filtering of the marketing content and advertisements if the user selects it as a filter. Attention needs to be focused around the relevance of the content and the timeliness and frequency of the delivery of marketing messages[6]. The history of transactions and the wish list can increase the "relevance" of the content.

OTN additionally provides the social information around the wish list, beyond just its content. These are context sensitive alerts that will be raised to the user depending on their location. If a user is entering a wish list item, it will query the server to see if there are any users with the same item on their wish list or have recently bought the same item. It does not pop up, but it is visible to the users after they immediately save a wish/shopping list item.

4 Evaluation

The system was used for approximately five months to collect over 600 trasactions from over 20 people. This represents on average about 15% of transactions of the participants. Volunteers were recruited to participate by providing them an iPhone and a data plan for the length of the study. About 50% were married and other details are described in Table 1. In this section we summarize the findings from the deployment and share the insights we gained from the collected data.

Table 1. Participant demographics

Parameter	Min	Max	Median
Age	23	55	38
No. of children	0	3	1
Total transactions made during the trial period	50	over 400	250

4.1 User Interface

The original goal was to collect itemized level data that is usually not available in credit card statements or online banking. However, the data we collected did not have desired details due to people's difficulty with manual keying on the mobile phone. For example, for Italian food, users simply recorded 'spaghetti' instead of the full dish name. For groceries, many people entered 'groceries' instead of specific items that they bought.

The application would have been more effective if data entry was minimal. A more user friendly design would be for a user to take a photo of their receipts with details filled in through a connection with the merchant or via crowd sourcing. We are attempting this design for our current version.

The collected data contained geo-information for location based filtering. As shown in Figure 4, the purchases were geo-coded so that they would naturally be useful information for those specific locations. Our initial design did not proactively make other people's purchases visible at pre-sale. One could see the list of people who had purchased the same item or had the same item on their wish list after they have logged the item. People found it most helpful to know this information when they were buying gifts for others, or when they had to choose wine at a restaurant. The main advantage beyond a Yelp like service was that a recommendation was available from *known* people and the system could be generalized beyond just restaurants.

At the time of the trial, 3G was not universally available and many indoor locations did not have adequate cell phone coverage or WiFi reception. Therefore, it was necessary to implement persistence on the device so that items logged were stored and uploaded the next time the application acquired a network

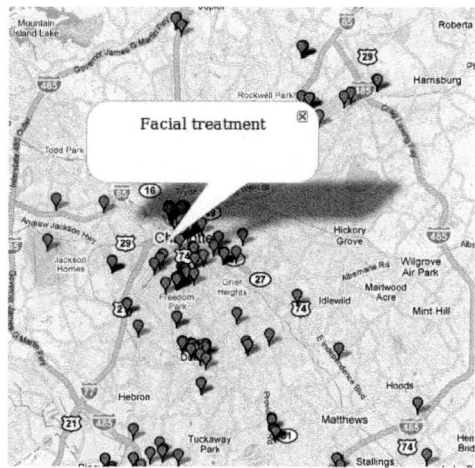

Fig. 4. Geo-coded purchases

connection. Since locations were not tagged due to lack of GPS coordinates when there were no GPS, WiFi or cell signals, it was necessary to estimate coordinates from the location name or previous entries. This constraint is a limitation with using the iPhone, since on Google G1 (Android phone), GPS coordinates using background process can be regularly tracked and used to estimate locations.

4.2 Individual and Group Behaviors

The primary benefit of OTN is that it networks people with similar transactions and also informs of the social distance. Individual profiling can be performed from participants' open transactions. Each individual in the study was represented as a vector of 16 values representing normalized fraction of purchases in each category. The basic use of this information is to create a signature of financial behavior. The transactions recorded in OTN are shared with friends and the public depending on the user's sharing preferences at the transaction level. Not only does this approach allow for comparisons with friends but also with other individuals in different dimensions. Participants in second degree and third degree friends are readily viewable through OTN.

Participants indicated it would be most useful to compare their information with people of similar income, similar family size, or lifestyle since these people would have useful information about stores of similar interest or sales of interest. In contrast to anonymous recommendations on the web, by adding social distance information, it has the added advantage of judging the reputation of the recommenders.

In order to understand how purchase behaviors relate to social relationships, we calculated the mean squared distance between the transaction vectors. We found that the behaviors of individuals compared to their social network of different distance was divergent on average (Figure 5). By comparing the vector distances, we found that the social network relationship does not provide any information about how similar friend's purchase behaviors are. One could explain this from the fact that they are coworkers and each have very different life styles. However, this also means that new product/service related information could be easily obtained from the social network due to the social network having knowledge about products and services that one might not be familiar with.

We refer to the "second degree social network" of an individual user as the collection of users who are either friends of friends. This second degree friends are particularly special since they can potentially provide significantly more recommendations. Compared to a randomly generated social network, the empirical data shows that the second degree friends can supplement the most amount of information (Figure 5). By comparing the empirical data with randomly generated social networks, we find that the purchase behavior of second degree friends in the OTN trial are statistically different (p < 0.01) from random social networks. Therefore being able to reach the second degree network more easily may provide more valuable information for the consumers. This is due to the real world social networks having hubs that bridge different groups in contrast to random networks.

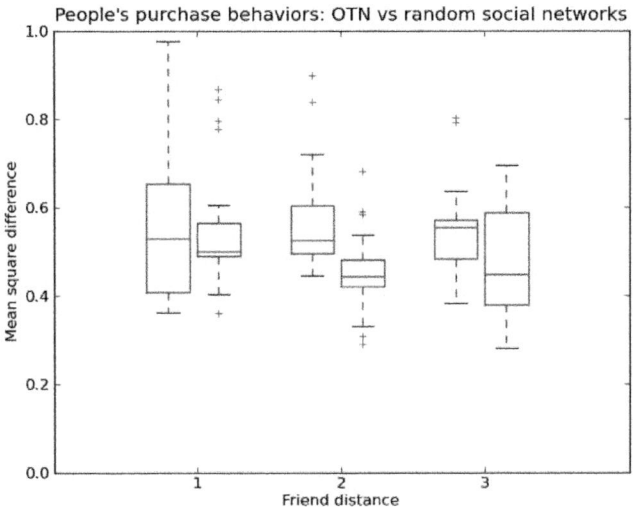

Fig. 5. 2nd degree friends can provide most information. Transaction behaviors related to social distance (0 very similar, 1 very different). First set of box plots are empirical data from real world social network of OTN participants. Second set of box plots are from participants connected through random networks. Second degree friends are different from random friends (p < 0.01).

The Open Transaction Network can also be used to analyze group behavior and identify the shared experiences from people's consumption data. The sequence of events also form a behavioral story of individuals and of the community. The transaction timeline (Figure 6) shows much more activity as the December holiday season was approaching. The 2 week cycle of large spending spikes aligned with the bi-weekly pay cycle of the company in which the participants were employed, illustrating temporal group pattern in consumption.

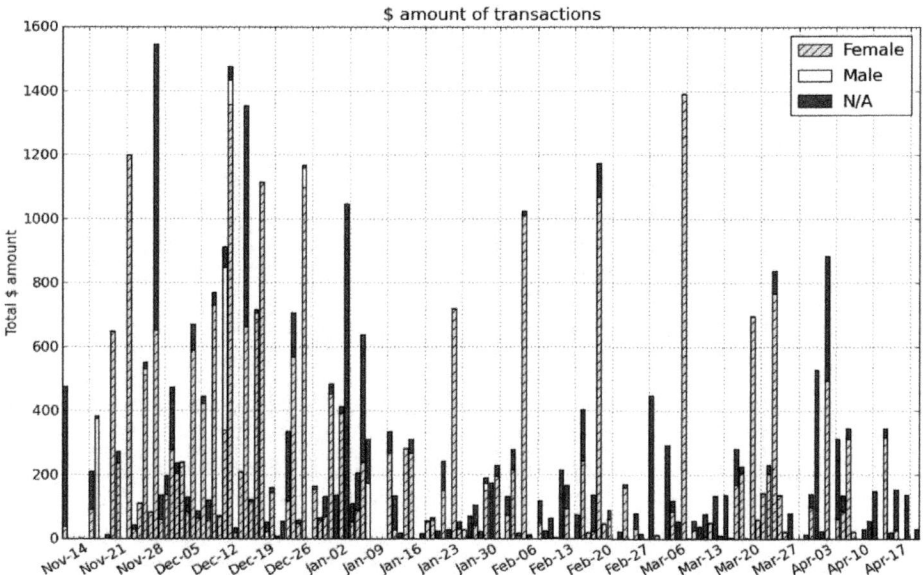

Fig. 6. Transaction time line from November 2007 to April 2008

4.3 Social Network of Recommenders

Figure 7 illustrates the social network of participants in the OTN. These were identified by participants adding people they knew through the OTN website. The color gradient shows the different number of friends with darker circles having larger number of friends. Subjects 15 and 19 form the hub of the social network.

The OTN community shows the benefits of open (shared) transactions versus closed (not shared or shared only with direct friends) transactions. Normally we are only connected by 1st degree friends in the real world or with a lot of random reviewers on the Internet and somewhat through word of mouth to 2nd and 3rd degree friends. By connecting people via transactions, we can see larger community (friends of friends) of people that can share their experiences for common category of purchases. Through OTN, people have trusted visibility to other participants' transactions in the social network and the OTN community.

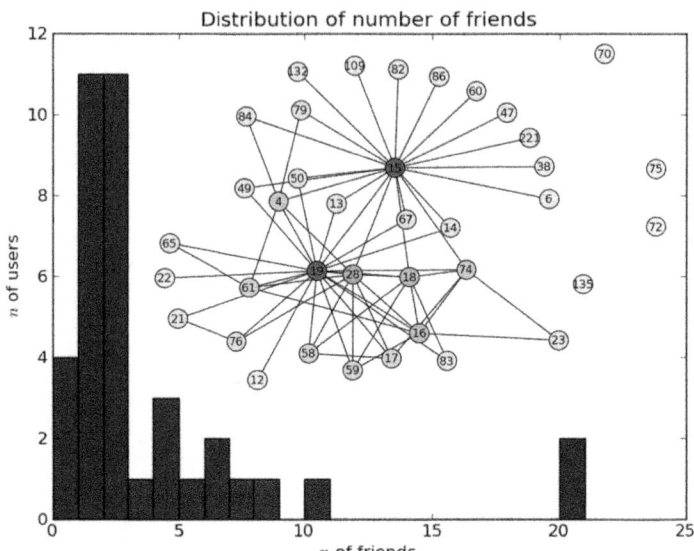

Fig. 7. Social network of participants and degrees of friends in OTN

In the small community of participants in current study, a range of 4 to 14 people purchased items in common categories (Figure 8). We define the *potential recommenders* for an individual to be the people in OTN who have a purchase experience in the same category before the individual's purchase date. The greater potential recommendations from 2nd degree relationships is especially noteworthy. If one were to seek information in a typical social network, one would have to ask a friend to refer a friend for information. However in OTN, the openness of the platform provides a direct link to a community of known, trusted recommenders and reviewers through similar transactions, making information equally accessible even to those who are socially disadvantaged.

The actual utility of these potential recommenders may vary depending on the product category. Those product categories that have greatest uncertainty about the quality would benefit the most from these recommenders who have past purchase experiences. Categories of products and services that require experience, denoted as experience goods[11], such as food, auto mechanics, medical services, plumbers and clothing would benefit from greater number of trusted second opinions. The benefit of OTN is that these are people who have actually purchased these goods or services and are potential friends or acquaintances. Also, if one has a high degree of friendship, each individual's opinion might not carry the same weight. As a result, those that have lower number of friends might actually benefit more from OTN by being able to filter information through first and second degree friendships.

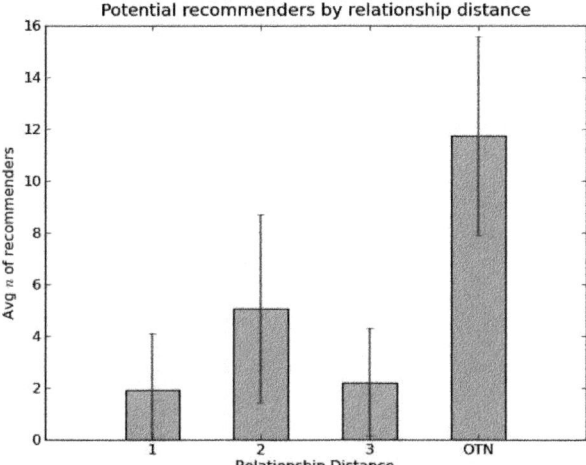

Fig. 8. Mean potential recommendations from different social network distances (1st, 2nd, 3rd degree) compared with OTN (when connected through common transaction categories). Whiskers indicate standard deviation.

4.4 Sociability Threshold

A key research question that results from OTN is about dimensions of people's willingness to share: if users are willing to contribute their data, then for which categories and under what conditions are they willing to reveal to others? Though the experimental trial was done with a reduced group of people in an organization, it revealed some insights to these questions.

OTN has four categories of sociability: private, friends, community and public. In the case of the mobile scenario OTN attempts to allow users to control sharing of their transactions in different contexts since, unlike the online context, offline context is dynamic. Users decide whether certain purchases were shared between friends, community (people who purchase in similar categories) or the public.

OTN users had an increased willingness to share with friends (20% increase) and public (30% increase) about their purchase information after the OTN trial as compared to the start of the trial (Figure 9). Through online surveys we investigated the general public's willingness to share purchases on 10 items between friends and the general public. The survey also questioned people's willingness to share the name, brand, location and price. Consistently price came up as the least willing to share attribute. The same survey was completed by OTN participants. We cannot directly compare the results since the total population that was surveyed without experience of OTN was much larger (~250 people) while the OTN participants were only 20 people. However, there was a consistent trend that the OTN participants were more willing to share with both friends and the public. The least difference between the survey groups was observed in pizza and the most difference was observed in expensive items such as "Ralph Lauren trench coat on sale" and a MacBook.

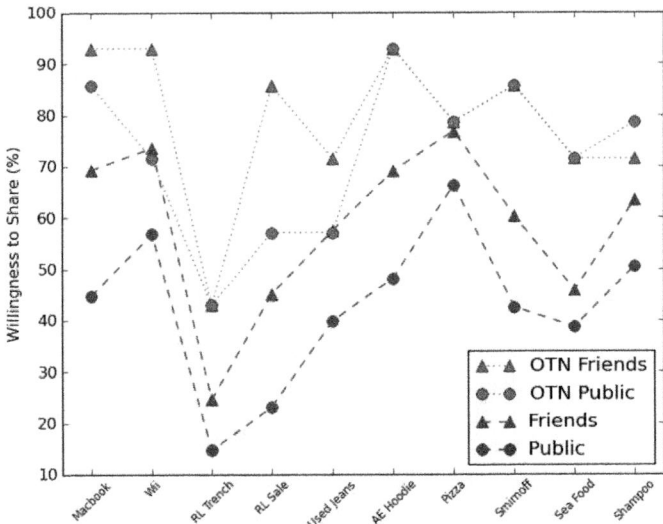

Fig. 9. Online Survey: public's willingness to share (dashed) compared with OTN participants (dotted)

The sociability threshold is a measure that indicates that there is a price threshold for each category of products that users are open to reveal and accept recommendations from friends. Dining, coffee, groceries, entertainment, fashion were the top categories that were openly shared by people (Figure 10). Categories purchased less frequent, for example electronics, were less likely to be open shared. From interviews we found that people were not willing to share their

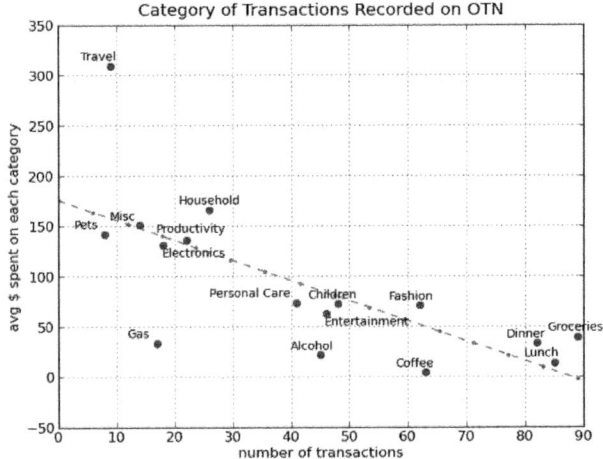

Fig. 10. Sociability threshold of categories correlation to price (Pearson r=-0.75)

pharmacy purchases since these purchases may reveal person health issues. Post interviews revealed that people would like to be able to find other friends and community participants that are actively purchasing specialized items such as electronics, pet accessories or wine. Another subject mentioned that she wanted to purchase a gift for her 12 year old niece and it would have been extremely helpful to obtain recommendations from their friends who have experience with gifts for that age group. Finally, one participant commented about her avid wine consumption which she did not publish on OTN due to potential misinterpretation by her colleagues.

5 Discussion

5.1 Applications to Advertising

There is no variable in current advertising models to capture a user's appetite to particular advertisements. OTN provides a social metric for products that can be used as a proxy for relevant advertising in those product categories. The frequency of transaction categories could be used to limit the frequency, timing, and amount of advertisements that a user receives.

A user's behavioral propensity to save or spend should be the basis of advertisements. We believe the new measure of success for advertising will be the ability to enhance closing rates instead of click-through rates and impressions. By presenting transactions socially, advertisements can become socially aware (i.e. your friend Mark likes/has purchased these items). We believe this targeted social information can make users more informed and more comfortable with their product purchases.

5.2 Limitations on Collecting Data

Sites like Amazon, BestBuy and others are collecting information about people's wish lists so people can share and manage them while the merchants may use it to make recommendations to increase purchase behavior. Purchase data have always been collected by offline retail, but only sold to third parties and rarely shared with the customers. The mobile phone allows such data to be accessible to customers and provide contextual recommendations when needed. In the case of OTN, one can search at the location whether others that they know may have purchased a similar item. By making this data more available for application developers and community, a more consumer centric data gathering and application environment can be fostered while increasing the number of applications that can benefit consumers.

Transaction information from credit and debit cards are readily available through sites like Buxfer that provide APIs to access transactions from multiple banks. However, the data provided through payment networks and card transactions are difficult to decipher due to the truncated text and numeric strings that do not provide much meaning. For example, if there are multiple

CVS locations in Cambridge area, it is not a trivial task to identify from the transaction string (CVS 1002 01002 Cambridge) which specific store was used. For smaller stores, the merchant names do not match the actual business names, making it difficult to identify the actual location of purchase. Categorization is also not readily available through these APIs. The time of the day of the transaction can provide a lot of information about the daily behavior of users, but this information is not available in current transaction information.

OTN overcomes these problems through user participation. However, similar to many user contribution based systems, it will require a much larger number of participants to contribute to cover the tremendous number of products. Participation by merchants and a more open system beyond existing point of sale systems could be a great way to seed the system.

5.3 Privacy Concerns

Stalking is the behavior of following someone incessantly by utilizing information that can be obtained publicly. When purchase behavior transactions are made widely available in the public domain especially in timely manner through mobile devices, stalking problems could arise. A person may not want a boyfriend or girlfriend to know where they shop or when and where they met another person for coffee. Similarly, college students may not want their parents to know that they are spending a majority of their available funds on alcoholic beverages. OTN provides the opportunity for resolution of disputes over certain events and/or the time of their occurrence.

However, with the prevalence of social network services (i.e. people following on Twitter), it is imperative to understand how we can manage such open information rather than continue the existing tradition of keeping the information vaulted by the banks and merchants. There is greater benefit to be gained for the greater community as we understand how socialization of transactions affects people's commerce behavior. The community may know the best price; the community may know the pitfalls for certain services; the community may know when new products are available and selling in what quantities they are selling.

OTN is currently an experimental platform to understand control given to users to manage their privacy. Making the purchases shareable to a community of people that have similar purchases is one way of limiting the information diffusion. For example, people who actually buy flowers will only be notified about others who bought flowers. The friendship degrees can be used as a way to control diffusion. Initially people will only see their first degree neighbors, but as they contribute more they will be able to access second degree friends that might contain more valuable information.

6 Conclusion

This paper presented the OTN, an extensible platform for mobile social transactions that is built on the principle of encouraging greater access to your financial data and community awareness of information. Despite privacy problems that were created by Beacon for publishing Amazon purchases on Facebook, startups like Blippy[19] and tweetwhatyouspend.com have reformulated the idea to allow people to share transaction data for online purchases in more practical ways. Companies like Square are also pioneering digital receipts that make it easier for sharing purchases. As the industry is moving towards providing the tools and APIs to access the transaction information, it will become important to consider what it means for transactions to be open and the possibilities of new applications and systems that can be built in such an environment[16].

An open transaction is not about just making the information public, but it's also about creating an environment where people feel safe to share their purchases and inform each other. When it is open, it is also open for verifying, annotating, and disputing for refunds[4]. Accounting is made open for users and comparisons can be made about each other. This trend has become prevalent in Wesabe and Mint like services where they aggregate many people's transaction data. This approach presents opportunities to socially influence people for behavioral change in-situ. However, we believe viewing this information on a website is less persuasive than viewing it during the purchase decision or at the point of sale. Mobile phones are the natural means for such interventions. The effectiveness of these interventions is currently being tested through a digital menu and a digital receipt applications.

References

1. Andrade, E.B., Kaltcheva, V., Weitz, B.: Self-disclosure on the web: The impact of privacy policy, reward, and company reputation. Advances in Consumer Research 29(1), 350–353 (2002)
2. Arora, N.N., Dreze, X., Ghose, A., Hess, J.D., Iyengar, R., Jing, B., Joshi, Y.V., Kumar, V., Lurie, N.H., Neslin, S., Sajeesh, S., Su, M., Syam, N.B., Thomas, J., Zhang, Z.J.: Putting One-to-One Marketing to Work: Personalization, Customization and Choice. Marketing Letters (2008)
3. Bohnenberger, T., Jameson, A., Krüger, A., Butz, A.: Location-Aware Shopping Assistance: Evaluation of a Decision-Theoretic Approach. In: Paternó, F. (ed.) Mobile HCI 2002. LNCS, vol. 2411, pp. 155–169. Springer, Heidelberg (2002)
4. Brin, D.: The Transparent Society. Basic Books (1999)
5. Bernheim Brush, A.J., Combs Turner, T., Smith, M.A., Gupta, N.: Scanning Objects in the Wild: Assessing an Object Triggered Information System. In: Beigl, M., Intille, S.S., Rekimoto, J., Tokuda, H. (eds.) UbiComp 2005. LNCS, vol. 3660, pp. 305–322. Springer, Heidelberg (2005)
6. Carroll, A., Barnes, S.J., Scornavacca, E.: Consumers perceptions and attitudes towards SMS mobile marketing in new zealand. In: ICMB 2005: Proceedings of the International Conference on Mobile Business, pp. 434–440. IEEE Computer Society, Washington, DC (2005)
7. Frenzen, J.K., Davis, H.L.: Purchasing behavior in embedded markets. The Journal of Consumer Research 17(1), 1–12 (1990)

8. Froehlich, J., Chen, M.Y., Consolvo, S., Harrison, B., Landay, J.A.: MyExperience: a system for in situ tracing and capturing of user feedback on mobile phones. In: MobiSys 2007: Proceedings of the 5th International Conference on Mobile Systems, Applications and Services, pp. 57–70. ACM, New York (2007)
9. Girardin, F., Calabrese, F., Fiore, F.D., Ratti, C., Blat, J.: Digital footprinting: Uncovering tourists with user-generated content. IEEE Pervasive Computing 7(4), 36–43 (2008)
10. Gross, R., Acquisti, A., Heinz III, H.J.: Information revelation and privacy in online social networks. In: WPES 2005: Proceedings of the 2005 ACM Workshop on Privacy in the Electronic Society, pp. 71–80. ACM, New York (2005)
11. Huang, P., Lurie, N.H., Mitra, S.: Searching for Experience on the Web: An Empirical Examination of Consumer Behavior for Search and Experience Goods. Journal of Marketing (forthcoming)
12. Jiang, X., Hong, J.I., Landay, J.A.: Approximate Information Flows: Socially-Based Modeling of Privacy in Ubiquitous Computing. In: Borriello, G., Holmquist, L.E. (eds.) UbiComp 2002. LNCS, vol. 2498, pp. 176–193. Springer, Heidelberg (2002)
13. Kondo, F.N., Nakahara, M.: Differences in customers' responsiveness to mobile direct mail coupon promotions. International Journal of Mobile Marketing 2(2), 68–74 (2007)
14. Meschtscherjakov, A., Reitberger, W., Lankes, M., Tscheligi, M.: Enhanced shopping: a dynamic map in a retail store. In: UbiComp 2008: Proceedings of the 10th International Conference on Ubiquitous Computing, pp. 336–339. ACM, New York (2008)
15. Moon, Y.: Intimate exchanges: Using computers to elicit self-disclosure from consumers. Journal of Consumer Research 26(4), 323–339 (2000)
16. Salmon, F.: Might the consumer banking revolution be coming? Reuters (December 2009)
17. Siau, K., Shen, Z.: Building customer trust in mobile commerce. Commun. ACM 46(4), 91–94 (2003)
18. Smith, M.A., Davenport, D., Hwa, H., Turner, T.: Object auras: a mobile retail and product annotation system. In: EC 2004: Proceedings of the 5th ACM Conference on Electronic Commerce, pp. 240–241. ACM, New York (2004)
19. Steinberg, D.: Introducing a twitter for credit card purchases. The New York Times (December 2009)
20. Weiss, A.M., Lurie, N.H., Macinnis, D.J.: Listening to strangers: Whose responses are valuable, how valuable are they, and why? Journal of Marketing Research (2008)

Mobile Lifelogger – Recording, Indexing, and Understanding a Mobile User's Life

Snehal Chennuru, Peng-Wen Chen, Jiang Zhu, and Joy Ying Zhang

Carnegie Mellon University
Moffett Field, CA 94035
{snehal.chennuru,pengwen.chen,jiang.zhu,joy.zhang}@sv.cmu.edu

Abstract. Lifelog system involves capturing personal experiences in the form of digital multimedia during an entire lifespan. Recent advancements in mobile sensor technologies have helped to develop these systems using commercial smart phones. These systems have the potential to act as a secondary memory and also aid people who struggle with episodic memory impairment (EMI). Despite their huge potential, there are major challenges that need to be addressed to make them useful. One of them is how to index the inherently large lifelog data so that the person can efficiently retrieve the log segments that interest him / her most. In this paper, we present an ongoing research of using mobile phones to record and index lifelogs using activity language. By converting sensory data such as accelerometer and GPS readings into activity language, we are able to apply statistical natural language processing techniques to index, recognize, segment, cluster, retrieve, and infer high-level semantic meanings of the collected lifelogs. Based on this indexing approach, our lifelog system supports easy retrieval of log segments representing past similar activities and automatic lifelog segmentation for efficient browsing and activity summarization.

Keywords: Lifelogger, activity language, mobile computing, indexing heterogenous data.

1 Introduction

Memory, our ability to store, retain and recall information, is crucial to day to day life. But, our memory fades and we tend to forget intricate details of our experiences. Memory-related problems can be very serious for people who suffer from brain injuries or have memory diseases like the Alzheimer's. As of September 2009, more than 35 million people around the world are living with Alzheimer's disease for which episodic memory impairment (EMI) is the main symptom [12]. And according to [3], the prevalence of Alzheimer's is thought to reach approximately 107 million people by 2050.

In 1945, Vannevar Bush proposed a prototype computer system named MEMEX, whose main functionality is to share people's burden in memorizing things. Such system has great potential in a variety of applications and is particularly useful for people who suffered from EMI [10]. To assist people with this ever growing population, Bush's MEMEX concept seems to be a promising solution and thus gives birth to the personal lifelog research area. To fulfill Bush's vision, a personal lifelog system must be able to 1) store a large volume of personal multimedia data and 2) efficiently retrieve the relevant data based on users' requests.

Technologies today make it possible to capture one's life experience in digital format. The advancement in mobile sensors and ubiquitous computing allows lifeLog systems to record almost every aspects of one's life to provide a digital memory [6]. While recording and storing all sensor information in a database poses some engineering challenges, indexing them is the key to make lifelog system useful. We need the index so that we can retrieve important pieces of memory from the lifelogs. We can not expect users to annotate everything in the lifelog and automatic extracting semantics from images, audio and video is still an open research problem. Without a convenient method to index and retrieve the recorded data, such logged memory is of little use.

In this paper, we present a new approach to indexing lifelog data using the activity language. In this ongoing research, we convert the ambulatory sensor inputs such as accelerometers' readings into the so called *activity language* and use the activity language as the main index of the multimedia lifelogs. The activity language approach enables us to use statistical natural language processing methods to index, retrieve, cluster, and summarize lifelogs which are not easy or not possible for images, audio, and video information.

The rest of the paper is organized as follows. We first describe our lifeLogger system in Section 2. Section 3 introduces the concept of "Activity as Language" where we quantize the sensory input and convert it into a text representation in order to interpret the meaning of lifelogs. In Section 3.2, 3.3 and 3.5, we present algorithms and preliminary results on activity recognition, similar activity retrieval, and automatic lifelog segmentation. Finally, we conclude our findings and discuss the future works.

2 System Implementation

Our lifelogger system consists of three major parts, namely, the Lifelogger Mobile Client, the Lifelogger Application Server and the Lifelogger Web Interface (http://www.lifelogger.info). Figure 2 depicts its overall software architecture.

2.1 Lifelogger Mobile Client

Thanks to the rapid advancement of commercial mobile devices, we are able to build our Lifelogger client devices directly from off-the-shelf products. Our LifeLogger Mobile Client is a helmet mounted with one Nokia N95 phone (Figure 1).

The software for the mobile client is written using the PyS60 SDK for the Symbian platform. The client records various types of sensory data including i) accelerometer for motion, ii) GPS coordinates for outdoor locations, iii) camera view finder for pictures, iv) microphone recordings for sound, v) rotation sensor readings for rotation, and vi) WiFi signal strength for indoor locationing. These sensory data are all collected together with their corresponding timestamps. The data is captured in the JSON format and transmitted to the LifeLogger Application Server using the HTTP protocol via wireless connections.

Fig. 1. Sensor helmet for collecting activity data

2.2 Lifelogger Application Server

The LifeLogger Application Server is responsible for storing, pre-processing, modeling recorded data as an activity language and finally retrieving similar activities of the user. It is also in charge of supplying the user interface for users to interact with their lifelogs.

The Server consists of mainly 4 components, namely - i)Indexing Service, ii) Similar Activities Retrieval Service, iii) Hierarchical Segmentation Service and iv) Activity Language Corpus Database. Once the mobile client transmits the sensor data to the server, the server pre-processes and converts the accelerometer readings into the activity language and stores the data to the datastore. The Indexing Service then indexes the activity language corpus. The Similar Activities Retrieval Service allows the user to select a portion of the lifelog and show similar activities performed by the user in the past. This feature allows the system to automatically label activities that are similar in nature to the one's labeled manually in the past by the user.

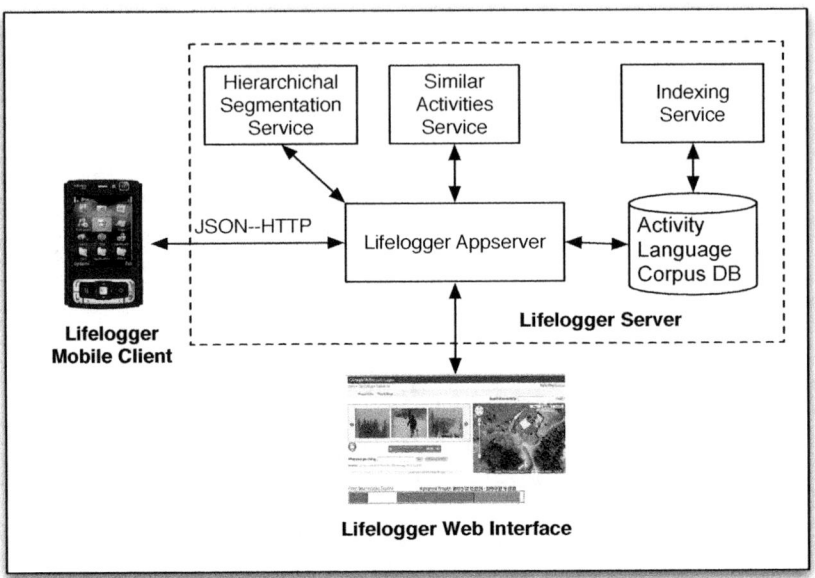

Fig. 2. System Architecture of the Mobile Lifelogger System

2.3 Lifelogger Web Interface

The server-side module is implemented as a web application so that users can access it easily through their favorite web browsers. The application provides an easy-to-use interface for end users to browse, annotate, and search their lifelogs. A personal calender based timeline is provided for all the recorded Lifelogs (Figure 3). The user can browse through the calendar and select a particular session that he's interested in revisiting his memory.

To make it easy for users to recall their past living experiences, we fit the collected images, audio, and GPS location data into one screen (Figure 4) to let users intuitively combine these memory clues. Moreover, since all sensory data are associated with synchronized time-stamps, users can navigate through the data set by simply dragging the timeline at the center of the screen, and all the three types of data would be updated simultaneously. The users are also provided with an option to "play" their lifelogs. This feature makes the images to automatically scrolls through, show the incremental trail of GPS on the map and play the audio at the same time. This acts as a memory cue so that the user can better recollect the experiences associated with that lifelog.

Users can also annotate a selected segment of lifelog by providing a short text description. Such text description will be used to learn the association between natural language query and the stored lifelog data. If the user is interested in a specific part of the lifelog and would like to find all his / her lifelog segments that are similar to it, all the user need to do is to first select that specific lifelog part using the timeline control and then click the "Find Similar Activities" button

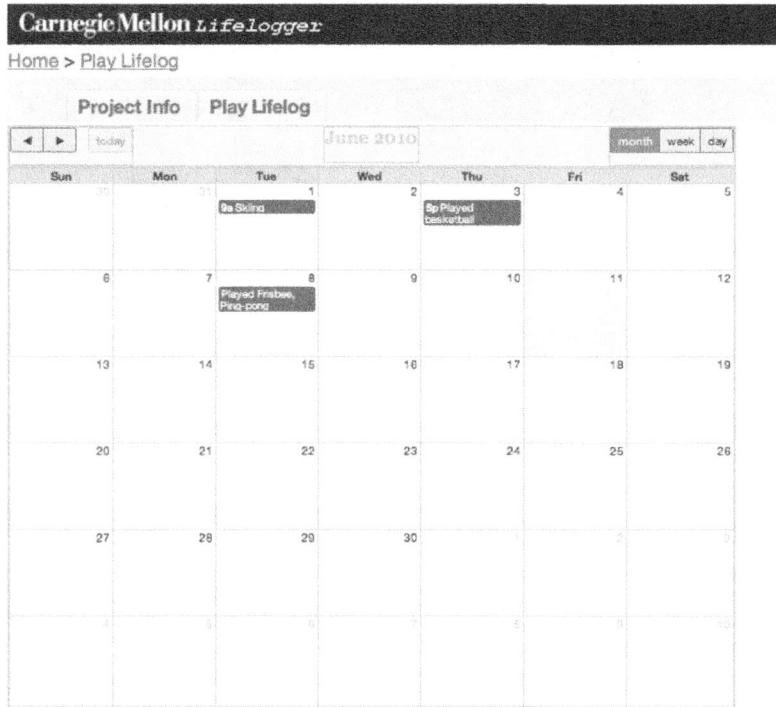

Fig. 3. Personal calendar based timeline

beside it. The similar lifelog segments will be returned and listed in a table at the bottom of the screen (Figure 4) and are already sorted based on their relevance scores ranging from 0 to 1.

Users can also view hierarchical segmentation of different activities that they have performed. A color code is assigned to each activity that the user has performed over the period of the lifelog, so that he/she can easily distinguish between different activities. Similar activities have the same color code, making it easier for the user to identify what activities he has repeatedly performed.

3 Language-Based Indexing

3.1 Main Idea

One of challenges in indexing, retrieving and interpreting lifelogs is that a lifelog is a collection of heterogenous sensory information and each sensory data type requires a special method to process the raw input. In most existing lifelog applications, raw input from sensors is classified into predefined classes by trained classifiers for further processing. This usually limits the scope of lifelog applications to those predefined activities.

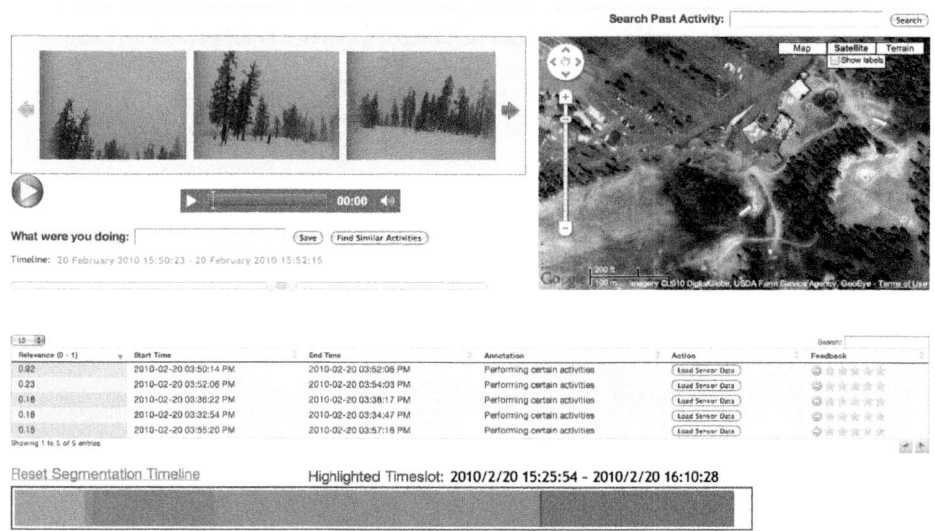

Fig. 4. Web interface of browsing/searching/annotating life logs

In this paper, we propose a novel method of representing the sensory input as "activity language" through quantizing the raw sensory input. Here we use *motion* information as an example. To record users' motion, we use 3-axis accelerometers to measures the acceleration at the X, Y, Z direction at the time of sampling. For the built-in accelerometers used in our experiments, the raw readings for each axis ranges from -360 to 360 which translates into 373,248,000 different (a_x, a_y, a_z) combinations. We quantize the raw accelerometer reading into V groups using K-Means clustering algorithm. Once the K-means clustering algorithm converges, it results in V cluster centroids and we give each cluster a label such as "D", "GC" and "DFR". We can then convert all the training and testing accelerometer data to their nearest cluster's label and thus convert the ambulatory activity into "activity text" (Figure 5).

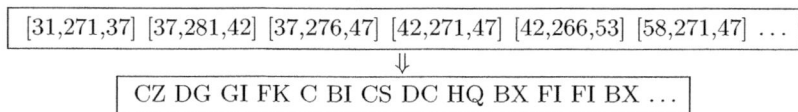

Fig. 5. An example of quantizing accelerometer readings to activity language representation

There are three benefits from quantizing the raw sensory input into " activity language" representation:

- Dimension reduction of sensory input. High dimension input data is reduced into one dimension to reduce the computation complexity. In the case of accelerometer readings, the original 3-dimension input of (a_x, a_y, a_z) is now reduced to one dimension.
- Efficient indexing and searching of lifelogs. Searching an indexed text corpus is much easier than searching a database of real numbers. Compared to the infinite real number space, the limited "vocabulary" size of the text representation allows the search algorithm to be much more efficient. Index and search algorithms such as the Inverted Index and Suffix Arrays [11] developed for strings can be applied on the " activity language" representation of lifelogs. This is more straight forward than searching the lifelogs based on the cosine or Euclidean distances between the query and the logged activities.
- Uniformed representation of heterogenous sensory data. By converting different sensory input into the same type of " activity language" representation, we can develop and apply the same " activity language processing" algorithms on different types of data. Although the activity language discussed in this paper was constructed only from accelerometer readings, other types of sensory data can be used to generate activity languages as well. For example, in [5] we had demonstrated our preliminary results of using GPS recordings to generate activity languages.

We call this representation "activity language" based on the analogy between human activity and natural languages. The similarity between human activity and language had been articulated by Burke [4] and Wertsh [18]. Based on the "principle of language as action", natural languages and human activities indeed share some important properties. For instance, they are both "mediational means" or tools by which we achieve our ends. Additionally, they both exhibit structure and satisfy "grammars".

Table 1. Activity as language at different levels

Natural Language	activity language	Example
Word	Atomic Movement	Turn upper body left
Phrase	Movement	Stand up
Sentence	Action	Climb up stairs
Paragraph	Activity	Enter building, climb up stairs and walk into office
Document	Event	Left home and ride bicycle to campus arrived at my office at 2nd floor

Table 1 illustrates that people's ambulatory activities share a lot in common with natural languages at all levels. The anatomy of human bodies allows us to perform certain *atomic movements* such as "turn upper body left" whereas "jump up at 10g acceleration" is not possible. Such atomic movements form the vocabulary of the activity language. A sequence of atomic movements performed in a meaningful order creates a *movement* such as an *action* of "standing up". *Actions* such as "climbing up stairs" are created by performing actions in a right order similar to create a "sentence". A sequence of actions builds up an *activity*. The higher level concept *event* is composed of a series of activities in a similar way as a *document*.

This " activity language" concept serves as the foundation for our approach to efficient lifelog retrieval. To empirically evaluate the similarity between the ambulatory activity and natural languages, we check if the activity language corpus follows the Zipf's law. Zipf's law states that given some corpus of natural language utterances, the frequency of any word is inversely proportional to its rank in the frequency table. In other words, the logarithm of a word's frequency is linear to its rank in a natural language corpus. Figure 6 plots the logarithm frequency of word types in the activity language corpus for word type ranked 1, 2, 3, etc. Though not exactly linear, it does plot a line similar to Zipf's distribution.

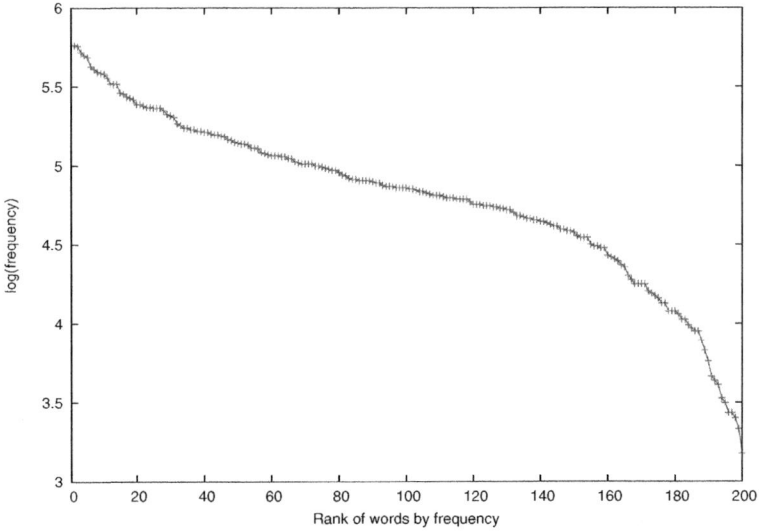

Fig. 6. log(freq) vs. rank of frequency of word types in a lifelog converted as activity language

3.2 Activity Recognition

Before applying the "activity language" concept in our lifelog system, we performed a series of predefined-activity recognition experiments to justify the benefits of modeling human activities as a language.

In our approach, we view the labeled data for each activity a_i as the training corpus and train a smoothed n-gram language model over the converted activity language text using the SRI language model toolkit [16]. For each testing "activity sentence" t, we input it to all the pre-built language models to calculate the probability of t being generated by activity a_i and predicts the activity of the testing sentence to be i^* such that

$$i^* = \arg\max_i P(t|a_i) \qquad (1)$$

One issue of using language models for activity recognition is that language model probabilities are not directly comparable if their respective training data have different vocabulary sizes. To solve this problem, each training data set is augmented with a universal vocabulary list built from all training data sets. As a result, all our activity language models have the same vocabulary size, and thus their generated probabilities are comparable.

Table 2. Classification accuracy on corpus with vocabulary=100

	Predicted Activity		
	walking	running	cycling
walking	94%	3%	3%
running	6%	92%	2%
cycling	8%	0%	92%

Table 3. Classification accuracy on corpus with vocabulary=200

	Predicted Activity		
	walking	running	cycling
walking	95%	1%	4%
running	4%	94%	2%
cycling	2%	0%	98%

Our preliminary results of using smoothed n-gram language model for activity recognition demonstrated an average accuracy rate of 94% in distinguishing among basic activities such as walking, running, and cycling. Table 2 and 3 compare the recognition accuracy of language models trained over a corpus of vocabulary size 100 vs. the one with 200 word types. With a larger vocabulary size, i.e., more atomic movement types, the activity language has more discriminative power to differentiate human activities.

Figure 7 shows the average activity recognition accuracy vs. the order of n in language model training. Overall, for this basic activity recognition task, the order of history does not play a significant role here.

The promising results of these experiments increase our confidence in using the activity language to improve lifelog systems' indexability.

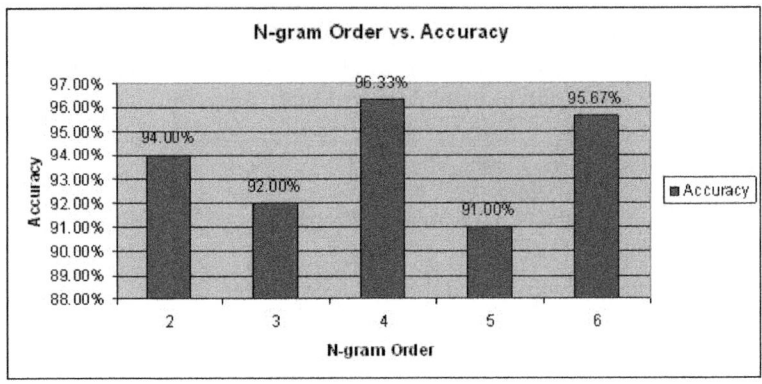

Fig. 7. Recognition accuracy vs. n-gram order

3.3 Similar Activity Retrieval

In many cases, we want to find out information about activities that are not predefined such as "how many tennis games did I play in the past two months?" or "how much time did I spend sitting in front of TV last week?" It is not possible to enumerate all possible activities and train Hidden Markov Models ahead of time in order to answer such questions. In our approach, we convert the lifelogs, in particular, the main indexing sensory information into a text representation. This allows us to apply Information Retrieval techniques to "retrieve' relevant activities from the past logs to answer users' queries.

In our implementation, a user can selects a segment from his/her lifelogs on the web interface and indicate that he/she may want to find similar activities from the past logs. The highlighted segment does not need to be annotated by natural language descriptions such as "playing tennis". The system will search from the past life logs and return most relevant segments for user to review.

The key here is to calculate "similarity" between two segments of lifelogs. Inspired by the BLEU metric [14] where averaged n-gram precision is used to measure the similarity between a machine translation hypothesis and human generated reference translations, we use *averaged n-gram precision* to estimate the similarity between two lifelog segments.

Assuming that P and Q are two activity language sentences of the same length l. P is the sequence of $P_1, P_2, \ldots, P_L$ and Q is the sequence of $Q_1, Q_2, \ldots, Q_L$. Denote the *similarity* between P and Q as $S(P, Q)$. Define the n-gram precision between P and Q as $\text{Prec}_n(P, Q) =$

$$\frac{\sum_{\tilde{p} \in \{\text{All } n\text{-gram types in } P\}} \min(freq(\tilde{p}, P), freq(\tilde{p}, Q))}{\sum_{\tilde{p} \in \{\text{All } n\text{-gram types in } P\}} freq(\tilde{p}, P)}, \quad (2)$$

and the similarity between P and Q is defined as:

$$S(P,Q) = \frac{1}{N} \sum_{n=1}^{N} \text{Prec}_n(P,Q) \qquad (3)$$

$\text{Prec}_n(P,Q)$ calculates the percentage of n-grams in P that can also be found in Q and $S(P,Q)$ averages the precision over 1-gram, 2-gram and up to N-gram. In our experiments, we empirically set $N = 5$.

Table 4 shows an example of calculating the similarity between activity sentence P ("NB NB P P P P P P NB NB") and Q ("NB P P NB NB P NB P P P").

Table 4. Calculating the similarity between two activity sentences using averaged n-gram precision

n	n-gram	$freq_P$	$freq_Q$	min	Prec_n
1	NB	4	4	4	
	P	6	6	6	
					10/10=1.0
2	NB NB	2	1	1	
	NB P	1	3	1	
	P P	5	3	3	
	P NB	1	2	1	
					$6/9 = 0.67$
3	NB NB P	1	1	1	
	NB P P	1	2	1	
	P NB NB	1	1	1	
	P P NB	1	1	1	
	P P P	4	1	1	
					$5/8 = 0.63$
4	NB NB P P	1	0	0	
	NB P P P	1	1	1	
	P P NB NB	1	1	1	
	P P P NB	1	0	0	
	P P P P	3	0	0	
					$2/7 = 0.29$
5	NB NB P P P	1	0	0	
	NB P P P P	1	0	0	
	P P P NB NB	1	0	0	
	P P P P NB	1	0	0	
	P P P P P	2	0	0	
					$0/6 = 0.0$
					$S(P,Q) = 0.52$

Given a query sentence of l words, we assume that similar activities in the lifelog should also be of length l. This assumption makes the retrieval algorithm easier to implement as varied length activity retrieval would require activity segmentation. For a lifelog with G words, there are $G - l$ different strings of l

words long. In our current setting, a 24 hours lifelog contains about 200 million activity words. Calculating the similarity between each of the $G-l$ strings with the query can be computationally expensive. To speed up the retrieval, we use suffix arrays to pre-select strings in the corpus that have high order n-gram matches with the query and calculate $S(P,Q)$ scores for those strings only. The observation is that if a string in the lifelog is similar to the query, then it should have many high order n-grams matched with those in the query string.

Top R similar activity segments is returned to the user on the web interface (as shown in lower panel in Figure 4). User can load each segment to "play" the corresponding lifelogs and for our ongoing experiments evaluate if the segment is truly "similar" to the query.

3.4 Hierarchical Segmentation of Lifelogs

Lifelog records a user's daily life as a continuous sequence of sensory data. After converting the sensory data to activity language text, a lifelog is now a long string of text. Just as we need punctuations, sentence boundaries and paragraph boundaries in written text, it would make lifelogs more readable if we could automatically segment the data based on user's activities.

Table 5. Configuration of the automatic segmentation experiments

Configuration	Value
Activity Type	Playing frisbee, Playing basketball
	Playing table tennis, Playing tennis
Activity Duration	5 to 10 minutes each
Lifelog Length	40 to 50 minutes each
Accuracy Measure	F1 score

The underlining assumption of our segmentation algorithm is that when a user switches his/her activity at time t, the similarity between string $[t-w, t-1]$ and $[t, t+w]$ should be much lower than if t is inside the same activity for a window of size w. For a window size w, define the "change of activity" at time t as:

$$H(t, w) = -log(0.00001 + S([t-w, t-1], [t, t+w-1])). \qquad (4)$$

The higher the value of $H(t,w)$, the more likely user changed his/her activity at time t. Figure 8 shows the H value at each data points given different window sizes for a segment of lifelog.

It can be noticed that: (1) peaks of activity change identified by larger windows are also peaks identified by smaller windows but not vice versa; and (2)activity changes over larger windows are smoother than smaller windows. Intuitively, larger window size captures changes of larger-scale activities whereas smaller window captures changes of smaller activities. Based on this finding, we first segment the lifelog data using large window size and then recursively segment

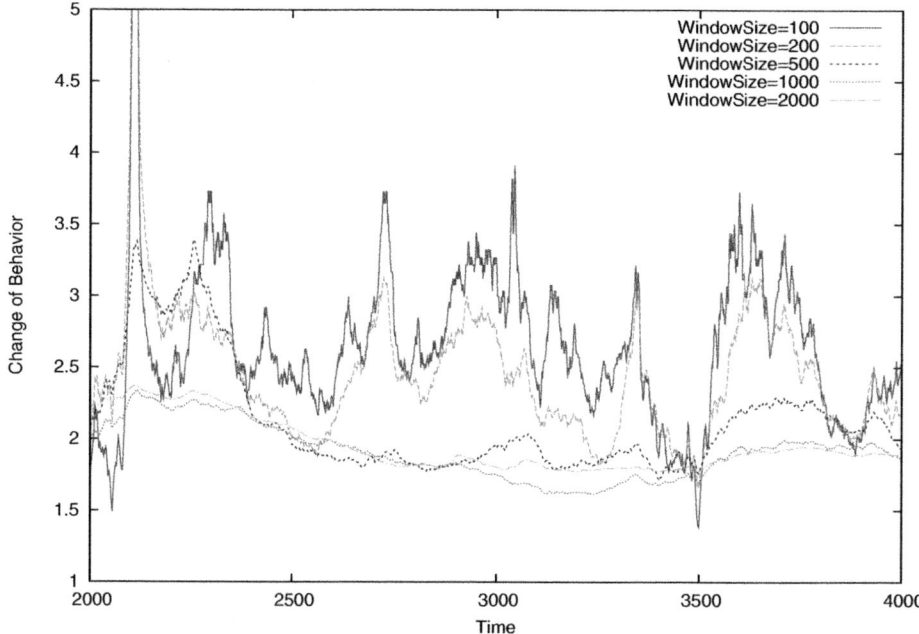

Fig. 8. Activity changes calculated by different size of sliding windows

the data using smaller windows. This results in a hierarchical segmentation of lifelogs which allows user to efficiently browse through the lifelog instead of playing the whole lifelog (Figure 10).

To evaluate the quality of this automatic segmentation method, experiments were performed with the configuration shown in Table 5. For every experiment, the participant would wear the sensor helmet to perform a series of different activities and there would be an observer who was in charge of recording the timestamps at which the activity changes occur. The collected human annotations would then serve as ground truth to help us evaluate the accuracy of the automatic segmentation. Only the first level of the segmentation would be evaluated and the metric we used is the F1 score calculated as

$$F = 2 * precision * recall/(precision + recall). \quad (5)$$

Table 6 shows the F1 scores of our automatic segmentation method according to different window sizes. Overall, our segmentation method achieved an F1 score of 60%.

3.5 Segmentation Clustering of Lifelogs

Hierarchical segmentation of the lifelogs shows different activities performed by the user over time. However, it would be further useful if the system groups

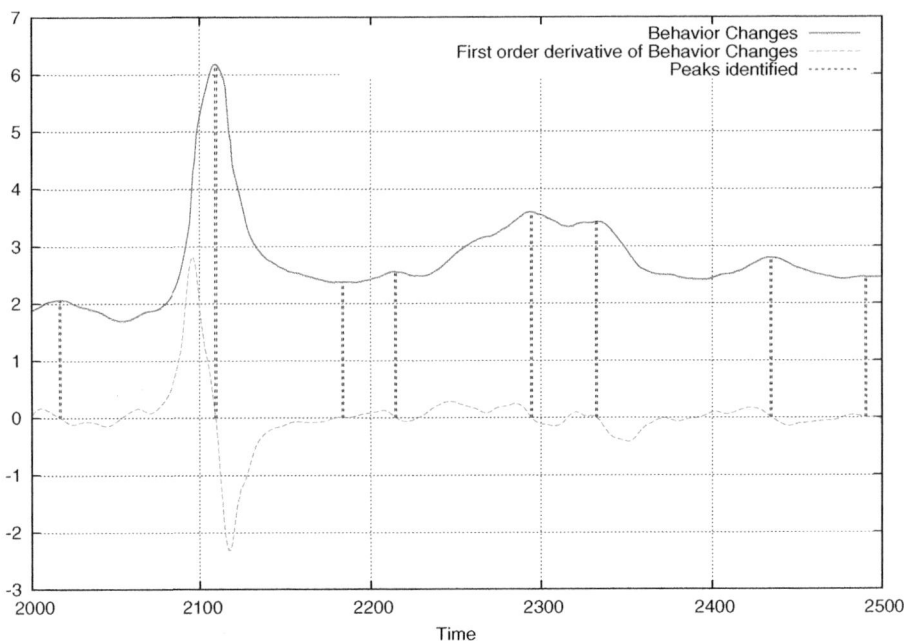

Fig. 9. Identifying peaks in the behavior-change-curve

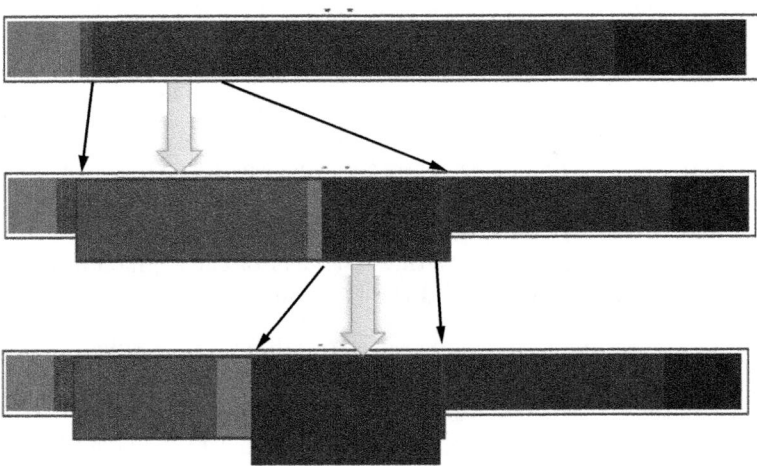

Fig. 10. Hierarchically segmented lifelog. Activity boundaries detected automatically by the system and descriptions are added by the user.

similar segments so that user perceives similar activities he/she has performed. By performing unsupervised clustering on the Autosegmentation ouput, we can group similar activities.

K-means clustering algorithm is chosen for its simplicity and performance [7]. As it requires similarity or distance between two segments, we use a similarity measure analogous to Section 3.3 - *averaged n-gram precision*. But, the key difference is that the segments need not be of the same length.

Let P and Q be two activity language sentences, with lengths l and m respectively. P is the sequence of $P_1, P_2, \ldots, P_L$ and Q is the sequence of $Q_1, Q_2, \ldots, Q_M$. Denote the *similarity* between P and Q as $S(P,Q)$. Define the n-gram precision between P and Q as $\text{Prec}_n(P,Q) =$

$$\frac{1}{2} \sum_{\tilde{p} \in \{\text{All } n\text{-gram types in } P \text{ or } Q\}} \min(freq(\tilde{p},P), freq(\tilde{p},Q)) \left(\frac{1}{\sum_{\tilde{p} \in \{\text{All } n\text{-gram types in } P\}} freq(\tilde{p},P)} + \frac{1}{\sum_{\tilde{q} \in \{\text{All } n\text{-gram types in } Q\}} freq(\tilde{q},Q)} \right) \quad (6)$$

and the similarity between P and Q is defined as:

$$S(P,Q) = \frac{1}{N} \sum_{n=1}^{N} \text{Prec}_n(P,Q) \quad (7)$$

Each sentence is vectorized using the frequencies of all its n-grams. Each dimension of the vector represents one n-gram type. Using the same example in Section 3.3, an activity sentence P ("NB NB P P P P P P NB NB") and Q ("NB P P NB NB P NB P P P"). Therefore, the vectors for sentence P and Q are constructed as

$$V_P = (4, 6, 2, 1, 5, 1, 1, 1, 1, 1, 4, 1, 1, 1, 1, 3, 1, 1, 1, 1, 2)$$
$$V_Q = (4, 6, 1, 3, 3, 2, 1, 2, 1, 1, 1, 0, 1, 1, 0, 0, 0, 0, 0, 0, 0)$$

Each component of the these vectors is the frequency of one of the n-gram types in the n-gram column of Table 4. Note that if a particular n-gram does not appear in a sentence, the component of that n-gram is set to zero. e.q. n-gram ("NB NB P P P") in sentence Q.

The centroid of cluster is constructed by calculating the **average vector** of all the vectors in the given cluster.

We have applied this method to the output of hierarchical segmentation to generate clustering results at different levels. The preliminary results show that the k-means algorithm can converge quickly, within 3-4 iterations, when running with a small number of clusters as 5. Due to the limited volume of annotated data, we could not evaluate the results in a systemical manner. However, manual

Table 6. F1 Score vs. Window Size

Window Size	100	200	300	400	500	600	700	800	900	1000
Precision (%)	26.09	40.00	43.75	46.67	41.67	44.44	44.44	50.00	50.00	66.67
Recall (%)	60.00	80.00	70.00	70.00	50.00	40.00	40.00	50.00	40.00	60.00
F1 Score (%)	36.36	53.33	53.84	56.00	45.45	42.11	42.11	50.00	44.44	63.16

cross-checking between the clustering results and the user annotations at the top level shows some level of relevance. This is outside of the scope of this paper and we will leave it for future work.

4 Related Work

4.1 Activity Recognition

There have been several techniques for recognizing or distinguishing basic human activities. They can be categorized into two flavors: heuristic threshold-based classifiers and pattern recognition techniques such as decision trees, nearest neighbor, Naive Bayes, support vector machines (SVM), neural networks, Hidden Markov Models (HMM) and Gaussian mixture models [13]. For recognizing high-level human activities, several attempts had been made in [1,15]. Among these techniques, the most popular ones we see so far are those based on HMM. These HMM-based approaches classify the input sensory information into one of the predefined activities such as walking, running, and standing. However, since HMM assumes the first order Markov chain in the state space and usually does not consider the inherent "grammar" or "structure" of human activities, the activities that can be recognized by HMM are limited to those pre-defined in the training data, which as a result limits HMM's application in people-centric computing.

4.2 Lifelog System

Different approaches have been used to implement a lifelog system. The MyLifeBits system [6] is designed to store and manage everything in a person's lifetime that can be captured in digital format. Its initial goal was to store all personal information found in PCs such as articles, video, office documents, email, keystrokes, and screen mouse clicks, etc. It then evolved into storing all ambient information of a person's daily life via a specialized camera device named Sense-Cam. MyLifeBits supports capture, storage, management, and retrieval of many media types, and its sophisticated database design is capable of storing a large volume of multimedia data. However, MyLifeBits only applies a basic metadata-based indexing approach which requires users to manually annotate most of the collected data in order to have meaningful search results. Our works address this issue well by providing a more effective indexing scheme which requires less user involvement and provides more meaningful search results by taking the "meaning" of the collected media into account. Another lifelog system implementation is discussed in [8]. This work focuses on realtime storage and retrieval of lifelog in a ubiquitous environment. The developed system supports semi-automatic activity analysis and provides an intuitive graphical interface for users to browse their lifelogs that correlates the space and temporal information of the displayed sensory data.

In addition to our language approach to indexing lifelog, Kim et al presented a multimodal sensor fusion technique which supports automatic generation of lifeline's metadata [9]. The key idea is to combine the analysis results of different kinds of low-level sensory data to better infer higher-level context information about the collected lifelog. For example, by combining the analysis of audio, GPS, and accelerometer readings, the system is able to better identify the environment in which the lifelog was taken. Machine learning techniques such as decision tree and Gaussian Mixture Model (GMM) are used to analyze the collected low-level sensory data. Similar techniques to this sensor fusion approach are explored in [2,17]. The former uses video key frame summarization and conversation scene detection to fulfill efficient lifelog retrieval. The latter proposes an integrated technique to process lifelog data using correlations between different types of the captured data from multiple sensors.

Kyoko et al. [19] have developed a wearable lifelogging device to recognize the experiences and activities of cats and post them as tweets on Twitter. C4.5 decision tree is used to recognize and classify different activities performed by cats such as eating, sleeping running.

5 Conclusions and Future Work

In this paper, we present our Mobile Lifelogging system to record, index and understand the life experiences of a mobile user. We discuss several functions of the system such Similar Activity Retrieval (Section 3.3) and Automatic Hierarchical Segmentation (Section 3.5) for efficiently retrieving the lifelogs and help the user visualize them. We present a novel yet straightforward approach of processing lifelogs of heterogenous sensory data. We verify the similarity between activity and language by demonstrating Zipf's distribution over our activity language corpus. The experimental results presented in Section 3.2 demonstrate high accuracy of using language models for human activity recognition. Unlike the traditional HMM approach which is limited to activity recognition task, modeling activity as language enables many other applications such as similar activity retrieval, and hierarchical activity segmentation. Besides, this language-based modeling approach can be applied to other types of sensory input such as geo-locations. Extracting high level semantic information from primitive activities that are recognized from the sensory data still remains a challenging problem. We will conduct user study to evaluate the effectiveness of our similar activity retrieval service. We will also carry out more experiments to generate sufficient annotated data. With these data, we hope to extend our work of hierarchical activity segmentation by grouping atomic sensory events in the sequence progressively to build abstraction hierarchy of HHMM towards automatic lifelog summarization.

References

1. Aipperspach, R., Cohen, E., Canny, J.: Modeling human behavior from simple sensors in the home. In: Proceedings of IEEE Conf. on Pervasive Computing, Dublin, Ireland, pp. 337–348 (April 2006)
2. Aizawa, K., Tancharoen, D., Kawasaki, S., Yamasaki, T.: Efficient retrieval of life log based on context and content. In: CARPE 2004: Proceedings of the the 1st ACM Workshop on Continuous Archival and Retrieval of Personal Experiences, pp. 22–31. ACM, New York (2004)
3. Brookmeyer, R., Johnson, E., Ziegler-Graham, K., Arrighi, H.M.: Forecasting the global burden of alzheimer's disease. Alzheimer's and Dementia 3(3), 186–191 (2007); predicted 107 million people will suffer from Alzheimer by 2050
4. Burke, K.: Language as Symbolic Action. University of California Press (1966)
5. Chen, P., Chennuru, S., Buthpitiya, S., Zhang, Y.: A language-based approach to indexing heterogeneous multimedia lifelog. In: Proceedings of 12th International Conference on Multimodal Interfaces (2010)
6. Gemmell, J., Bell, G., Lueder, R., Drucker, S., Wong, C.: Mylifebits: fulfilling the memex vision. In: MULTIMEDIA 2002: Proceedings of the Tenth ACM International Conference on Multimedia, pp. 235–238. ACM, New York (2002)
7. Hartigan, J.A.: Clustering Algorithms. Wiley (1975) ISBN 0-471-35645-X
8. Kim, I.-J., Ahn, S.C., Kim, H.-G.: Personalized Life Log Media System in Ubiquitous Environment. In: Stajano, F., Kim, H.-J., Chae, J.-S., Kim, S.-D. (eds.) ICUCT 2006. LNCS, vol. 4412, pp. 20–29. Springer, Heidelberg (2007)
9. Kim, I.-J., Ahn, S.C., Ko, H., Kim, H.G.: Automatic lifelog media annotation based on heterogeneous sensor fusion. In: Proceedings of IEEE International Conference on Multi Sensor Fusion and Integration for Intelligent Systems, Seoul, Korea, August 20-22 (2008)
10. Lee, M.L., Dey, A.K.: Lifelogging memory appliance for people with episodic memory impairment. In: UbiComp 2008: Proceedings of the 10th International Conference on Ubiquitous Computing, pp. 44–53. ACM, New York (2008)
11. Manber, U., Myers, G.: Suffix arrays: a new method for on-line string searches. SIAM J. Comput. 22(5), 935–948 (1993)
12. Neergaard, L.: Report: 35 million-plus worldwide have dementia. Associate Press (September 21, 2009)
13. Nguyen, A., Moore, D., McCowan, I.: Unsupervised clustering of free-living human activities using ambulatory accelerometry. In: Proceedings of the 29th Annual International Conference of the IEEE Engineering in Medicine and Biology Society (EMBS), Lyon, France, August 22-26, pp. 4895–4898 (2007)
14. Papineni, K., Roukos, S., Ward, T., Zhu, W.: Bleu: a method for automatic evaluation of machine translation. Technical Report RC22176(W0109-022), IBM Research Division, Thomas J. Watson Research Center (2001)
15. Patterson, D.J., Liao, L., Fox, D., Kautz, H.: Inferring High-Level Behavior from Low-Level Sensors. In: Dey, A.K., Schmidt, A., McCarthy, J.F. (eds.) UbiComp 2003. LNCS, vol. 2864, pp. 73–89. Springer, Heidelberg (2003)

16. Stolcke, A.: Srilm – an extensible language modeling toolkit. In: Proc. Intl. Conf. on Spoken Language Processing, Denver, CO, vol. 2, pp. 901–904 (2002)
17. Takata, K., Ma, J., Apduhan, B.O., Huang, R., Jin, Q.: Modeling and analyzing individual's daily activities using lifelog. In: ICESS 2008: Proceedings of the 2008 International Conference on Embedded Software and Systems, pp. 503–510. IEEE Computer Society, Washington, DC (2008)
18. Wertsch, J.V.: Mind As Action. Oxford University Press, USA (1998)
19. Yonezawa, K., Miyaki, T., Rekimoto, J.: Cat@log: sensing device attachable to pet cats for supporting human-pet interaction. In: ACE 2009: Proceedings of the International Conference on Advances in Computer Enterntainment Technology. ACM (2009)

Activity-Aware Mental Stress Detection Using Physiological Sensors

Feng-Tso Sun[1], Cynthia Kuo[1,2], Heng-Tze Cheng[1], Senaka Buthpitiya[1], Patricia Collins[1], and Martin Griss[1]

[1] Carnegie Mellon University
{lucas.sun,hengtze.cheng,senaka.buthpitiya,
patricia.collins,martin.griss}@sv.cmu.edu
[2] Nokia Research Center
cynthia.kuo@nokia.com

Abstract. Continuous stress monitoring may help users better understand their stress patterns and provide physicians with more reliable data for interventions. Previously, studies on mental stress detection were limited to a laboratory environment where participants generally rested in a sedentary position. However, it is impractical to exclude the effects of physical activity while developing a pervasive stress monitoring application for everyday use. The physiological responses caused by mental stress can be masked by variations due to physical activity.

We present an activity-aware mental stress detection scheme. Electrocar- diogram (ECG), galvanic skin response (GSR), and accelerometer data were gathered from 20 participants across three activities: sitting, stand- ing, and walking. For each activity, we gathered baseline physiological measurements and measurements while users were subjected to mental stressors. The activity information derived from the accelerometer en- abled us to achieve 92.4% accuracy of mental stress classification for 10-fold cross validation and 80.9% accuracy for between-subjects classi- fication.

Keywords: Mental stress, electrocardiogram, galvanic skin response, physical activity, heart rate variability, decision tress, Bayes net, support vector machine, stress classifier.

1 Introduction

Stress is a physiological response to the mental, emotional, or physical chal- lenges that we encounter. Immediate threats provoke the body's "fight or flight" response, or acute stress response [5]. The body secretes hormones, such as adrenaline, into the bloodstream to intensify concentration. There are also many physical changes, such as increased heart rate and quickened reflexes. Under healthy conditions, the body returns to its normal state after dealing with acute stressors.

Unfortunately, many of the stressors in modern life are ongoing. Chronic stress can be detrimental to both physical and mental health. It is a risk fac-tor for hypertension and coronary artery disease [22, 12]. Other physical disor- ders, including irritable bowel syndrome (IBS), gastroesophageal reflux disease (GERD), and back pain, may be caused or exacerbated by stress [16]. Chronic stress also plays a role in mental illnesses, such as generalized anxiety disorder and depression [11].

Chronic stress is difficult to manage because it cannot be measured in a consistent and timely way. One current method to characterize an individual's stress level is to conduct an interview or to administer a questionnaire during a visit with a physician or psychologist. This method provides only a momentary snapshot of the individual's stress level, as most individuals cannot accurately recall the history of the ebb and flow of their stress symptoms [3].

Continuous monitoring of an individual's stress levels is essential for under- standing and managing personal stress. A number of physiological markers are widely used for stress assessment, including: galvanic skin response, several fea- tures of heart beat patterns, blood pressure, and respiration activity [31, 15]. Fortunately, miniaturized wireless devices are available to monitor these physio- logical markers. By using these devices, individuals can closely track changes in their vital signs in order to maintain better health.

Measuring physiological signals during everyday activity is more difficult than in a rigorous laboratory environment. First, the physiological responses caused by mental stress can be masked by variations due to physical activity [1]. For example, people may have higher heart rate when standing than when sitting. Heart rate may also increase when people are mentally stressed. Hence, using heart rate alone as an indicator to detect mental stress may lead to misclas- sification. Second, signal artifacts caused by motion, electrode placement, or respiratory movement affect the accuracy of measured recordings. Third, it is also difficult to determine the ground truth of a user's stress level when label- ing training data in mobile environment. These factors increase the difficulty of developing a pervasive mental stress detection application for everyday use.

We introduce an activity-aware, multi-modal system that combines ac- celerometer, ECG, and GSR information to differentiate between physical ac- tivity and mental stress. We conducted a user study with 20 participants across three different physical activities: sitting, standing, and walking. With activ- ity information derived from the accelerometer, we achieved 92.4% accuracy for 10-fold cross validation and 80.9% accuracy for between-subject's classification.

In the next section, we describe how we can measure the body's responses to mental stress. Next, we discuss prior work on stress detection. Section 4 describes our experimental protocol and our physiological feature extraction and classification methods. Experimental results are presented in Section 7.

2 Background

The autonomic nervous system (ANS) regulates the body's major physiolog- ical activities, including the heart's electrical activity, gland secretion, blood pressure, and respiration. The ANS has two branches: the sympathetic nervous system (SNS) and the parasympathetic nervous system (PNS). The SNS mobi- lizes the body's resources for action under stressful conditions. In contrast to the SNS, the PNS relaxes the body and stabilizes the body into steady state.

2.1 Heart Rate Variability (HRV) and Stress

Under acute stress, the SNS increases heart rate, respiration activity, sweat gland activity, etc. After the stress has passed, the PNS reverses the stress re- sponse [17]. Since the ANS controls the heart, measuring cardiac activity is an ideal, non-invasive means for evaluating the state of the ANS.

An ECG is a recorded tracing of the electrical activity generated by the heart. Figure 1 shows a P wave, a QRS complex, and a T wave in the ECG. The P wave represents atrial depolarization, the QRS represents ventricular depolarization, and the T wave reflects the rapid repolarization of the ventricles [8]. The R-R interval is the time interval between two R peaks and is used to calculate heart rate.

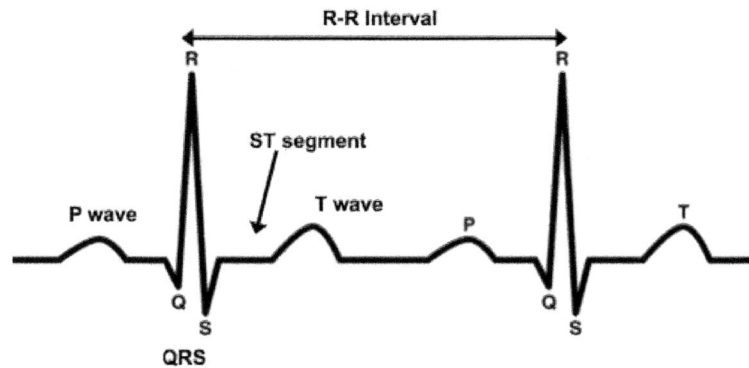

Fig. 1. Electrocardiogram sample

Heart rate variability (HRV) refers to the beat-to-beat variation in the R-R interval. HRV analysis can be categorized into time-domain and spectral-domain analysis. Several time-domain parameters include:

- mean HR: mean heart rate (beats per minute);
- mean RR: mean heartbeat interval (ms);
- SDNN: standard deviation of RR-intervals between normal beats;
- RMSSD: root mean square of the difference between successive RR-intervals; and
- pNN50: the percentage of heartbeat intervals with a difference in successive heartbeat intervals greater than 50 ms.

Three widely used components can be found in HRV power spectrum:

- LF (0.04-0.15 Hz): a low-frequency component that is mediated by both the SNS and PNS;
- HF (0.15-0.4Hz): a high-frequency component mediated by the PNS; and
- LF/HF: LF to HF ratio that is used as an index of autonomic balance.

2.2 Galvanic Skin Response (GSR) and Stress

GSR is a measure of the electrical resistance of the skin. A transient increase in skin conductance is proportional to sweat secretion[6]. When an individual is under mental stress, sweat gland activity is activated and increases skin conductance. Since the sweat glands are also controlled by the SNS, skin conductance acts as an indicator for sympathetic activation due to the stress reaction.

The hands and feet, where the density of sweat glands is highest, are usually used to measure GSR. There are two major components for GSR analysis. Skin conductance level (SCL) is a slowly changing part of the GSR signal, and it can be computed as the mean value of skin conductance over a window of data. A fast changing part of the GSR signal is called skin conductance response (SCR), which occurs in relation to a single stimulus. Widely used parameters for GSR include the amplitude and latency of SCR and average SCL value[2].

3 Related Work

The validity of using ECG and GSR measurements in mental stress monitor- ing has been demonstrated in both psychophysiology and bio-engineering. HRV analysis based on ECG measurement is commonly used as a quantitative marker describing the activity of the autonomic nervous system during stress. For example, Sloten et al. conclude that the mean RR is significantly lower (i.e., the heart rate is higher) with a mental task than in the control condition while pNN50 is significantly higher in the control condition than with a mental task [26].

Also, conventional short-term HRV features (e.g., a 5-minute sample win- dow) may not capture the onset of acute mental stress for a mobile subject. Salahuddin et al. noted that HR and RR-intervals within 10 sec, RMSSD and pNN50 within 30 sec, high frequency band (HF: 0.15 to 0.4 Hz) within 40 sec, LF/HF, normalized low frequency band (LF: 0.04 to 0.15 Hz), and normalized HF within 50 sec can be reliably used for monitoring mental stress in mobile settings [23]. Hence, mental stress can be recognized with most HRV features calculated within one minute.

Boucsein provided an extensive coverage of early research of GSR related to stress [2]. He showed that slowly changing SCL and SCR aroused by specific stimulus are sensitive and valid indicators for the course of a stress reaction. Setz et al. demonstrated the discriminative power of GSR in distinguishing stress caused by a cognitive load and psychosocial stress by using a wearable GSR device in an office environment [25]. In this study, analysis of the data showed that the distributions of the SCL peak height and the SCR peak rate carry information about the stress level of a person.

Some research has used multiple physiological features to determine the existence of the subject's stress response. Zhai and Barreto applied an interactive "Paced Stroop Test," a psychological test of the subject's mental attention and flexibility, as a stimulus to elicit emotional stress in the subject [33]. The Paced Stroop Test requires the subject to select the font color of a word shown on the screen. The word itself names a potentially different color. The authors proposed to extract features from the subject's physiological response (blood volume pulse, galvanic skin response, skin temperature and pupil diameter) during both the congruent phase (matching color name and font color) and incongruent phase (mismatching color name and font color). An example of the incongruent phase is shown in Figure 3. Three learning algorithms, Naive Bayes, Decision Tree, and Support Vector Machine (SVM), are used to classify relaxed and stressed states. The SVM classifier reached an accuracy of 90.1% with 20-fold cross validation.

Some experiments have been conducted in the real world. For instance, Healey and Picard measured drivers' stress reactions by monitoring multiple physiological signals, such as ECG, GSR, electromyogram (EMG), and respira- tion in a prescheduled route setting [10]. They used 5-minute intervals of data during the rest, highway, and city driving conditions to distinguish between three levels of driver stress. Heart rate and skin conductance provided the highest over- all correlations with drivers' stress level across multiple drivers and driving days, reaching an accuracy of over 97%.

Most previous research considers distinguishing the physiological response to mental stress from subjects at rest. While developing mental stress monitoring algorithms in real-life ambulatory situations, it is crucial to take physical activity (e.g.,walking) and posture (e.g., sitting or standing) into account. Cardiovascu- lar variability is highly affected by changes in body posture and physical activ- ity [30]. In Van Steenis et al.'s sample, subjects' mean HR increased significantly from a supine to sitting posture (from 66 to 77 bpm), from a sitting to standing posture (86 bpm), and from a standing posture to dynamic body movements (92 bpm). A major obstacle for ambulatory monitoring is that physiological dysregulation or emotion effects can be confounded by physical activity. Many physiological parameters, including heart rate, respiratory sinus arrhythmia, and skin conductance level, are strongly affected by both anxiety and exercise [32]. The daily routines involve different psychophysiological body activation charac- teristics. Kusserow et al.'s study showed that the physical-related and mental-related routines that are correlated with heart activity can be characterized and visualized as different activation patterns using RR-intervals and accelerometer data [13]. Schumm et al.'s work validated that it is possible to provoke and mea- sure GSR with the startle event during different walking speeds [24]. However, the faster a person is walking, the more the peak distribution of GSR approaches a uniform distribution. Activity information is also helpful in ECG-based iden- tity authentication area. The perturbation of the ECG signal due to physical activity is a major obstacle in applying the technology in real-world situations. Sriram et al.'s work presented a novel ECG- and accelerometer-based system that can authenticate individuals in an ongoing manner under various activity conditions [27].

Thus, physical activity distorts the result of mental stress detection in a mobile health monitoring scenario. In this paper, we compensate for the effects of physical activities by extracting a set of accelerometer features that characterize different physical activities along with ECG and GSR features. We hypothesize that the accelerometer features provide the necessary auxiliary information for differentiate physical activity and mental.

4 Methodology

In this section, we describe the components of the wireless sensor system we used, the procedure of the experimental environment, and the segmentation of experimental dataset.

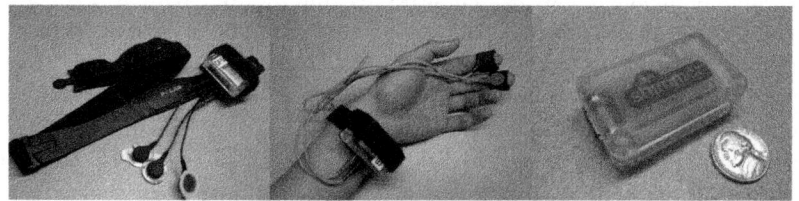

(a) ECG sensor and chest strap   (b) GSR sensor   (c) Accelerometer

Fig. 2. SHIMMER sensors including ECG, GSR, and accelerometer

4.1 Wireless Sensor Network

We used the SHIMMER platform developed by Intel's Digital Health Group. SHIMMER is a small wireless sensor platform with an integrated 3-axis accelerometer designed to support wearable applications. We also used SHIM- MER's ECG and GSR daughter boards for data acquisition. The sensor data from the ECG sensor and accelerometer were sampled at 100 Hz, and the data from the GSR sensor were sampled at 32 Hz. Data were transmitted to a PC via Bluetooth connectivity and saved to binary and comma-separated value files. We used three sensor nodes for the wireless sensor network configuration. Pho- tos of the sensors are shown in Figure 2. The ECG sensor node was strapped to an elastic chest belt and three electrodes were placed on the body to form lead II and lead III[1] recording configurations. The GSR sensor was attached on a wrist band. Then, skin conductance was measured at the base of two fingers by measuring the electrical current that flowed as a result of applying a constant voltage. The third sensor node which was placed on the waist belt was used to collect accelerometer data.

---

[1] (Lead II is the voltage between the left leg (LL) electrode and the right arm (RA) electrode), and Lead III is the voltage between the (positive) left leg (LL) electrode and the left arm (LA) electrode.

4.2 Experimental Protocol

20 participants were monitored, 13 men and 7 women. Participants were students, faculty, and staff at our university. A computer application that randomly presented Stroop Color-word interference tests and mental arithmetic problems was provided as stressor. The Stroop test has been widely utilized as a psy- chological or cognitive stressor to introduce emotional response and autonomic reactivity [28]. Because participants would answer so many questions during the study, we added a variant of the Stroop test to prevent habituation, where participants were asked to select either color-name or font-color. The mental arithmetic is based on the Montreal Imaging Stress Task (MIST), consisting of two levels of difficulty under time pressure [7]. The mental arithmetic will adapt to participants' level or adjust the time limit in order to maintain an appropriate level of stress. The participants completed the mental tasks by interacting with a 19-inch touchscreen. Examples of two mental tasks are shown in Figure 3. When the participant provided an answer before the end of time limit, the feed- back "correct" or "wrong" was displayed. The interface also shows the elapsed time and the participant's accuracy rate.

Participants were confronted with mental stress in each of three different conditions: sitting, standing, or walking. Each condition consisted of a baseline measurement with no stressor, measurement during the mental tasks, and a recovery segment:

1. Baseline segment (10 minutes): Listen to meditation music (in seated, stand- ing, or walking position).
2. Mental task segment (10 minutes): Complete Stroop test and mental arithmetic under time pressure while seated, standing, or walking.
3. Recovery segment (10 minutes): Sit in a chair with closed eyes and listen to meditation music.

All participants completed three conditions in random order. For the baseline and mental task segments, the participant had to complete the physical activity simultaneously. For example, when the participant was in the mental task segment of walking session, the participant was required to walk on the treadmill at 3 mi/hr and complete the mental task using the touch screen at the same time.

4.3 Data Collection

We collected sensor data from each participant for three physical activity conditions. The data were separated into six datasets. Hence, for each participant, we collected six data sets shown in Figure 4. The dataset for each segment contains 19200 GSR samples and 60000 ECG and accelerometer samples. The dataset for all 20 participants, including the six segments, is around 45 Megabytes.

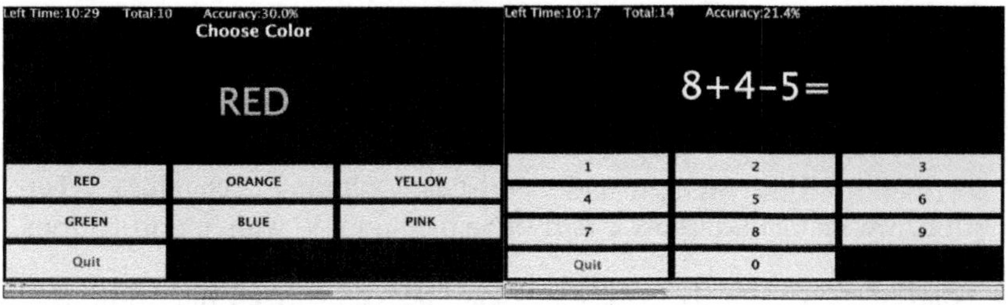

(a) Stroop Test User Interface   (b) Mental Arithmetics User Interface

Fig. 3. Screenshots of the stressor application

| Baseline Segment (Sitting Only) 10 min | Stressed Segment (Mental Task + Sitting) 10 min | Recovery Segment (Sitting) 10 min |

(a) Sitting condition

| Baseline Segment (Standing Only) 10 min | Stressed Segment (Mental Task + Standing) 10 min | Recovery Segment (Sitting) 10 min |

(b) Standing condition

| Baseline Segment (Walking Only) 10 min | Stressed Segment (Mental Task + Walking) 10 min | Recovery Segment (Sitting) 10 min |

(c) Walking condition

Fig. 4. Experimental conditions

5 Data Analysis

5.1 Feature Extraction

For each participant's 60 minutes of data, we segment each channel of data into a 60-second window to obtain the data windows $\omega_1, \omega_2, ..., \omega_{60}$. We denote F_i as the feature vector extracted from the data window ω_i and $F_i(j)$ is the j_{th} feature in the feature vector F_i. We create a set of feature vectors F for each participant's data set. Each segment in the experiment protocol has 10 feature vectors (e.g. $F_1 - F_{10}$ for SitBase, $F_{11} - F_{20}$ for SitStress,..., $F_{51} - F_{60}$ for WalkStress).

We chose a 60-second window for two reasons. First, the HRV features that we used in this study can distinguish between stressed and baseline segments using

60-second windows [23]. Second, the 60-second feature window reduces the impact of misclassified R-peaks by averaging HRV features within the window. All feature extraction algorithms are implemented in MATLAB. To eliminate the artifacts caused by variations in electrode contact and physical motion, we applied both moving average and band-pass filtering techniques. For R-peak detection, we mainly adapted a derivative method [19] with modifications.

HRV analysis: HRV analysis methods can be categorized into time domain and spectral domain analysis. Time domain analysis is calculated directly from RR-intervals over the feature window. Examples of time domain features include mean value of the RR-interval (mean RR), standard deviation of the RR-interval (Std RR), mean value of the HR (mean HR), standard deviation of the HR (Std HR), RMSSD, and pNN50. Moreover, in the spectral domain methods, a power spectrum density (PSD) estimate is calculated for the RR interval series. Fre- quently used spectral measures are the very low frequency (VLF, 00.04 Hz), low frequency band (LF) and high frequency band (HF), and the ratio LF/HF. These spectral domain features are often interpreted as a measure of sympathovagal balance (autonomic state influence by the sympathetic and parasympathetic nervous system). We first calculated six time-domain features of HRV including mean RR, Std RR, mean HR, Std HR, RMSSD, and pNN50. Then, we applied a Fast Fourier Transform (FFT) to convert the time-domain RR-interval sequence to the power spectrum. The frequency components are used to calculate three spectral-domain features of HRV for each window: LF, HF, and LF/HF ratio.

GSR analysis: Due to the startle response (the physiological response of body to a sudden stimulus), the resistance of the skin can vary. The GSR can measure these subtle differences [29]. All GSR signals were filtered with a 256-point low- pass filter with 3Hz cutoff frequency to reduce noise. We calculated three GSR features: the total number of the startle responses in the segment, the sum of the response magnitude, and the sum of the response duration. These three features characterize the startle response, and Healey and Picard demonstrated their reliability [10]. Two additional features, mean and standard deviation of skin conductance level, are calculated over the feature window. Figure 5 shows the R-R interval and skin conductance recordings of a subject over six experimental segments.

Accelerometer analysis: Olguin and Pentland's work indicated that an accelerometer placed on hip significantly helped classify activities such as sitting, running, crawling, and lying down [18]. Therefore, we placed one accelerometer on the waist belt close to the hip in order to maximize the difference of signal among sitting, standing, and walking activities. For each of the three axial dimensions, we calculated twelve features: mean value, standard deviation, energy, and correlation of each two axes. Table 1 lists the features derived from the ECG, GSR, and accelerometer data.

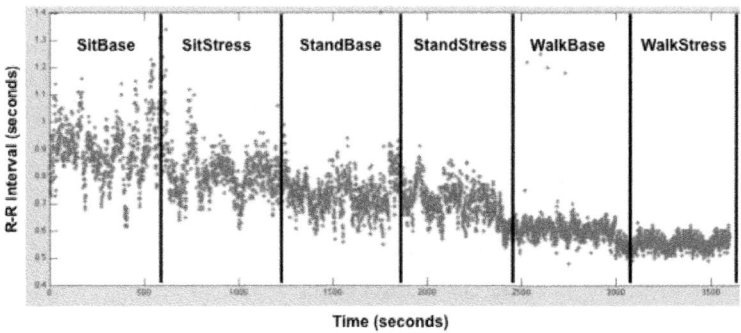

(a) RR interval data of a subject

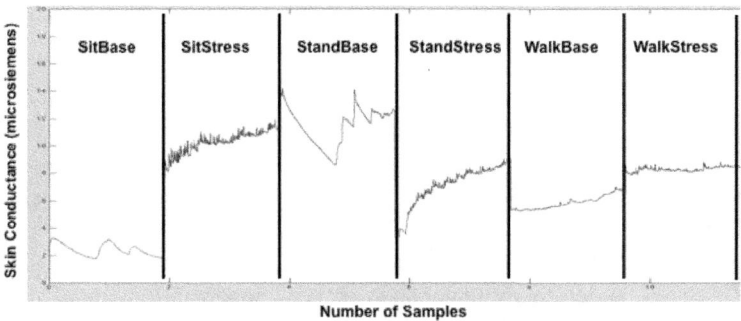

(b) Skin conductance of a subject

Fig. 5. RR interval and GSR data in six experiemental segments

Table 1. Feature Vectors

	Sensor	Features
f 1-f 9	ECG	Mean RR, Std RR, Mean HR, Std HR RMSSD, pNN50, LF, HF, LF/HF ratio
f 10-f 14	GSR	Mean SCL, Std SCL, Total magnitude, Duration, and Number of startle responses
f 15-f 26	Accel	Mean of X, Y and Z axis Standard deviation of X, Y, and Z axis Energy of X, Y, and Z axis Correlation coefficient of XY, YZ, and ZX

5.2 Feature Normalization

Skin conductivity and heart rate signals are dependent on each individual's initial physiological level. Even when the GSR or HR baseline level is measured from the same individual, these signals are likely day-dependent due to variations in physiology caused by diet or sleep, variations in mental state affected by mood, or variations in the sensor's connectivity with skin [20]. Hence, to eliminate the intra-individual factor, we applied Equation 1 to each feature in feature vector set F. Since we conducted a short recording interval (one hour) for each participant, day-to-day variation caused by mood fluctuations is not considered in this study.

$$F(j)_{norm} = \frac{F(j) - Z_{min}(j)}{Z_{max}(j) - Z_{min}(j)}, \text{ where}$$
$$Z_{min}(j) = \min\{F_i(j)\}, \forall i \in |F|$$
$$Z_{max}(j) = \max\{F_i(j)\}, \forall i \in |F| \quad (1)$$

Equation 1 describes the normalization process for each feature. The first step is to subtract the minimum value from each feature such that the feature with minimum value becomes 0. Then, the feature values are divided by the overall range in six segments to make all the features lie between 0 to 1. The normalized feature values are fed to the classifiers described in the next section.

6 Stress Classification

We used the WEKA machine learning engine to train classifiers using various learning methods, including the J48 Decision Tree, Bayes Net, and support vector machine (SVM) for stress inference [9]. We divided the training data into two different sets in order to evaluate how activity information may influence the results of stress inference. One set of training data only includes the ECG- and GSR-related features while the second set also includes the accelerometer information. We also evaluated classification performance for between-subjects datasets and within-subject datasets.

6.1 Decision Tree

Decision Tree is a commonly used machine learning technique that uses a divide-and-conquer approach to classify testing data. During the learning stage, the tree structure is constructed. The tree structure has internal nodes and leaves. Internal nodes represent the test conditions while the leaves represent the clas- sification results. We used a J48 Decision Tree for mental stress classification. Again, since we are interested in observing how the accelerometer information affects the accuracy of mental stress classification, we separated the training data into two sets. One dataset includes features extracted from accelerometer, ECG, and GSR recordings. The other dataset only consists of ECG and GSR features. When the Decision Tree is being constructed, the most informative feature (with a higher information gain) will be used near the root. Therefore, we are interested in observing the constructed decision tree structure to see if the accelerometer information is used in the test conditions and provides higher information gain.

6.2 Bayesian Network

We are also interested in using a Bayesian network structure to model the probabilistic relationships among physical and mental stress. Figure 6 shows two Bayesian network structures with and without considering the activity information to predict the existence of mental stress.

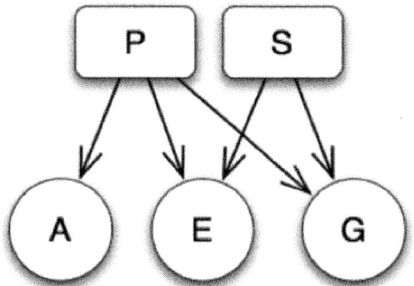 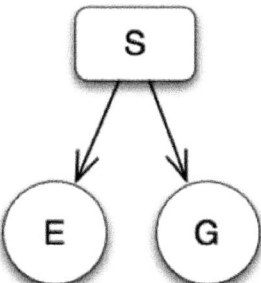

(a) Bayesian network with activity information

(b) Bayesian network without activity informa- tion

Fig. 6. Variable S represents a binary stress state and variable P represents the three physical activities. Variables A, E, and G are the accelerometer, ECG, and GSR features, respectively.

By comparing the inference results of the models in Figure 6a and in Figure 6b, we can investigate the effects of activity information in a mobile stress detection scenario.

In Figure 6a, S ∈ {baseline, stressed} represents a binary stress state, and P ∈ {sitting, standing, walking} represents the three physical activities. A, E, and G are the subsets of features defined in Table 1. A = (a_1,...,a_{12}) is a 12-feature vector corresponding to the accelerometer measure. E = (e_1,...,e_9) is a 9-feature vector related to the HRV parameters. G = (g_1,...,g_5) represents a set of 5 GSR-related features. Equation 2 shows the joint probability distribution encoded by the Bayesian network structure shown in Figure 6a. The Bayesian network structure shown in Figure 6b only uses physiological signals from ECG and GSR to infer the probability of the mental stress state.

$$P(S, P, A, E, G) = P(S) \cdot P(P) \cdot P(A|P) \cdot P(E|P, S) \cdot P(G|P, S) \quad (2)$$

We ran the K-Means clustering algorithm with the Euclidean distance metric on accelerometer data to automatically label variable P. Because we have three types of activities (sitting, standing, and walking), we set K = 3. Figure 7a shows an example of accelerometer raw data we collected from the experiment.

Figure 7b shows the class of activity derived from Figure 7a using the K-Means algorithm.

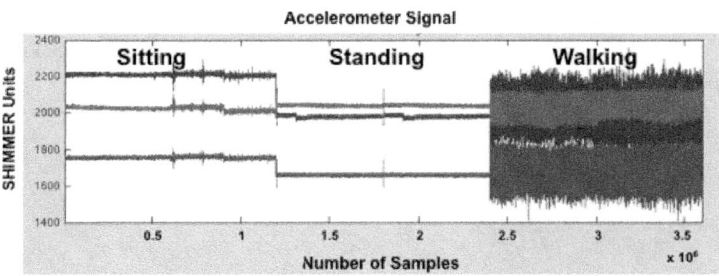

(a) Accelerometer data from one subject (red:x-axis, green:y-axis, and blue:z-axis)

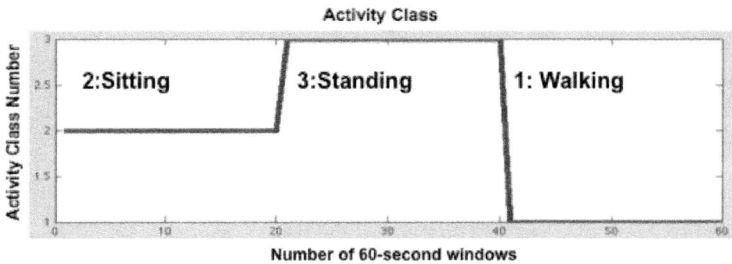

(b) Activity classes derived from accelerometer data using the K- means algorithm

Fig. 7. Accelerometer data of a subject during three activities and the activity class derived by the K-Means algorithm

The probability distribution of S conditioned on the three observed variables ($A = a$, $E = e$, and $G = g$) can be obtained by marginalizing the activity variable P as shown in Equation 3.

$$P(S|A = a, E = e, G = g) = \sum_{i=0}^{|p|} P(S, P = p_i | A = a, E = e, G = g) \tag{3}$$

The binary class of the stress state is estimated by maximizing the posterior probability in Equation 4.

$$S_{estimated} = \underset{S}{\mathrm{argmax}} \sum_{i=0}^{|p|} P(S, P = p_i | A = a, E = e, G = g) \tag{4}$$

An evaluation of cross validation for these two Bayesian networks is presented in Section 7.

6.3 Support Vector Machine

Support Vector Machine (SVM) is a classifier that performs classification by constructing a high-dimensional hyper-plane [4]. The constructed high-dimensional hyper-plane is optimized to separate the testing data into two classes. SVM also allows different types of kernel functions to transform testing data points into a higher dimensional space and make the transformed data easier to be classi- fied. Since SVM has recently become a popular machine learning technique for classification, we are interested in investigating its performance with our testing dataset.

7 Experimental Results

In this section, we present the results from two experiments and a comparison of ECG and GSR feature efficacy. First, to investigate how a combination of features affect stress classification accuracy, we design four feature combinations from the measured accelerometer, ECG, and GSR data. Second, we test if the classifiers generalize across subjects by training our classifier on subset of subject data and testing our classifier on the remainder. Finally, we analyze the ECG and GSR features in six conditions across 20 participants.

7.1 Cross Validation with Different Feature Combinations

The first type of feature combination includes data measured from the accelerometer, ECG, and GSR. In each of the other three types of feature combinations, one feature is excluded. For each of the three classifiers described in Section 6, we evaluated its classification accuracy using these four types of feature combinations. Figure 8 plots the results of using 10-fold cross validation. For all of the three types of classifiers, excluding data recorded from the accelerometers degrades in the classification accuracy. The experimental results provide evidence to support our hypothesis that accelerometer data help the classification accu- racy in a mobile stress detection scenario; physiological signals are both affected by physical activities and mental stress levels.

Furthermore, since heart rate is highly affected by the intensity of physical activities in our experiment, the classification results are even better for Bayesian network and SVM classifiers without including ECG features compared to the all-feature combination. Unlike ECG, the GSR features are good indicators to identify the presence of mental stress. When GSR features are excluded, the accuracy of each classifier decreases compared to the all-feature combination. We also found that the best classification accuracy (92.4%) is obtained from using the decision tree classifier with the all-feature combination. Moreover, the structure of the decision tree uses the energy of the x-axis from the accelerometer data as the root test condition. Several accelerometer features are also used as test conditions close to the root of the tree. It proves that activity information provides higher information gain in the decision tree learning stage. Table 2 shows more detail of the cross validation results on 1200 samples. The grey cells highlight correctly recognized instances (true-positives).

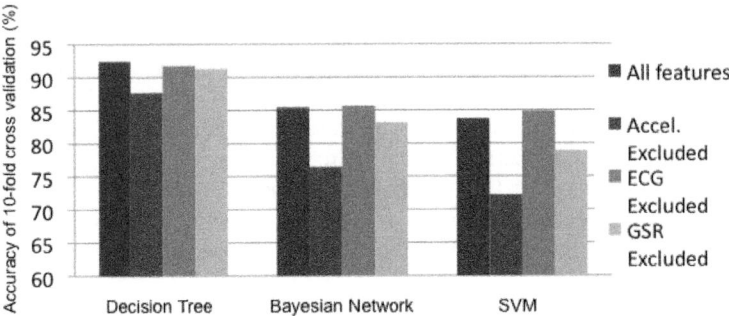

Fig. 8. Accuracy of the three classifiers using different feature combinations

Table 2. Confusion matrix for the combination of three classifiers and different feature combinations

Features		Classification schemes					
		Decision Tree		Bayesian Network		SVM	
		Baseline	Stressed	Baseline	Stressed	Baseline	Stressed
All features	Baseline	553	47	475	125	509	91
	Stressed	44	556	50	550	104	496
Accel. Excluded	Baseline	530	70	412	188	455	145
	Stressed	78	522	95	505	189	411
GSR Excluded	Baseline	539	61	431	169	461	139
	Stressed	44	556	33	567	115	485
ECG Excluded	Baseline	552	48	483	117	510	90
	Stressed	52	543	55	545	92	508

7.2 Between-subjects Experiment

We randomly selected a subset of our twenty subjects and used their data to train our classifier. The, we tested the classifier on the remainder. For a given between-subjects

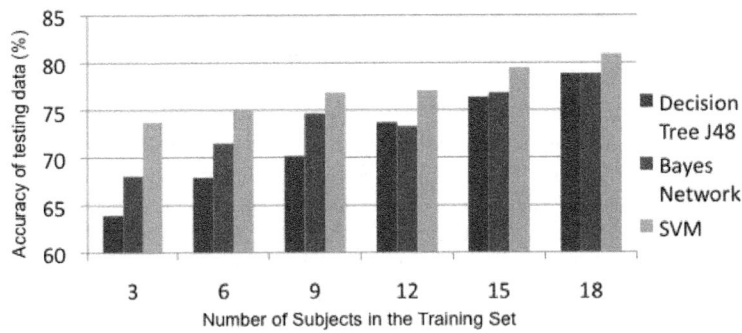

Fig. 9. Classification accuracy between subjects with the three classifiers

classification setting, we repeated 10 times and calculated the average accuracy. To observe the effect on the size of the subset data, we changed the number of subjects in the training data set from 3, 6, ...to 18. The results shown in Figure 9 demonstrate that the SVM classifier outperforms the other two classifiers for between-subjects classification.

7.3 Comparison of HRV and GSR Parameters

Table 3 compares HRV features for the six experimental segments: (SitBase, Sit-Stress, StandBase, StandStress, WalkBase, and WalkStress). Each HRV parameter is calculated by the average value across 20 participants for each segment. The mean RR interval decreases in all of the three mental task segments while the mean HR increases. The trend of mean RR and HR proves their efficacy in distinguishing mental stress across the three physical activities. The pNN50 feature also has a tendency to decrease from the baseline to the mental task segment. In the walking condition, pNN50 significantly decreases compared to the lower-intensity activities (sitting and standing). The standard deviation of RR is relatively high in walking condition and decreases the percentage of heartbeat intervals with difference in successive heartbeat intervals greater than 50 ms.

Table 3. Comparison of HRV parameters in six conditions

HRV Parameters	Sit Base	Sit Stress	Stand Base	Stand Stress	Walk Base	Walk Stress
*Mean RR (ms)	887.59	814	752.07	722.43	586.03	562.94
Std RR (ms)	70.88	85.39	82.44	68.35	92.47	98.94
*Mean HR (bmp)	69.53	75.59	82.84	85.66	107.09	110.79
Std HR (bmp)	5.54	7.56	8.00	9.50	18.98	16.21
*pNN50 (%)	19.54	15.69	12.09	11.38	4.49	4.23
LF (%)	7.04	8.45	7.49	7.77	9.43	9.45
HF (%)	6.25	6.51	6.33	6.73	13.95	15.64
LH Ratio	1.34	1.51	1.45	1.48	0.67	0.71

The last three rows in Table 3 are three spectral-domain parameters. The LF component is an indicator for both sympathetic modulations and cardiac vagal activity. It slightly increases from the baseline during all three activities. The HF component is considered to reflect parasympathetic modulations. There is a large increase in the HF component when the participants are walking. The imbalance of increase on LF and HF causes the LH ratio (LF / HF), a widely adapted index of sympathetic modulations, to decreases from sitting to walking. The HF increases during exercise has also been noted by other researchers [21].

From our analysis of all HRV parameters, we found that mean HR and RR are the most reliable features to recognize mental stress across three physical activities. The standard deviation of RR and HR did not demonstrate a coherent relation to the baseline and stressed segments. Spectral-domain parameters are sensitive to the physical activity conditions. Hence, this explains why excluding HRV features even increases in accuracy compared to the all-feature combination as shown in Figure 8.

Table 4 lists five GSR parameters for each segment. For each startle response, we can indicate its duration and magnitude. The total duration was calculated by accumulating the total elapsed time of the responses in the window. The total magnitude was measured by summing up the difference of the onset and the peak of each startle response in window. The number of response occurrences over the one minute window was also recorded. Total duration, total magnitude, total occurrence of the responses, and mean GSR level illustrate an obvious increase from baseline to stressed segment. However, the standard deviation does not provide significant change between conditions.

8 Discussion and Conclusion

Previous mental stress studies were conducted in the laboratory with seden- tary subjects. However, the controlled setting in a laboratory is not suitable for mobile mental stress monitoring because physical activity affects the measured physiological signals. The main goal of this study was to determine whether activity information can compensate for the interactive effects of mental stress and physical activity, which affect the accuracy of mental stress detection. Therefore, we conducted a user study in which participants completed baseline and mental task segments across three physical activities (sitting, standing, and walking).

Table 4. Comparison of GSR parameters in six conditions

GSR Parameters	Sit Base	Sit Stress	Stand Base	Stand Stress	Walk Base	Walk Stress
*Total duration(second)	3.17	14.30	4.16	13.15	13.72	16.32
*Total magnitude(μSiemens)	0.79	2.04	0.75	3.32	1.69	1.97
*Total occurrence	1.09	6.58	3.13	6.37	5.63	7.47
*Mean GSR(μSiemens)	4.69	4.83	6.19	6.97	6.42	7.22
Std GSR(μSiemens)	0.62	0.53	0.62	0.71	0.63	0.52

This paper presented a multimodal approach to model the mental stress activation affected by physical activities using accelerometers, ECG, and GSR sensors. Our analysis showed that accelerometer data is necessary to improve mental stress detection in a mobile environment. We also noticed that the Decision Tree classifier has the best performance in our experiments using 10-fold cross validation. Decision Tree is recognized as one of the classification methods with low computational complexity [14]. Therefore, the performance along with the low complexity of the Decision Tree classifier makes it a practical design choice for stress detection on mobile devices.

Furthermore, we also compared how physical activities and mental stress affect HRV and GSR parameters. We found that the GSR features are relatively independent of the three activities, even when participants were walking at a 3mi/hr. The between-subjects experiment demonstrated that we need to use up to 90% of subjects' data to achieve the classification accuracy of around 80%. It indicates that physiological signals tend to be user-dependent; hence, mental stress monitoring applications should also rely on personalized data in the training stage. We plan to further investigate the user-dependent attribute with more participants in the future. This study was limited to three specific activities and a relatively short recording time. The next step is to design a mobile platform enabling participants to wear sensors on a daily basis.

Our activity-aware scheme for mental stress detection can facilitate the de- velopment of many affective mobile applications using physiological signals (e.g., stress management, affective tutoring, and emotion-aware human computer in- terfaces). Including activity recognition techniques to interpret users' emotional states helps produce more feasible wearable sensors in everyday life.

References

1. Bernardi, L., et al.: Physical activity influences heart rate variability and very-low-frequency components in holster electrocardiograms. Cardiovascular Research 32(2), 234 (1996)
2. Boucsein, H.: Electrodermal Activity. Plenum, New York (1992)
3. Breslau, N., Kessler, R., Peterson, E.L.: Post-traumatic stress disorder assess-ment with a structured interview: reliability and concordance with a standardized clinical interview. International Journal of Methods in Psychiatric Research 7(3), 121–127 (1998)
4. Burges, C.J.: A Tutorial on Support Vector Machines for Pattern Recognition. Data Mining and Knowledge Discovery 2, 121–167 (1998)
5. Cannon, W.: Bodily Change in Pain, Hunger, Fear and Rage: An Account of Recent Research into the Function of Emotional Excitement. Appleton, New York (1915)
6. Darrow, C.: The rationale for treating the change in galvanic skin response as a change in conductance. Psychophysiology 1, 31–38 (1964)
7. Dedovic, K., Renwick, R., Mahani, N.K., Engert, V., Lupien, S.J., Pruessner, J.C.: The Montreal imaging stress task: Using functional imaging to investigate the effects of perceiving and processing psychosocial stress in the human brain. Journal Psychiatry Neuroscience 30, 319–325 (2005)
8. Dubin, D.: Rapid Interpretation of EKG's. Cover Publishing (2000)

9. Hall, M., Frank, E., Holmes, G., Pfahringer, B., Reutemann, P., Witten, I.H.: The WEKA Data Mining Software: An Update. SIGKDD Explorations 11 (2009)
10. Healey, J.A., Picard, R.W.: Detecting stress during real-world driving tasks using physiological sensors. IEEE Transactions on Intelligent Transportation Systems 6(2), 156–166 (2005)
11. Herbert, J.: Fortnightly review: Stress, the brain, and mental illness. British Medical Journal 315, 530–535 (1997)
12. Holmes, S., Krantz, D.S., Rogers, H., Gottdiener, J., Contrad, R.J.: Mental stress and coronary artery disease: A multidisciplinary guide. Progress in Cardiovascular Disease 49, 106–122 (2006)
13. Kusserow, M., Amft, O., Troster, G.: Psychophysiological body activation char- acteristics in daily routines. In: IEEE International Symposium on Wearable Computers, pp. 155–156 (2009)
14. Lim, T.-S., Loh, W.-Y., Cohen, W.: A comparison of prediction accuracy, complexity, and training time of thirty-three old and new classification algorithms. Machine Learning 39 (2000)
15. Lundberg, U., et al.: Psychophysiological stress and EMG activity of the trapezius muscle. International Journal of Behavioral Medicine 1(4), 354–370 (1994)
16. Monnikes, H., Tebbe, J., Hildebrandt, M., et al.: Role of stress in functional gastrointestinal disorders. Digestive Diseases 19, 201–211 (2001)
17. T.F. of the European Society of Cardiology, the North American Society of Pacing, and Electrophysiology. Heart rate variability: Standards of measurement, physio- logical interpretation, and clinical use. European Heart Journal 17(2), 1043–1065 (1996)
18. Olguin, D., Pentland, Y.: Human activity recognition: Accuracy across common locations for wearable sensors. In: Proc. 10th International Symposium Wearable Computer, pp. 11–13 (2006)
19. Pan, J., Tompkins, W.J.: A real-time QRS detection algorithm. IEEE Transactions on Biomedical Engineering BME- 32(3), 230–236 (1985)
20. Picard, R.W., Vyzas, E., Healey, J.: Toward machine emotional intelligence: Analysis of affective physiological state. IEEE Transactions on Pattern Analysis and Machine Intelligence 23(10), 1175–1191 (2001)
21. Pichon, A., Bisschop, C., Roulaud, M., et al.: Spectral analysis of heart rate variability during exercise in trained subjects. Medicine and Science in Sports and Exercise 36, 1702–1708 (2004)
22. Pickering, T.: Mental stress as a causal factor in the development of hypertension and cardiovascular disease. Current Hypertension Report 3(3), 249–254 (2001)
23. Salahuddin, L., Cho, J., Jeong, M.G., Kim, D.: Ultra short term analysis of heart rate variability for monitoring mental stress in mobile settings. In: 29th Annual International Conference of the IEEE, Engineering in Medicine and Biology Society, EMBS 2007, pp. 4656–4659 (August 2007)
24. Schumm, J., Bächlin, M., Setz, C., Arnrich, B., Roggen, D., Tröster, G.: Effect of movements on the electrodermal response after a startle event. In: Proceedings of 2nd International Conference on Pervasive Computing Technologies for Healthcare, Pervasive Health (2008)
25. Setz, C., Arnrich, B., Schumm, J., La Marca, R., Troster, G., Ehlert, U.: Discriminating stress from cognitive load using a wearable EDA device. IEEE Transactions on Information Technology in Biomedicine 14(2), 410–417 (2010)

26. Sloten, J.V., Verdonck, P., Nyssen, M., Haueisen, J.: Influence of mental stress on heart rate and heart rate variability. In: International Federation for Medical and Biological Engineering Proceedings, pp. 1366–1369 (2008)
27. Sriram, J.C., Shin, M., Choudhury, T., Kotz, D.: Activity-aware ECG-based patient authentication for remote health monitoring. In: ICMI-MLMI 2009: Proceedings of the 2009 International Conference on Multimodal Interfaces, pp. 297–304. ACM, New York (2009)
28. Stroop, J.: Studies of interference in serial verbal reactions. Journal of Experimental Psychology 18, 643–661 (1935)
29. Tarvainen, M., Koistinen, A., Valkonen-Korhonen, M., Partanen, J., Karjalainen, P.: Analysis of galvanic skin responses with principal components and clustering techniques. IEEE Transactions on Biomedical Engineering 48(10), 1071–1079 (2001)
30. Van Steenis, H., Tulen, J.: The effects of physical activities on cardiovascular variability in ambulatory situations. In: Proceedings of the 19th Annual International Conference of the IEEE, Engineering in Medicine and Biology Society, vol. 1, pp. 105–108 (November 1997)
31. Vrijkotte, T., et al.: Effects of work stress on ambulatory blood pressure, heart rate, and heart rate variability. Hypertension 35(4), 880–886 (2000)
32. Wilhelm, F.H., Pfaltz, M.C., Grossman, P., Roth, W.T.: Distinguishing emotional from physical activation in ambulatory psychophysiological monitoring. Biomedical Sciences Instrumentation 42, 458–463 (2006)
33. Zhai, J., Barreto, A.: Stress detection in computer users ased on digital signal processing of noninvasive physiological variables. In: 28th Annual International Conference of the IEEE, Engineering in Medicine and Biology Society, EMBS 2006, pp. 1355–1358 (September 2006)

A Decentralized Decision Support System for Mobile Devices

Gert Scholten, Nicholas Palmer, Roelof Kemp, Thilo Kielmann, and Henri Bal

Vrije Universiteit
De Boelelaan 1081A, 1081 HV, Amsterdam, The Netherlands
gscholt@gmail.com,
{palmer,rkemp,kielmann,bal}@cs.vu.nl

Abstract. With this demonstration we present *Decisionlib*, a Distributed Decision Support System that runs on mobile devices. Decisionlib is a flexible, durable, and robust voting system which places emphasis on communication and error handling in order to provide reliable and easy to program group-based decisions on distributed mobile devices. Written on top of the *Ibis* [4] distributed communication system, Decisionlib represents the state of the art for group decision making on mobile devices.

Group based decisions generally involve reaching some form of agreement between a number of different participants. This kind of decision is vital in the disaster management field where leadership roles are often not well defined[2] and quick decisions need to be made by people in the field based on varying conditions they inevitably encounter in order to provide the agility[1] required for an effective response.

To illustrate the problems that arise in mobile, distributed group based decisions we present the following scenario from the disaster management literature[3]; A major disaster such as an earthquake or a hurricane landfall and there is destruction on a massive scale leaving the remaining survivors stranded. Rescue workers and non-injured effected individuals search the area, collect groups of survivors and provide immediate first aid. For this process to work smoothly, people need to be grouped so that efficient aid can be provided, and people can be easily transported out of the disaster area. Rescue workers and effected people find each other by use of GPS locations, using ad-hoc networking via Wi-Fi and Bluetooth and mark their preference for a gathering point. The gathering points are evaluated and voted upon among those in the area while some participants lose connectivity. Nonetheless, a reachable location is selected to bring everyone together for faster evacuation. Shortly after the decision is made users who lost connectivity during the decision process reconnect to the network and receive notification of the chosen gathering point.

We have identified the following requirements for a distributed decision support system for disaster situations:

Flexible: The system should be highly configurable to allow users to specify what kind of decisions to make, and how to make them.

Durable: The system should provide durable results. In the scenario above, some participants lose network connectivity. This should not have a permanent impact on the availability of the result of a decision. Optionally, entities not actively participating in the decision should be made aware of the result. This allows for partial participation in decisions, and public versus private decisions.

Ubiquitous: The system should allow for decisions to be easily made while out in the field and preferably already available on devices which are normally carried by people likely to be effected by disasters.

Robust: The system should manage failures in a configurable manner and should not rely on central components, because these are prone to failure, especially in disaster scenarios.

To summarize, we need a flexible system on mobile phone which handles failures well, provides durable results and runs without relying on centralized components.

In this demo we present the Decisionlib Distributed Decision Support System which satisfies all of these requirements. It is configurable to support any kind of decision process and to provide the information required for group based decision making. It deals with connectivity issues in ways that do not hinder the reaching of a decision. A decision is considered final and possibly universally available after it has been made, meeting all of the above requirements.

The demonstration of this system will be run using a number of Android powered Smartphones. No additional materials are required from the Venue. One of the phones will be used in Ad-Hoc mode to provide a network connection

Fig. 1. List of Decisions

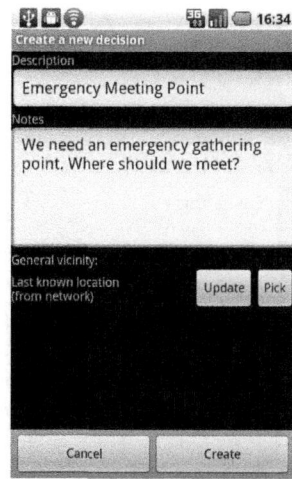

Fig. 2. Create a New Decision

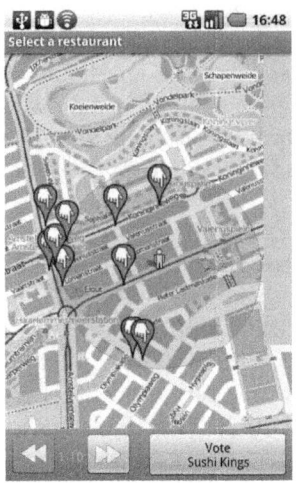

Fig. 3. Join a Decision   **Fig. 4.** Free Vote On A Map

between the phones demonstrating that our system can be used in a fully decentralized, ad-hoc environment. During the demonstration participants will be invited to use one of our phones to participate in a decision making process. For this demonstration participants will be invited to select a restaurant near the conference venue at which to meet other participants. Note that this process is completely analogous to the situation presented in our disaster management example above but is of more direct interest to conference participants. Users will be invited to create a decision, as shown in Figure 2 and select the participants,

Fig. 5. Vote List   **Fig. 6.** View Completed Decision

to view a past decision as shown in Figure 1, or join a currently running decision as shown in Figure 3. In the first round of voting users will be allowed to select a restaurant from a map with local venues as shown in Figure 4. Voting then proceeds in rounds using a list of selected restaurants as seen in Figure 5. Once a decision is reached it is available for viewing as shown in Figure 6. Note that the fact that our application can be used for a number of every day scenarios means that it is much more likely to be deployed and users will already know how to use it when a disaster strikes.

Conclusion

This demonstration features the *Decisionlib* Distributed Decision Support System. Our analysis of disaster management literature led us to the following requirements for a distributed decision support library: flexible, durable, ubiquitous and robust. The Decisionlib decision support system we have built meets these requirements by supporting reliable communication, membership management, recovery in case of connection failures and persistence in decisions and their results as long as a participating node is alive. It supports many types of decisions, and any round-based voting algorithm. It comes with several pre-defined algorithms and application designers are free to add their own or tailor an existing algorithm to meet the needs of the application. It runs on smartphones which people are likely to carry with them everywhere.

Our Restaurant Decision test application was used to evaluate the utility of the Decisionlib system. The application is a reliable application for deciding on temporary gathering points, and uses only a few lines of code to use the Decisionlib system. It illustrates the ease of writing an application on top of the Decisionlib, and furthermore demonstrates that it works reliably in hostile situations such as those encountered as part of disaster management.

References

1. Harrald, J.: Agility and Discipline: Critical Success Factors for Disaster Response. The ANNALS of the American Academy of Political and Social Science 604(1), 256 (2006)
2. Plotnick, L., Ocker, R., Hiltz, S., Rosson, M.B.: Leadership roles and communication issues in partially distributed emergency response software development teams: A pilot study. In: HICSS, vol. 29 (2008)
3. U.S.C.S.B.C. to Investigate the Preparation for, R. to Hurricane Katrina, T. Davis, and U. S. G. A. Office: A Failure of Initiative: Final Report of the Select Bipartisan Committee to Investigate the Preparation for and Response to Hurricane Katrina. Govt. Printing Office (2006)
4. van Nieuwpoort, R.V., Maassen, J., Wrzesinska, G., Hofman, R., Jacobs, C., Kielmann, T., Bal, H.E.: Ibis: a flexible and efficient Java based grid programming environment. Concurrency and Computation: Practice and Experience 17(7-8), 1079–1107 (2005)

Inferring Complex Human Behavior Using a Non-obtrusive Mobile Sensing Platform

Bruce DeBruhl[1], Michele Cossalter[1], Roy Want[2],
Ole Mengshoel[1], and Pei Zhang[1]

[1] Carnegie Mellon University
{bruce.debruhl,michele.cossalter,ole.mengshoel,pei.zhang}@sv.cmu.edu
[2] Intel Labs
roy.want@intel.com

1 Introduction

Thanks to the decreasing cost, increasing mobility, and wider use of sensors, a great number of possible applications have recently emerged, including applications that may impact the high-level goals in people's lives. Applications can be found in many areas ranging from medical devices to consumer devices. Information about activity and context can be inferred from sensors and be used to provide automated recommendations.

Previous research has suggested that, using simple sensors, devices can improve human life on an everyday basis. Interesting examples can be found in the area of fluid intake monitoring [2,7]. Also, activity and context recognition from wearable sensors have been investigated [8–11].

Still, many challenges remain open in this area, including how to effectively determine the sensor data that is valuable, how to select sensors and also how to best connect the low-level sensory data with high-level goals that are explicitly or implicitly monitored by a user. As demand for these devices increases it is desired to generalize sensor platforms to monitor a broad set of activities. Moreover, it is desirable to make devices non-obtrusive and more naturally fit into our daily life. This further motivates researchers to design simple devices able to complete complex sensor-supported tasks in real-time.

In this paper we:

- Demonstrate use of a mobile sensor platform for high-level inference, with an example about water consumption recommendation while hiking;
- Highlight the selection of simple, inexpensive sensors for a non-obtrusive device running real-time algorithms;
- Build a dehydration model for implementation on a sensor platform.

To further understand applications of sensors we look at a specific problem found in outdoor activity, namely water consumption recommendation. This problem is challenging as it demands data from many sensors to be properly fused for inference and recommendation.

Fig. 1. Intel Ubiquity Lab sensing platform providing the following features: triaxial accelerometer, 3D compass, triaxial gyroscope, capacitive touch sensors, light sensor, thermometer, barometer, ZigBee radio and switches

2 System Description

Towards the goal of demonstrating high-level inference with a mobile sensor platform, we design a system that recommends appropriate water intake during a hike. Hiking can be a physically demanding exercise that becomes dangerous if hikers get dehydrated. This makes this application useful in a practical sense because learning how to avoid dehydration can be extremely difficult. Although many guidelines exist in the literature [1, 5, 6], little effort has been made in integrating them in a unified model. More importantly, this also raises the issue of how to connect high-level goals (such as "maintain healthy water intake") to low-level sensory data (such as "your acceleration is currently [x, y, z]"). The amount of water that needs to be consumed is strictly related to the level of dehydration. It is advisable to drink at rates comparable to a person's sweating rate [1]. Different variables affect the dehydration rate, including intensity of activity and environment conditions.

Hardware: our prototype system is implemented using the Intel Ubiquity Lab mobile sensing platform, depicted in Figure 1, providing the following features: triaxial accelerometer, 3D compass, triaxial gyroscope, capacitive touch sensors, light sensor, thermometer, barometer, ZigBee radio and switches. This hardware allows for a non-obtrusive implementation, where the device can be carried anywhere over the outerwear or on the user's backpack. The most relevant sensors to activity and environment inferencing were chosen as a subset of the available features, as described in the following.

Sensor selection: the accelerometer is used to achieve information about the intensity of the user's activity, as physical activity recognition using acceleration data has been shown to be successful in the literature [3]. Dehydration rate is also affected by temperature, altitude and degree of exposure to sunlight [1, 5, 6]. For this reason, the light sensor, the barometer and the thermometer are employed to make inferences about the environment conditions in which the

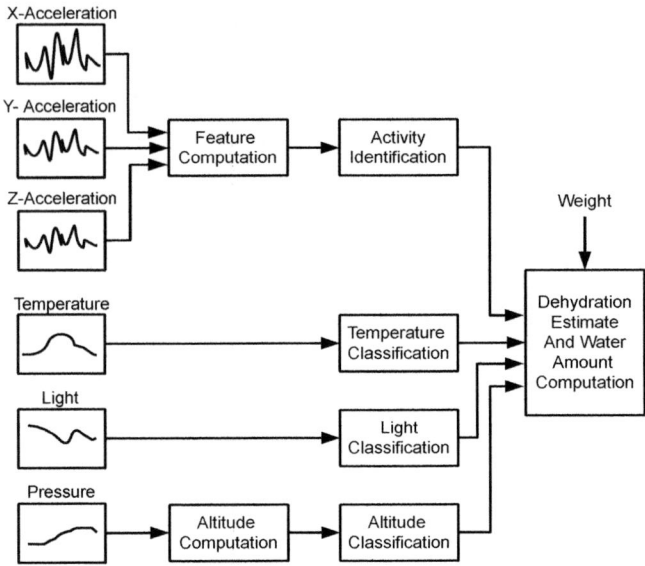

Fig. 2. Overview of the prototype system: acceleration data is used for activity identification while temperature, pressure and light intensity are used to make environmental inferences. Information about intensity of activity and environment conditions is then combined to estimate dehydration rate.

activity is performed. Altitude is computed from the pressure by inverting the barometric formula [4]. These sensors were preferred to perspiration and CO_2 sensors because of their non-obtrusive and inexpensive nature.

Classification: the system is depicted in Figure 2. We use classification as opposed to more complex interpolation to mitigate computational load. The following four common hiking activities are considered, listed in increasing intensity level: resting, walking level, walking downhill and walking uphill. This particular choice of actions is able to characterize a general hiking experience providing a reasonable estimate of the amount of undertaken work and, as a consequence, a recommendation about the amount of water that the user should drink. The model makes use of a well-known approach [3] to recognize activity using acceleration data.

We address the modeling of the external environment conditions by considering a number of temperature, light and altitude classes. Exploiting the guidelines present in the literature, a different dehydration rate is associated with each possible combination of classes. The following temperature classes are used: low (under 20°C), medium (over 20°C but below 30°C) and high (above 30°C) [1]. Altitude is classified as either low or high based on a threshold set to 6000 feet [5]. In a similar way, illumination is classified as low or high based on a threshold set to 800 lux, distinguishing between direct and indirect sunlight exposure [6].

Inference: starting from the identified activity and the temperature, altitude and light classes, the estimated dehydration rate d is computed according to the following equation:

$$d = (1 + k_a a)(1 + k_l l) \left[d_{00} + \frac{d_{23} - d_{03}}{2} t + \frac{d_{03} - d_{00}}{2} i \right]$$

where $a \in \{0,1\}$, $l \in \{0,1\}$, $t \in \{0,1,2\}$ and $i \in \{0,1,2,3\}$ represent altitude, light, temperature and activity intensity, respectively. The algorithm was designed to minimize processor load. The constants k_a, k_l, d_{00}, d_{23} and d_{03} were estimated from the literature [1,5,6]. Dehydration rate is then used to estimate the amount of water that should have been consumed over the hike to keep dehydration below 2% of body weight.

3 Conclusion

We demonstrated use of a mobile sensing platform for complex behavior inference, developing an application for recommending water intake when hiking. Our prototype system meets the design criteria including use of simple sensors and algorithms, non-obtrusive nature and real-time operation. Preliminary experiments provided results within reason of current sport medicine expectations. Upon completion of a 30 minute level walk in direct sunlight, low altitude and moderate temperature, a water amount of 0.22 liters was recommended for a 70 kg person, which is in the range of 0.25 liters expected.

With regards to the hiking application, more research has to be formalized on the effect of weather conditions and intensity of exercise on dehydration. Also, the use of additional sensors can be investigated to achieve a richer description of the environment and a more detailed activity classification. Possible improvements include using a flow rate sensor to detect amount of water consumed and microphones to monitor user breathing and conversation level.

In a more general frame, future work includes formalizing the design process to choose appropriate sensors and algorithms. More research is needed to bring machine learning algorithms to low computational power platforms.

References

1. Coyle, E.F.: Fluid and fuel intake during exercise. Journal of Sports Sciences 22, 39–55 (2004)
2. Chiu, M., Chang, S., Chang, Y., Chu, H., Chen, C.C., Hsiao, F., Ko, J.: Playful bottle: a mobile social persuasion system to motivate healthy water intake. In: Ubicomp, pp. 185–194. ACM, New York (2009)
3. Ravi, N., Nikhil, D., Mysore, P., Littman, M.L.: Activity recognition from accelerometer data. In: IAAI, pp. 1541–1546. AAAI Press (2005)
4. U.S. Standard Atmosphere. U.S. Government Printing Office, Washington, D.C. (1976)

5. Cook, C.J.: High Altitude Hydration (2008),
   http://ezinearticles.com/?High-Altitude-Hydration$&$id=1216328
6. Steadman, R.G.: The Assessment of Sultriness. Part II: Effects of Wind, Extra Radiation and Barometric Pressure on Apparent Temperature. Journal of Applied Meteorology 18(7), 874–885 (1979)
7. Lester, J., Tan, D., Patel, S., Brush, A.J.B.: Automatic Classification of Daily Fluid Intake. In: International Conference on Pervasive Computing Technologies for Healthcare (2010)
8. Sundaram, S., Cuevas, W.W.: High level activity recognition using low resolution wearable vision. In: IEEE CVPR (2009)
9. Subramanya, A., Raj, A.: Recognizing activities and spatial context using wearable sensors. In: UAI (2006)
10. Klasnja, P., Harrison, B.L., LeGrand, L., LaMarca, A., Froehlich, J., Hudson, S.E.: Using wearable sensors and real time inference to understand human recall of routine activities. In: UbiComp, pp. 154–163. ACM, New York (2008)
11. Welbourne, E., Lester, J., LaMarca, A., Borriello, G.: Mobile Context Inference Using Low-Cost Sensors. In: Strang, T., Linnhoff-Popien, C. (eds.) LoCA 2005. LNCS, vol. 3479, pp. 254–263. Springer, Heidelberg (2005)

Magic Wand: A Framework for Developing Remote Controlled Web Applications

Vibhor Nanavati

Carnegie Mellon University, Silicon Valley
vibhor.nanavati@sv.cmu.edu

Abstract. This poster describes the MagicWand framework for developing web applications that can be controlled remotely by smart phone running on android OS.

Keywords: Human Computer Interfaces, Android, Smart Controllers, Mobile Applications, Web Applications.

1 Introduction

With the massive adoption of smart phones since 2007, the expectations of user interfaces have dramatically increased. Users are now comfortable with two-finger gestures such as pinch scaling or gravity based mobile applications. The landscape of desktop applications and web applications in particular, has largely remained unchanged. In particular, users have seen a dramatic change in the kinds of web applications they can interact with in the last decade. But the modes of interaction with these applications are still confined to conventional interfaces. The poster illustrates a revolutionary idea that can take the human computer interaction to a next level.

2 Background

The idea of using smart phone as a conventional mouse and/or keyboard has been shown in past. AirMouse [7] on iPhone allows users to control the mouse cursor on desktop over local Wi-Fi network using the touch screen. Remote-Droid [10], an open source framework allows similar user experience on Android devices. If we look beyond the conventional interfaces, Nintendo Wii [8] consoles provide a motion controlled gaming experience. Onomy Tilty Table [9] demonstrates a version of Google Earth [2] that can be navigated using a large tilting and rotating table. Google Earth for smart phones[3] itself presents a unique experience for user where one can change the viewing camera angle by simply rotating the phone around any of the 3 axes.

3 Design Goals and Constraints

An ideal solution for a smart controller of web applications shall address following goals:

- Fast response time on motion controlled gestures.
- Install once, use anytime and anywhere.
- Make multiple interface devices redundant.
- Allow web developers to define the interactions.

Due to the given environment, the following roadblocks have to be overcome:

- Only channel of communication in browser is http.[1]
- No Bluetooth API implemented by any major browser vendor.
- Browser does not allow listening on a socket.
- Browser connections to web application and smart phone are subject to cross-site scripting restrictions.

4 Solution

4.1 Framework

MagicWand framework comes bundled with a mobile application for Android [1] powered devices that provides the end user controller behavior, and a JavaScript library that can be included in the target web application. The JavaScript library hides the MagicWand protocol and provides convenient routines for a web application to connect/disconnect with an Android device and send messages to it. Messages are passed over HTTP in JSON format [6]. The application in browser can send configuration messages to control the UI and layout of MagicWand app in mobile. The mobile app sends the sensor data periodically. The period is determined by the sampling frequency which is also configurable. Apart from sending configuration requests, browser can also send a text-to-speech request or request the device to provide a haptic (vibration) feedback. The framework requires both the Android device and the desktop which runs the web application to be on the same local Wi-Fi network.

4.2 Architecture

In order to circumvent the limitations of working within the browsers, following design choices were made:

- Android application opens a listening TCP socket on start up. Web application that includes the MagicWand JavaScript makes a request to connect.

---

[1] With HTML5, browsers can also open websocket connections to websocket-enabled servers.

- The sensor data transfer takes place on a dedicated channel. Any configuration requests, haptic feedback or text-to-speech requests are made on a secondary channel.
- The communication with the phone takes place in a browser Iframe. A web application on different domain runs in the parent frame. The data between 2 frames is exchanged using HTML5 messages [5].
- In order to keep the response time low, the android application keeps the connection with application persistent[2], instead of browser making a new HTTP request for every sensor poll.

5 Results

To demonstrate this framework, I have created tiltymap.heroku.com. It uses Google Maps API [4] to support various mapping functionality. This sample application demonstrates how the MagicWand JavaScript is used in conjunction with the magic wand Android app. User of this application can use the phone to control the map displaying on a larger screen such as desktop. Some of the key interactions include gravity based map movement; voice-command enabled searching of locations and pinch scaling of the map.

6 Future Work

MagicWand framework can be extended to applications beyond browsers. A Java based http client library is under development and shall expose an API similar to its JavaScript equivalent. Another interesting application of the framework that will involve work in future is allowing applications to be controlled by more than one Android device.

NOTE: The project is in process of being open-sourced. The project page is hosted at http://code.google.com/p/magicwand/

Acknowledgements. My sincere thanks to: Ted Selker, faculty at CMU Silicon Valley for inspiring a project of this impact, Martin Griss, Director CMU Silicon Valley for proposing several possible applications of the framework and Neha Mohan, my wife for constantly contributing with valuable user feedback.

References

[1] Android, http://developer.android.com/index.html
[2] Google earth, http://earth.google.com/
[3] Google earth for mobile, http://www.google.com/mobile/earth/

---

[2] Using HTTP response header transfer-encoding: chunked, a server can keep a long-lived connection.

[4] Google Maps API,
    http://code.google.com/apis/maps/documentation/javascript/
[5] Html5 inter-frame communication,
    https://developer.mozilla.org/en/DOM/window.postMessage
[6] JSON, http://en.wikipedia.org/wiki/JSON
[7] Mobile AirMouse, http://www.mobileairmouse.com
[8] Nintendo Wii, http://www.nintendo.com/wii/console
[9] Onomy Tilty Table, http://www.onomy.com/video/Tilty.mov
[10] Remote-Droid, http://www.remotedroid.net

Bringing the Cloud Down to Earth: Transient PCs Everywhere

Mahadev Satyanarayanan[1], Stephen Smaldone[2], Benjamin Gilbert[1], Jan Harkes[1], and Liviu Iftode[2]

[1] Carnegie Mellon University
[2] Rutgers University

Abstract. The convergence of cloud computing with mobile computing opens the door to the creation of new applications and services that can be delivered to users at any time and any place. At the heart of this convergence lies a delicate balance between centralization and decentralization. We explore the forces underlying this balance, and examine the role of virtual machine (VM) technology. We observe that a VM-based model of cloud computing called a *Transient PC* offers an approach to "carry-nothing" mobile computing that harnesses the full power of local hardware at the edges of the Internet. In particular, we show how a zero-install Transient PC implementation can safely use local storage.

1 The Roots of Cloud Computing

The intersection of cloud computing with mobile computing is a hot topic today. There is a growing sense that at their nexus lies immense potential for the creation of new applications and services that can be delivered to users at any time and any place. Large dollar signs dance before the eyes of investors and entrepreneurs. What is the basis for this excitement? What are the aspects of this technological convergence that are truly new, and what is merely old wine in new bottles? To fully understand the tectonic forces that are driving mobile computing and cloud computing towards each other, it is important to briefly revisit the past.

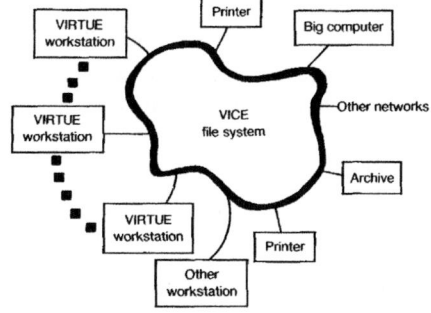

(a) Original Image   (b) Original Caption

Fig. 1. Cloud Computing *circa* 1986 (Source: Morris et al, 1986 [1])

At the heart of this convergence lies a fundamental tension between *autonomy* and *interdependence* in distributed systems. This can equally well be characterized as a tension between *centralization* and *decentralization*. On the one hand, mobility and decentralization are all about freedom: the ability to do anything, anywhere, anytime in a completely unconstrained and untethered manner. On the other hand, total isolation is rarely desired by a user. Communication across space and time is essential for collaboration and knowledge sharing across users. Even with respect to a single user, there is often a need to access personal information resources (such as email or files) from the past and across many different mobile or static devices. Unless carefully constrained, the totality of these interrelationships can be overwhelming to a user. System-level mechanisms such as distributed file systems, databases, email clients and hypertext systems as well as design primitives such as cache consistency protocols, content-addressable storage, and atomic transactions can be viewed as order-preserving tools that try to simplify the complexity of information space-time for a user. Centralization of storage in a cloud can be also viewed as an order-preserving tool. Centralization is simpler, less expensive and more orderly than decentralization from the viewpoint of system management.

How to preserve order and hence reduce human-visible complexity (both for users and system administrators), while minimally constraining mobility and flexibility (in their broadest sense) across time, space and multiple computing devices is a fundamental challenge. Finding the "sweet spots" in this tradeoff space has challenged the system designers for over a quarter of a century.

One of the earliest efforts to reconcile this tension took place at the dawn of personal computing in the early 1980s. The goal of this effort, called *the Andrew project,* was to create a system in which users enjoyed the load-invariant, high-quality, feature-rich interactive environment of personal computing while preserving the ease of information sharing that had emerged as a valuable side effect of timesharing. Contemporary accounts of this project [1,2] discuss the design considerations that led to its architecture. The image and caption shown in Figure 1 are reproduced from one of those accounts. It is interesting to note the similarity to modern cloud architectures, even though the term "cloud computing" was nearly 20 years in the future. An Andrew user did not know or care where his data was stored: it was just "somewhere in the cloud" and magically appeared at his current location when it was accessed. An Andrew system administrator could move data across servers for load balancing or add new servers without disrupting users. We associate these attributes of mobility and serviceability with cloud computing today. Usage experience in Andrew revealed the importance of these attributes as early as 1990, as shown by the quotation in Figure 2.

User mobility is supported: A user can walk to any workstation in the system and access any file in the shared name space. A user's workstation is personal only in the sense that he owns it.

System administration is easier: Operations staff can focus on the relatively small number of servers, ignoring the more numerous and physically dispersed clients. Adding a new workstation involves merely connecting it to the network and assigning it an address.

Fig. 2. Cloud-like Attributes *circa* 1990 (Source: Satyanarayanan, 1990 [3])

2 VM-Based Cloud Computing Today

Fast forward to the present day. The buzz around cloud computing directly reflects renewed interest in the attributes shown in Figure 2, in the hope of finding a new "sweet spot" in the centralization-decentralization tradeoff space that reflects current economic reality. As hardware costs continue to plummet, the people cost of system adiminstration looms ever larger. The attention now being paid to the metric called *total cost of ownership* reflects this rise in people cost relative to hardware cost. There is pressure towards centralization because it tends to lower system administration costs.

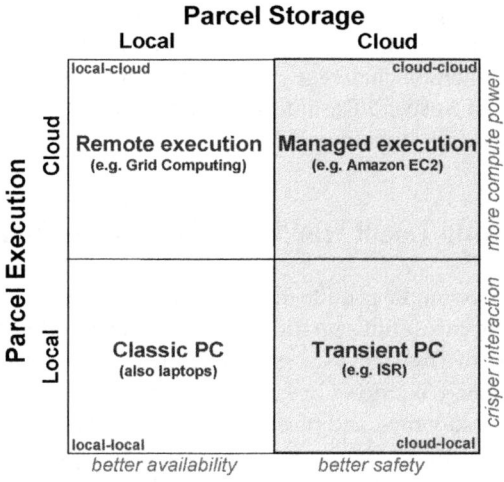

Fig. 3. Taxonomy of VM-based Cloud Computing

Virtual machine (VM) technology leads to new "sweet spots" in the tradeoff space that provide greater centralization without loss of user mobility. Figure 3 identifies these "sweet spots" as quadrants of a two-dimensional space. The vendor-neutral and format-independent term *parcel* in this figure refers to encapsulated VM state. A parcel may contain both volatile state (a memory image) and persistent state (a disk image), or just persistent state. The "Parcel Storage" dimension shows where parcels are stored when not in active use; the "Parcel Execution" dimension shows where they run. The top right quadrant, labeled *Managed Execution,* contains approaches such as Amazon's EC2 [4], IBM's Research Compute Cloud (RC2) [5], and the open source Eucalyptus infrastructure [6], where parcel storage and execution both happen within the cloud. The top left quadrant corresponds to *VM-based grid computing,* where a locally-created parcel is remotely executed. The bottom left quadrant, labeled *Classic PC,* is the degenerate case of a standalone PC, laptop or mobile device.

The bottom right quadrant of Figure 3, labeled *Transient PC,* is of particular interest to mobile computing because it makes possible a "carry-nothing" model of mobility. In this case, parcels are stored in the cloud but execute on a computer that is close to the user. That computer could, of course, be a mobile device. Alternatively, mobile computing can be realized by the user taking advantage of whatever hardware is nearby and letting the system magically deliver a relevant parcel upon use. The latter approach directly corresponds to the Andrew architecture of Figure 1, with the important difference that the granularity of location-transparent access is now an entire parcel rather than user files. In other words, the Transient PC model improves upon the Andrew model by ensuring that the operating system, applications and user files are all delivered from the cloud as a cohesive unit. The technique of *hoarding,* originally developed in the context of a distributed file system [7], can be also applied to parcel state. This enables the Transient PC model to support disconnected operation, a critical capability at the intersection of cloud computing and mobile computing.

There are many technical challenges in efficiently implementing the Transient PC model, the most obvious of these being efficient handling of large parcel size. Fortunately, these implementation challenges can be overcome, as shown by our work in the context of the Internet Suspend/Resume® system (ISR) [8,9] and by related work in the Collective [10,11] and MokaFive [12].

3 Fully Exploiting Local Hardware Resources in Transient PCs

The Transient PC usage model is quite different from the ubiquitous email, Web access, and social networking capabilities provided by BlackBerries, iPhones, and other mobile devices. The strength of Transient PC systems lies in their ability to precisely, safely and rapidly re-create a user's Windows or Linux desktop environment as a thick client on borrowed hardware at any time and place. When a user chooses to use a Transient PC, he or she implicitly confirms the importance of its strengths over the mobile device alternatives mentioned above. It is therefore critical to take full advantage of the local hardware resources of a Transient PC in order to provide a satisfactory user experience.

The most important property of these local resources is their proximity to the user. For interactive applications, executing the application in a distant cloud and viewing the output locally (as would be the case with the top right quadrant of Figure 3) results in an unsatisfactory user experience. The long WAN latencies to the cloud hurt the crisp interaction that is so critical for smooth and non-disruptive interactions. Humans are acutely sensitive to delay and jitter, and it is very difficult to control these parameters at WAN scale. The work by Lagar-Cavilla et al [13] has shown that latency can negatively impact interactive response even when bandwidth is adequate. Networks with high delay-bandwidth products are therefore disastrous for remote execution of interactive applications; in contrast, local execution on a Transient PC does not suffer from this problem. This is particularly true for graphics-intensive applications such as scientific visualization and games, which can further take advantage of graphics acceleration provided by local hardware. Fitting such applications into the Transient PC model is predicated on the ability to virtualize graphics hardware. This was a questionable assumption for many years because of the lack of standardization in such hardware, but VMM-independent virtualization of graphics hardware is now feasible [14].

Another performance-critical resource is the local disk of a Transient PC. To achieve truly ubiquitous availability of Transient PCs, a zero-install solution would be ideal.

Table 1. Portable Storage Device Characteristics

Storage Device	Label	Type	Size (GB)	Speed (RPM)	Transfer Rate (MB/sec) Read	Write
PNY Attache USB Flash Drive	SanDrive	USB	16	Flash	30.51 (1.01)	6.65 (0.29)
SanDisk MicroSD Card	MSD	USB	8	Flash	16.03 (0.23)	11.9 (0.2)
Apple iPod	IPOD	USB	20	4200	12.63 (0.18)	12.38 (0.17)
Internal SATA Drive	TransPart	SATA	250	7200	41.78 (2.48)	32.05 (1.14)

Transfer rate results are the mean of 5 measurements. Standard deviations are in parentheses.

This would obviate the need to require pre-installed software on the Transient PC. Instead, the user boots the hardware from a USB storage device to establish the Transient PC environment, and then accesses his cloud-based parcel. The user can thus transform any available hardware with an Internet connection into a Transient PC. Although conceptually simple, this approach runs into a serious performance obstacle. Portable storage devices sacrifice I/O performance in order to obtain the highest capacity and robustness at the lowest cost, size, and weight. As Table 1 shows, the I/O read and write performance of typical USB-attached storage devices is substantially slower that of an internal disk. This severely impacts operating system performance, including basic functionality such as swapping and application launch. Upgrading the interconnect to USB 3.0 will not eliminate this problem, since it is due to internal storage limitations.

To solve this problem, we have developed a mechanism called *TransPart* that constructs a transient virtual disk out of the free disk blocks of local hardware. This mechanism requires no modifications to the software or hardware of a borrowed computer, and thus preserves the zero-install attribute. TransPart currently supports discovery of free disk blocks from ext2/3/4 and NTFS file systems as well as Linux swap partitions.

Figure 4 illustrates the operation of TransPart. During the host boot process from a USB device, TransPart constructs the virtual disk in two phases. In the first phase, TransPart enumerates local devices and then discovers individual storage volumes stored

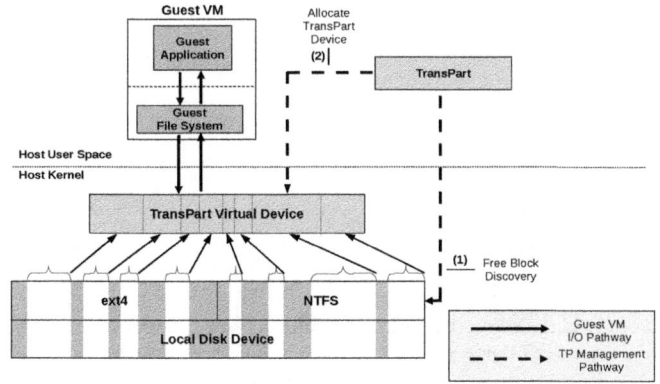

Fig. 4. TransPart Implementation

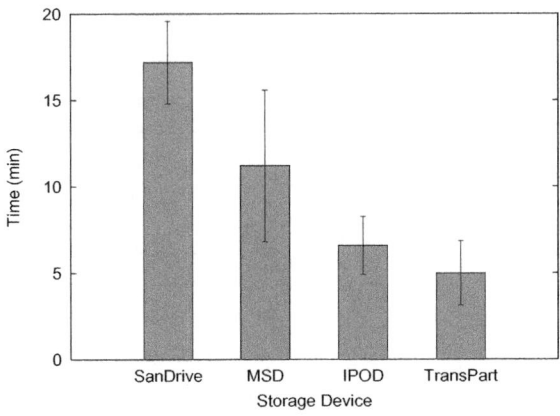

Fig. 5. Completion time of the Postmark v1.51 benchmark (minutes)

on these devices. Such volumes include physical disk partitions as well as aggregate volumes such as software RAID and Logical Volume Manager volumes. Each storage volume usually contains a single file system or swap partition. In the second phase, TransPart searches through each storage volume to discover the available free blocks. Most modern file systems maintain a set of block allocation tables as meta-data on disk. TransPart utilizes file system on-disk semantics to properly parse the file system metadata and to discover free disk blocks. Once TransPart has discovered free disk blocks, it allocates a TransPart device for the guest VM. An ext4 filesystem is created inside the TransPart device, and this filesystem is used to store parcel data demand-fetched from the cloud. All data written to the TransPart device is encrypted, so that a user never leaves behind residual state that could compromise his privacy.

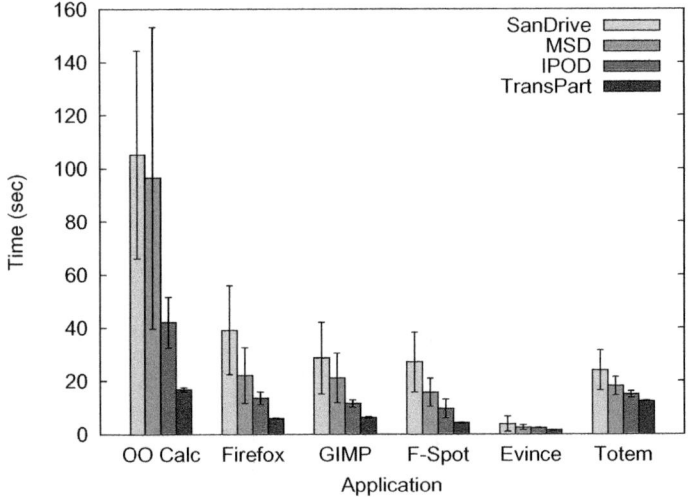

Fig. 6. Application launch latency time for 6 common desktop applications (seconds)

Our experiments show that TransPart can significantly improve the I/O performance of a Transient PC. For example, Figure 5 shows the performance of the well-known Postmark benchmark [15] on typical desktop hardware. The graph shows that TransPart offers significant performance improvement over the portable storage alternatives shown in Table 1. Figure 6 shows launch latency for a number of common Linux applications on the same Transient PC. Once again, the use of TransPart provides a significant improvement in user-visible performance. Experimental details of the results shown in Figures 5 and 6, as well as additional implementation details on TransPart are available in related technical reports [16,17].

4 Closing Thoughts

The convergence of cloud computing and mobile computing is the latest chapter in a long-running dialectic between centralization and decentralization in system design. Over the past 50 years, the pendulum has swung back and forth between these extremes. Today, cloud computing represents a thrust in which the forces of centralization are ascendant. Mobile computing, on the other hand, represents a thrust in which the forces of decentralization dominate. Where these tectonic forces meet, there will inevitably be a lot of heat generated and, hopefully, also some light. In this paper, we have put forth the view that use of virtual machine technology can lead to new "sweet spots" in the space of system architectures that try to reconcile the tradeoffs between centralization and decentralization. Notably, it yields the Transient PC computing model which preserves the centralization benefits of cloud computing without sacrificing mobility or usability.

Acknowledgements. This research was supported by the National Science Foundation (NSF) under grant numbers CNS-0509004, CNS-0833882 and CNS-0831268, and an IBM Open Collaborative Research grant. Any opinions, findings, conclusions or recommendations expressed in this material are those of the authors and do not necessarily represent the views of the NSF, IBM, Rutgers University, or Carnegie Mellon University.

References

1. Morris, J.H., Satyanarayanan, M., Conner, M.H., Howard, J.H., Rosenthal, D.S., Smith, F.D.: Andrew: a Distributed Personal Computing Environment. Commun. ACM 29(3), 184–201 (1986)
2. Satyanarayanan, M., Howard, J.H., Nichols, D.A., Sidebotham, R.N., Spector, A.Z., West, M.J.: The ITC Distributed File System: Principles and Design. In: Proceedings of the 10th ACM Symposium on Operating Systems Principles, Orcas Island, WA (December 1985)
3. Satyanarayanan, M.: Scalable, Secure, and Highly Available Distributed File Access. IEEE Computer 23(5), 9–18, 20–21 (1990)
4. Amazon Inc.: Amazon Web Services - Elastic Compute Cloud, http://aws.amazon.com/ec2/
5. Ammons, G., Bala, V., Berger, S., Da Silva, D.M., Doran, J., Franco, F., Karve, A., Lee, H., Lindeman, J.A., Mohindra, A., Oesterlin, B., Pacifici, G., Pendarakis, D., Reimer, D., Ryu, K.D., Sabath, M., Zhang, X.: RC2: A living lab for cloud computing. IBM Research Report RC24947, IBM (2010)

6. Nurmi, D., Wolski, R., Grzegorczyk, C., Obertelli, G., Soman, S., Youseff, L., Zagorodnov, D.: The Eucalyptus Open-Source Cloud-Computing System. In: CCGRID 2009: Proceedings of the 2009 9th IEEE/ACM International Symposium on Cluster Computing and the Grid (2009)
7. Kistler, J.J., Satyanarayanan, M.: Disconnected Operation in the Coda File System. ACM Transactions on Computer Systems 10(1) (February 1992)
8. Kozuch, M., Satyanarayanan, M.: Internet Suspend/Resume. In: Proceedings of the Fourth IEEE Workshop on Mobile Computing Systems and Applications, Callicoon, NY (June 2002)
9. Satyanarayanan, M., Gilbert, B., Toups, M., Tolia, N., Surie, A., O'Hallaron, D.R., Wolbach, A., Harkes, J., Perrig, A., Farber, D.J., Kozuch, M.A., Helfrich, C.J., Nath, P., Lagar-Cavilla, H.A.: Pervasive Personal Computing in an Internet Suspend/Resume System. IEEE Internet Computing 11(2) (2007)
10. Sapuntzakis, C., Chandra, R., Pfaff, B., Chow, J., Lam, M., Rosenblum, M.: Optimizing the Migration of Virtual Computers. In: Proceedings of the 5th Symposium on Operating Systems Design and Implementation, Boston, MA (December 2002)
11. Chandra, R., Zeldovich, N., Sapuntzakis, C., Lam, M.: The Collective: A Cache-Based System Management Architecture. In: Proceedings of the Second Symposium on Networked Systems Design and Implementation (May 2005)
12. MokaFive, Inc.: MokaFive Home Page, http://www.mokafive.com
13. Lagar-Cavilla, H.A., Tolia, N., de Lara, E., Satyanarayanan, M., O'Hallaron, D.R.: Interactive Resource-Intensive Applications Made Easy. In: Cerqueira, R., Pasquale, F. (eds.) Middleware 2007. LNCS, vol. 4834, pp. 143–163. Springer, Heidelberg (2007)
14. Lagar-Cavilla, H.A., Tolia, N., Satyanarayanan, M., de Lara, E.: VMM-Independent Graphics Acceleration. In: Proceedings of the 3rd ACM SIGPLAN/SIGOPS Conference on Virtual Execution Environments (VEE), San Diego, CA (2007)
15. Katcher, J.: PostMark: A New File System Benchmark. Technical Report TR3022, Network Appliance (1997)
16. Smaldone, S., Harkes, J., Iftode, L., Satyanaryanan, M.: Safe Transient Use of Local Storage for VM-based Mobility. Technical Report CMU-CS-10-110, School of Computer Science, Carnegie Mellon University (2010)
17. Gilbert, B., Goode, A., Satyanarayanan, M.: Pocket ISR: Virtual Machines Anywhere. Technical Report CMU-CS-10-112, School of Computer Science, Carnegie Mellon University (2010)

VStore++: Virtual Storage Services for Mobile Devices

Sudarsun Kannan, Karishma Babu, Ada Gavrilovska, and Karsten Schwan

Center for Experimental Research in Computer Systems
Georgia Institute of Technology
{sudarsun,karishma,ada,schwan}@cc.gatech.edu

Abstract. This paper addresses media sharing via an approach that offers 'fungible' storage, where storage services implement virtual stores that are dynamically mapped to suitable 'nearby' or otherwise available physical devices. In particular, the novel VStore++ system provides seamless and flexible data storage, access, and sharing services, by exploiting virtualization technology to aggregate and make use of both 'nearby' and private storage (e.g., in a mobile user's home), and public storage resources offered on remote cloud platforms.

Keywords: mobile virtualization, cloud computing.

1 Introduction

Mobile devices with their increased CPU speeds, core counts, memory sizes, and improved communication rates may well become the next generation personal computers. To meet the resulting increased end user demands for rich and diverse types of services on these platforms, however, industry must address constraints that include issues with battery life, processing capabilities dwarfed by those of server systems, limited storage, smaller display form factors, and others. In response, our research is exploiting the fact that mobile devices are often surrounded by and used in contexts where there are many other resources that could enhance their capabilities. Consider, for instance, the enormous aggregate processing power and storage capacity available in say, a soccer stadium in the forms of other spectators' devices, the server systems supporting broadcast and organizational functions, and devices engaged in ancillary tasks like security. Another example are users' homes where there may be home PCs, laptops, and computerized home entertainment systems. Further, often associated with such resources is locally captured state like home videos, security images, or the context information needed to distinguish important from less important content.

Locally available resources and context suggest solutions that use distributed, multi-device service implementations. We formulate the following simple principles for mobile service realization and delivery:

- *Fungibility for dynamic flexibility*: physical resources should be 'fungible', so as to create dynamic options in the mappings from the resources applications believe they are using – virtual resources – to the physical resources actually being used.
- *Explicit cooperation*: devices must agree to participate in service delivery, creating sets of cooperative devices operating in common domains (e.g., a user's home).
- *Guided active management*: since the 'best' mappings of virtual to physical resources depend on current context, user needs, and resource availabilities, active management of these mappings must have continuous inputs from methods that monitor these factors.
- *Automation and independence*: guided management should not require end user participation, i.e., it should be automated, and in addition, management should be independent of specific operating systems or application frameworks being present on mobile devices.
- *Universal operation*: managed services should function wherever mobile devices are used, which implies that there must also be ways to store and access global state across disconnected periods of operation. Access to Internet-based services, therefore, is a critical element of any solution for fungible mobile services.

Current solutions that 'simply use the Internet' are insufficient. Although they clearly enhance mobile devices via remote (e.g., for storage, DropBox, Gmail, etc.), they are lacking in terms of independence and universal operation. This is because of (1) disconnected operation – where sometimes, mobile devices will not be able to access Internet services like those offered by public cloud infrastructures,(2) undue communication overheads or costs – referring to the performance (or lack thereof) or the expense of reaching Internet resources in comparison to the lower costs of using local resources and exploiting locally available state, and (3) data privacy or security, for which it may be preferable to use and maintain local resources that cannot be accessed by others.

The VStore++ service described in this paper provides efficient, fungible storage for mobile devices. This is done by exploiting distributed storage via a 'cooperative' model in which devices choose to participate in joint data storage and access. Interactions may take place across wireless networks, as in the aforementioned case of the soccer stadium, across the Internet, when using Internet-based resources like cloud storage, or across wired links when the mobile device operates in a user's home. The outcome is a 'personal mobile cloud' comprised of a dynamically varying set of interacting devices that cooperate to provide end users with seamless storage services.

The implementation of VStore++ attains fungibility and independence by operating at the virtualization level, where it can use local disks, remote machines' stores, or even Internet-connected storage in ways that are transparent to and independent of end users, application frameworks, and even the operating systems running on mobile devices. At the same time, since the context in which the mobile device operates will change dynamically, as will end user requirements,

VStore++ will track resource availability in order to direct requests to whichever resource is currently accessible, using a global index maintained during its operation. This is done in ways that maintain user-defined data access controls. Further, VStore++ as a service can be used wherever there is connectivity to participating devices and using whichever connectivity methods are currently available, but automation in terms of making such choices or determining suitable storage targets remains subject of our future work. The outcome is a storage service accessible from a wide range of platforms, independent and potentially decoupled from their inherent constraints.

VStore++ has been implemented on Atom-based machines that are virtualized with the Xen open source hypervisor. Its evaluation uses a prototypical 'home' setup in which there is cooperation among mobile devices, tethered home machines like PCs, and remote services like Amazon's EC2 storage. Our ongoing work is exploring mobile services other than storage, is developing methods for automatically guided management, and is improving the ubiquity by inter-operating by use of other service delivery means, such via cable or satellite TV-based connections. Performance results reported in this paper clearly demonstrate the utility of fungibility, showing much improved performance when storage resources are local vs. in public clouds and showing advantages with local aggregation when there is substantial local state or when there are requirements for fast local response (as with home security systems, for instance).

2 VStore++ Architecture and Implementation

To provide end-users with mobile devices seamless access to state stored on diverse locally present storage resources, as well as remote, publicly available compute and storage cloud platforms, VStore++ must implement transparent enforcement of varying sharing policies on objects, and search for content belonging to other domains, both local and remote, via standard content access interfaces (e.g., the file system interface) In addition, there must be interfaces for dynamically associating with data accesses additional functionalities such as trust management, access control, location attributes, and methods for data manipulation or customization. VStore++ attains these implementation goals using as its service interface that of an object-based file system, virtualized for use by guest domains. VStore++'s basic object API is enhanced with additional object-level metadata (e.g., privacy attributes) and with 'activity specifications' that make it possible to associate data manipulation functions with object accesses. VStore++ is implemented in a trusted service domain, which is 'dom0' in its Xen-based prototype.

The main components of VStore++, illustrated in Figure 1, are:

1. *Virtualized object-based file system*: VStore++ is a virtualized storage service exposing an object-based file system interface. Internally, it uses a standard file system to represent objects, using a one-to-one mapping of objects to files. The current implementation is based on the PVFS object-based file system,

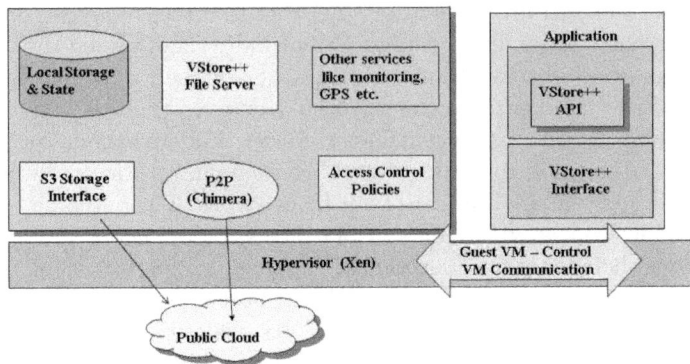

Fig. 1. VStore++ Architecture

but alternative, more lightweight implementations are currently under consideration (e.g., based on SyncFS).

2. Peer-to-peer service: this layer provides the facilities for establishing peer-to-peer overlays, for object searches, for request routing and for dynamic overlay management (e.g., participant discovery, etc.) in ad hoc mobile environments. The current implementation is based on the lightweight Chimera peer-to-peer overlay system. In order to support robust and efficient tag-based routing services, as needed by VStore++ to enable discovery of and access to remotely stored tagged data, a scalable DHT layered on top of Chimera's key based routing service provides file system-like semantics, i.e., insert and retrieve operations. The DHT layer also provides additional features, including caching, replication, node failure detection, and handling of new node arrivals.

3. Enhancement services: additional services associated with VStore++, at runtime, can provide functionality that includes (i) run-time trust mechanisms [2], (ii) methods that protect data via data manipulation functions that range from simple read/write permissions to content-dependent processing like watermarking for images, etc., or (iii) location mechanisms that operate with or without GPS hardware support, by using approximate GPS coordinates obtained via services such as Georgia Tech's WhereAmI or mechanisms such as WiFi triangulations, marker recognition, or others.

4. Interface to public/remote clouds: in order to seamlessly enable access to state or services available on 'nearby' vs. remote resources, such as those present in current public cloud platforms like Amazon's EC2, VStore++ transparently integrates corresponding cloud client components. The same metadata services which keep track of location of 'nearby' content, are used to encode operations and accompanying attributes needed for access to the remote cloud storage.

3 Experimental Evaluation

For brevity, this section outlines only some of the experimental results of our research, with the dual goals of illustrating the feasibility of supporting virtualized

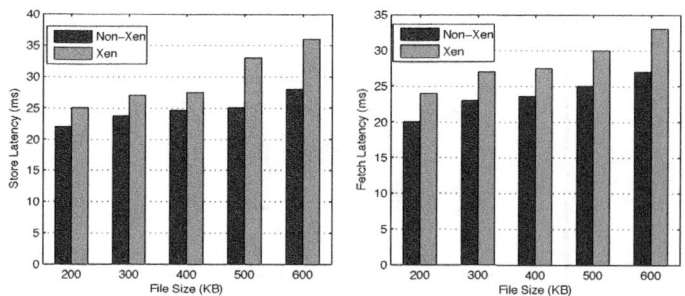

Fig. 2. VStore++ virtualization overheads

file sharing solutions such as VStore++ on mobile platforms and demonstrating the potential advantages of such a solution compared to pure 'Internet'-based methods. Experiments are conducted using the current VStore++ implementation on Xen-3.0.4 as the virtualization platform, with PVFS-2.6.3 as the object-based storage service, and Chimera 1.20 for DHT and peer-to-peer communication services. The experimental testbed consists of dual-core 1.66GHz Intel Atom N280 netbooks and multiple 3 GHZ 64-bit core-duo laptops, running Linux 2.6.16 on Xen. Access to Amazon S3 cloud storage is via a home-based broadband link, to better emulate future wireless bandwidths.

To determine the overheads due to virtualization, we compare the virtualized VStore++ implementation with one that runs on a non-virtualized system, where the client directly interacts with the 'back-end' server. The graphs in Figure 2 show that the use of Xen contributes from 4.5% to 15.74% to the data access latencies of VStore++'s operations. In large part, these are due to our use of TCP sockets for VM-dom0 communication. It is known that such costs can be reduced substantially by using shared memory based VM-VM communication mechanisms already available in the open source community. At the same time, the moderate costs of this un-optimized implementation enable functionality and flexibility that is otherwise not easily attainable – storage fungibility, independence, and universal operation, as well as the ability to transparently extend the data access service with additional functionality, such as location transparency, rich access control policies, useful data manipulations (e.g., for privacy protection), etc.

Additional experiments evaluate the impact of virtualization on sustainable throughput rates, with results indicating that for smaller file sizes (200 KB) virtualization impacts performance by no more than 10%. As file sizes increase, particularly as the distributed store becomes more full, overheads increase. Such costs are accompanied, however, by substantially increased total storage capacity and flexibility. Overheads can be reduced through additional optimizations of Xen for small form factor devices, and by using additional hardware-level virtualization support available on these (Atom-based) devices.

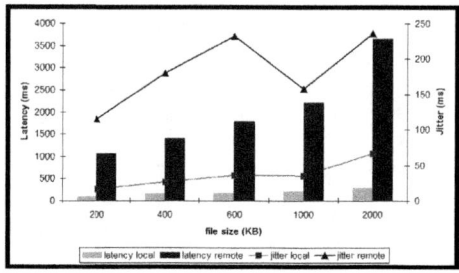

Fig. 3. Benefits of service delivery via local vs. remote entities

Finally, to demonstrate the importance of enabling access to local- vs. remote 'cloud' services, we compare between the latencies and response time variability between accesses to 'nearby' storage (i.e., in our prototypical 'home' environment) vs. to Amazon's EC2 storage service. The measurements in Figure 3 show that access to local storage can not only be seamlessly integrated under the VStore++ interface, but that it also results in superior access properties (e.g., response time and jitter), which may be critical for certain types of services.

4 Conclusions and Future Work

VStore++ is an experimental vehicle for understanding and exploring ways to construct 'personal clouds' comprised of dynamic sets of mobile and stationary computing platforms. By using virtualization technology to access storage service at the object level rather than at the block level, VStore++ can associate useful semantic data with storage objects, such as privacy or access control metadata, and can then use such information to enforce diverse end user requirements. Performance evaluations show that the VStore++ hybrid solution of using both local and remote resources for data storage can be superior to the purely Internet-based service offering now available to mobile platforms, with additional benefits derived from the ability to deal with trust and data privacy concerns.

Our ongoing and future work is exploring two avenues toward further enriching mobile services. One direction of research is considering other services, such as those needed for multimedia delivery and customization, and for universal service operation in lieu of certain dynamic resource deficiencies. Another direction is to automate virtual to physical resource mappings under changing conditions, such as when a mobile device's battery become depleted.

References

1. Seshasayee, B., Narasimhan, N., Biljani, A., Pai, A., Schwan, K.: VStore - Efficiently Storing Virtualized State Across Mobile Devices. In: MobiVirt 2008 (2008)
2. Kong, J., Schwan, K.: ProtectIt: Trusted Distributed Services Operating on Sensitive Data. In: EuroSys 2008 (2008)

On Economic Mobile Cloud Computing Model

Hongbin Liang[1,3], Dijiang Huang[2], and Daiyuan Peng[1]

[1] School of Information Science and Technology, Southwest Jiaotong University
[2] School of Computing Informatics and Decision Systems Engineering, Arizona State University
[3] Department of Electrical and Computer Engineering, University of Waterloo

Abstract. Cloud has become a promising service model for mobile devices. Using cloud services, mobile devices can outsource its computationally intensive operations to the cloud, such as searching, data mining, and multimedia processing. In this service computing model, how to build an economic service provisioning scheme is critical for mobile cloud service providers. Particularly when the mobile cloud resource is restricted. In this paper, we present an economic mobile cloud computing model using Semi-Markov Decision Process for mobile cloud resource allocation. Our model takes the considerations the cloud computing capacity, the overall cloud system gain, and expenses of mobile users using cloud services. Based on the best of our knowledge, our presented model is the first to address the economic service provisioning for mobile cloud services. In the performance evaluation, we showed that the presented economic mobile cloud computing model can produce the optimal system gain with a given cloud service inter-domain transfer probability.

Keywords: Mobile Cloud Computing, Semi-Markov Decision Process.

1 Introduction

With the development of wireless access technologies such as 3/4G, LTE, and WiMax, mobile devices can gain access to the network core over longer distances and larger bandwidths. This allows for very effective communication between mobile devices and the cloud infrastructure. A new service architecture is necessary to address the requirements of users in their unique operational environment and create new mobile applications. Cloud computing is a new business model focusing on resource-on-demand, pay-as-you-go, and utility-computing [1]. Cloud computing can be broadly classified as infrastructure-as-a-service (IaaS), platform-as-a-service (PaaS), and software-as-a-service (SaaS). Critical research issues for cloud computing such as computation offloading, remote execution, and dynamic composition have been extensively discussed in previous literature.

Recent research have been focused on cloud computing for mobile devices [4,6, 9]. Cloud computing for mobile devices has a major benefit in that it enables running applications between resource-constrained devices and Internet-based Clouds. Moreover, resource-constrained devices can outsource computation/communication/resource intensive operations to the cloud. CloneCloud [2] focuses on execution augmentation with less consideration on user preference or device status. Samsung has proposed the concept of elastic applications, which can offload components of applications from mobile devices to cloud [10]. We generalize mobile cloud services based on the MobiCloud computing model presented in [3], which is shown in Figure 1. Mobile cloud uses weblets (application components) to link the cloud services and mobile devices. A weblet can be platform independent such as using Java or .Net bytecode or Python script or platform dependent, using a native code. However, its execution location can be run on a mobile device or migrated to the cloud, i.e., run on one or more virtual nodes offered by an IaaS provider. In this way, an elastic application can dynamically augment the capabilities of a mobile device, including computation power, storage, and network bandwidth, based on the devices' status with respect to CPU load, battery power level, network connection quality, security, etc. One or more weblets are running in the Weblet Container (WC).

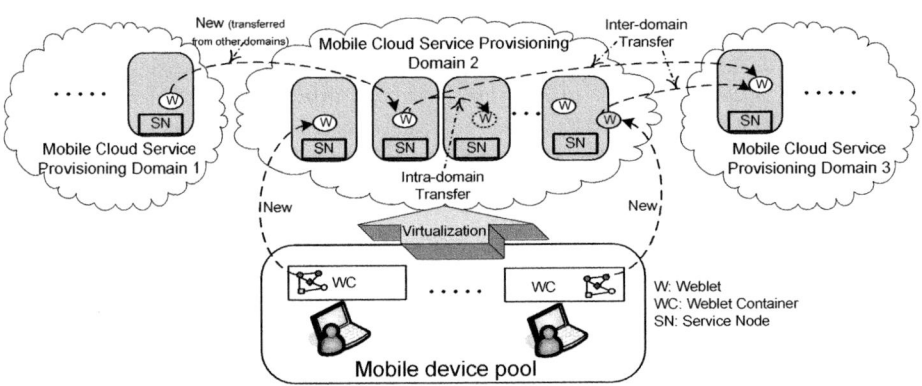

Fig. 1. Reference Model of Mobile Cloud Computing

In the cloud, a service node (SN) is responsible for managing the weblet's loading and unloading in the virtual image. An SN can only handle one weblet from either a new migrated weblet request or a transferred weblet request. Each virtual image has a capacity to hold one weblet at a time. The SN can handle three types of service requests: (i) *New*: a new weblet migration request received from a mobile device or transferred from other mobile cloud service provisioning domains, (ii) *Intra-domain transfer*: an existing weblet transferred from one SN to another within the same mobile cloud service provisioning domain, and (iii) *Inter-domain transfer*: a weblet transferred from current mobile cloud service provisioning domain to another one.

In our presented mobile cloud service model, the cloud can provide a large numbers of virtual images (one virtual image is associated with one CPU, and the CPU can only handle one weblet at a time), however, in reality, the number of virtual images is restricted by the capacity of the cloud hardware configuration. The inter-domain weblet transfer (the third type of service request) means the lose of revenue for the current mobile service provider. As a result, an economic mobile computing model is desired to maximally utilize the cloud resource and achieve the maximum benefit (economic gain) at the same time. The presented economic mobile computing model consists three types of weblet migrations. We differentiate these migrations based on their economic gains, in which a intra-domain weblet transfer migration from another SN usually generates higher economic gain than a new weblet migration from the mobile device or another mobile cloud service provisioning domain, and the inter-domain transfer migration means the lose of revenue. Besides the economic gain, the presented economic mobile computing model also needs to consider the cost due to CPU (or virtual image) occupation. Moreover, the model also needs to consider the trade-offs of the battery consumptions of mobile devices vs. the expenses of using cloud services. Thus, the total economic gain is determined by a comprehensive approach taking all the above mentioned considerations.

In this paper, we present an economic mobile computing model based on Semi-Markov Decision Process (SMDP) model. The contributions of our solutions are in three-fold:

- We firstly apply the Semi-Markov Decision Process to derive the optimal resource allocation policy for mobile cloud computing.
- Our model can take into the considerations both maximizing the system gain of the cloud and reducing the expenses for mobile users.
- Finally, our model can be used to achieve the maximum system gain with a given inter-domain transfer probability constraint.

The rest of this paper is arranged as follows: In Section 2, we present basic system models. In Section 3, the Semi-Markov Decision Process model for mobile cloud computing is presented. We present the inter-domain transfer probability in Section 4. The performance evaluation is presented in Section 5. Finally, we conclude our work in Section 6.

2 System Description

We consider that a mobile cloud consists of two types of nodes, virtual SNs and physical mobile devices (MDs). An MD is a wireless node with limited computing capability and energy supply. An MD can migrate its mobile codes (i.e., weblet) to the cloud. When the cloud receives a migration request, it will decide: wether the SN should accept the request or perform inter-domain transfer based on the consideration if the overall system gain of acceptance.

In the following, we present the system assumptions and states, and the reward model for the mobile cloud computing system.

2.1 System Assumptions

We assume that a service running at an MD or an SN in the cloud costs differently. For simplicity, we also assume the CPU in the cloud is single thread, thus, each weblet in processing occupies one CPU. There are K CPUs in the cloud system. We reserve $K - L$ ($L < K$) CPUs for the intra-domain transfer to ensure that the weblet transfers mostly occur within a mobile cloud service provisioning domain. The distribution for a new weblet migration and an intra-domain transfer weblet migration follows the Poisson distribution with mean rate λ_n and λ_t, respectively. The CPU occupation time of a new weblet and that of an intra-domain transfer weblet in an SN follow exponential distribution with mean rate μ_n and μ_t, respectively.

2.2 System States

The system states can be described based on the service events (including both arrival and leave events) and the service load. In mobile cloud computing system model, we can define three service events: 1) a new weblet request arrives from an MD or another mobile cloud service provisioning domain, denoted by A_n; 2) an intra-domain transfer weblet request arrives from one SN to another within the same mobile cloud service provisioning domain, denoted by A_t; and 3) a weblet leaves current mobile cloud domain, denoted by F. The service load can be represented as the numbers of new weblets and intra-domain transfer weblets in the mobile cloud, which are denoted as s_n and s_t, respectively. Therefore, the system state can be expressed as

$$S = \{\hat{s} | \hat{s} = (s_n, s_t, e)\},$$

where $0 \leq s_n + s_t \leq K$, $0 \leq s_n \leq L$, $L < K$, K is the number of CPUs, $K - L$ is the number of reserved CPUs for the intra-domain transfer weblet migration. Here, L is the maximal number of CPUs for the new weblet migration and $e \in \{A_n, A_t, F\}$.

2.3 Reward Model

For a system state with an incoming weblet migration service request (i.e., A_n or A_t), two actions can be adopted by the mobile cloud: *accept* or *transfer* (without speciall notice, the "transfer" means inter-domain transfer in the rest of this paper). We denote the action to accept the request as $a_{<s,e>} = 1$ and the action to transfer the request as $a_{<s,e>} = 0$, where $s = (s_n, s_t)$ and $e \in \{A_n, A_t\}$. On the other hand, for a system state with a weblet leave, there is no action to be performed and we define the action as $a_{<s,F>} = 0$. Then, the action space is defined as $Act_{\hat{s}}$, where

$$Act_{\hat{s}} = \begin{cases} 0 \text{ (no action)}, & e = F \\ 0 \text{ (transfer)}, & e \in \{A_n, A_t\} \\ 1 \text{ (accept)}, & e \in \{A_n, A_t\}. \end{cases} \quad (1)$$

We also simplify the action as a, where $a \in Act_{\hat{s}}$.

Based on the system state and its corresponding action, one can evaluate the reward to the cloud, which is computed based on the income and the cost as follows:

$$r_{<s,e>} = w_{<s,e>} + g_{<s,e>}, \ e \subseteq \{A_n, A_t, F\}, \tag{2}$$

where $w_{<s,e>}$ is the net lump sum income for the cloud and a mobile device, and it is computed as:

$$w_{<s,e>} = \begin{cases} 0, & a_{<s,e>} = 0, \ e \in \{A_n, A_t, F\} \\ (\alpha_s - \alpha_d) E_n + \gamma_d U_d, & a_{<s,A_n>} = 1 \\ (\alpha_s - \alpha_d) E_t + \gamma_d U_d, & a_{<s,A_t>} = 1. \end{cases} \tag{3}$$

Here, α_s and α_d are weight factors for cloud and mobile device, respectively. They satisfy $0 \leq \alpha_s, \alpha_d \leq 1$ and $\alpha_s + \alpha_d = 1$. E_n and E_t are the incomes of the cloud when it accepts a new weblet migration request from an MD, or an intra-domain transfer weblet migration request from a different SN. Here, U_d represents the income measured by the saved battery energy for the MD when the cloud accepts the weblet migration. γ_d is the weight factor that satisfies $0 \leq \gamma_d \leq 1$.

In (2), $g_{<s,e>}$ denotes the system cost and it is given by:

$$g_{<s,e>} = \tau_{<s,e>} o_{<s,e>}, \ a_{<s,e>} \in Act_{\hat{s}}. \tag{4}$$

In (4), $\tau_{<s,e>}$ is the average service time when the system state transfers from $<s,e>$ to the next potential state; $o_{<s,e>}$ is the cost rate of the service time, and it is defined as

$$o_{<s,e>} = \begin{cases} -f(s_n, s_t), & a_{<s,e>} = 0, \ e \in \{A_n, A_t, F\} \\ -f(s_n + 1, s_t), & a_{<s,A_n>} = 1 \\ -f(s_n, s_t + 1), & a_{<s,A_t>} = 1, \end{cases} \tag{5}$$

where $f(\cdot)$ is a linear function of s_n and s_t.

3 SMDP Based Mobile Computing Model

SMDP known as stochastic dynamic programming can be used to model and solve dynamic decision making problems. The SMDP model has the following elements: *system states, action sets, the events cause the decision, decision epochs, transition probabilities,* and *rewards.* We use standard notations and definitions as defined in [7] for our SMDP-based problem formulation.

Based on the SMDP model, to obtain the maximum long term reward, we need to calculate the transition probabilities between each system state. There are only three events in the cloud (i.e., a new weblet migration request arrival, an intra-domain transfer weblet migration request arrival, and a weblet leave). The next decision epoch occurs when any of the events takes place. T_{A_n} and T_{A_t} denote the time intervals from current state to the next weblet migration event, and T_F denotes the time interval from current state to the next weblet leave event. Then, the next decision epoch T satisfies $T = \min(T_{A_n}, T_{A_t}, T_F)$. T_{A_n}, T_{A_t}, and T_F follow exponential distributions with rate λ_n, λ_t, and $(s_n\mu_n + s_t\mu_t)$, respectively. Thus, T follows exponential distribution with rate $\lambda_n + \lambda_t + s_n\mu_n + s_t\mu_t$. Then, the expected time between current state and a new state can be expressed as:

$$\tau(\hat{s}, a) = \begin{cases} [s_n\mu_n + s_t\mu_t + \lambda_n + \lambda_t + a_{<s,A_n>}\mu_n]^{-1}, & e = A_n \\ [s_n\mu_n + s_t\mu_t + \lambda_n + \lambda_t + a_{<s,A_t>}\mu_t]^{-1}, & e = A_t \\ [s_n\mu_n + s_t\mu_t + \lambda_n + \lambda_t]^{-1}, & e = F. \end{cases} \quad (6)$$

$q(j|\hat{s}, a)$ denotes the state transition probability from the current state $\hat{s}$ to the next state j when action a is chosen. For a states $\hat{s} = <s, e>$ ($e \in \{A_n, A_t, F\}$) with action $a = 0$, $q(j|\hat{s}, a)$ can be obtained as follow:

$$q(j|\hat{s}, a) = \begin{cases} \lambda_n\tau(\hat{s}, a), & j = <s_n, s_t, A_n>, s_n \geq 0, s_t \geq 0 \\ \lambda_t\tau(\hat{s}, a), & j = <s_n, s_t, A_t>, s_n \geq 0, s_t \geq 0 \\ s_n\mu_n\tau(\hat{s}, a), & j = <s_n - 1, s_t, F>, s_n \geq 1, s_t \geq 0 \\ s_t\mu_t\tau(\hat{s}, a), & j = <s_n, s_t - 1, F>, s_n \geq 0, s_t \geq 1. \end{cases} \quad (7)$$

where $0 \leq s_n + s_t \leq K$, $0 \leq s_n \leq L$.

For a states $\hat{s} = <s_n, s_t, e>$ ($e \in \{A_n, A_t\}$) with action $a = 1$, $q(j|\hat{s}, a)$ can be obtained as follow:

$$q(j|\hat{s}, a) = \begin{cases} \lambda_n\tau(\hat{s}, a), & j = <s_n + 1, s_t, A_n>, s_n \geq 0, s_t \geq 0 \\ \lambda_t\tau(\hat{s}, a), & j = <s_n, s_t + 1, A_t>, s_n \geq 0, s_t \geq 0 \\ (s_n + 1)\mu_n\tau(\hat{s}, a), & j = <s_n - 1, s_t, F>, s_n \geq 1, s_t \geq 0 \\ (s_t + 1)\mu_t\tau(\hat{s}, a), & j = <s_n, s_t - 1, F>, s_n \geq 0, s_t \geq 1. \end{cases} \quad (8)$$

where $0 \leq s_n + s_t \leq K$, $0 \leq s_n \leq L$.

Figure 2 shows the state transition probabilities when there exists only one type of weblet migrations in the mobile cloud.

Since the time between two decision epochs can be regarded as exponentially distributed and the expected time between two decision epochs is $\tau(\hat{s}, a)$. Then the distribution of the time between two decision epochs is given as:

$$F(\bar{t}|\hat{s}, a) = 1 - e^{-\tau(\hat{s},a)^{-1}\bar{t}}, \bar{t} \geq 0. \quad (9)$$

Then we have

$$Q(\bar{t}, j|\hat{s}, a) = q(j|\hat{s}, a)F(\bar{t}|\hat{s}, a), \quad (10)$$

where (10) denotes if at a decision epoch the system occupies state $\hat{s} \in S$, after the cloud chooses an action a from the set of $Act_{\hat{s}}$ at state $\hat{s}$. The next decision epoch occurs at or before time $\bar{t}$, and the system state at that decision epoch equals j with probability $Q(\bar{t}, j|\hat{s}, a)$. We use $Q(d\bar{t}, j|\hat{s}, a)$ and $F(d\bar{t}|\hat{s}, a)$ to represent the time-differential.

Then we can get the reward of the system when an event (arrival or leave) occurs. To incorporate the action into the notations, we let $r(\hat{s}, a)$ denote $r_{<s,e>}$, $h(\hat{s}, a)$ denote $h_{<s,e>}$, and $o(\hat{s}, a)$ denote $o_{<s,e>}$. As the system state does not

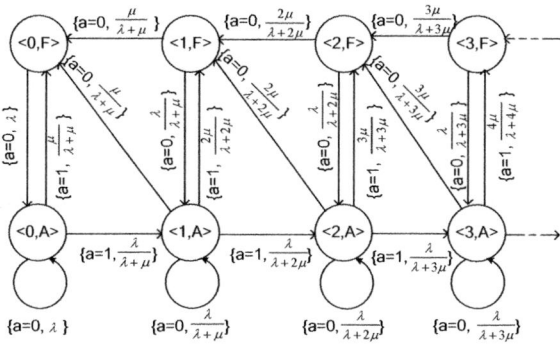

Fig. 2. An example of state transition probabilities for only one type of weblet migrations. The first item in the brackets is the action and the second item in the brackets is the state transition probability.

change between two decision epochs, the expected discounted reward during $\tau(\hat{s}, a)$ satisfies:

$$r(\hat{s}, a) = h(\hat{s}, a) + o(\hat{s}, a) E_{\hat{s}}^a \left\{ \int_0^{\tau_1} e^{-\alpha \bar{t}} d\bar{t} \right\} = h(\hat{s}, a) + o(\hat{s}, a) E_{\hat{s}}^a \left\{ \frac{[1 - e^{-\alpha \tau_1}]}{\alpha} \right\}$$
$$= h(\hat{s}, a) + \frac{o(\hat{s}, a) \tau(\hat{s}, a)}{1 + \alpha \tau(\hat{s}, a)}, \qquad (11)$$

where α is the discounted rate.

Let D denote the class of deterministic Markovian decision rules and d denote each deterministic Markovian decision rule. There exists a stationary deterministic optimal policy denoted by d^∞. At the current decision epoch, state $\hat{s}$ is occupied and the cloud makes the decision to choose action $d(\hat{s}) \in Act_{\hat{s}}$ under the deterministic Markovian decision rule d. Then, when the system occupies state j at the next decision epoch, for each d, $q(j|\hat{s}, d(\hat{s}))$, $r(j|\hat{s}, d(\hat{s}))$, and $\tau(\hat{s}, d(\hat{s}))$ denote the state transition probability, reward, and expected occupation time between two states, respectively, in which they are defined as follows:

$$q_d(j|\hat{s}) = q(j|\hat{s}, d(\hat{s})); \quad r_d(j|\hat{s}) = r(j|\hat{s}, d(\hat{s})); \quad \tau_d(\hat{s}) = \tau(\hat{s}, d(\hat{s})).$$

Thus, under the decision rule d, we define the distribution of the time between two decision epochs as $F_d(\bar{t}|\hat{s}) = F(\bar{t}|\hat{s}, d(\hat{s}))$, and then we can rewrite (10) as:

$$Q_d(\bar{t}, j|\hat{s}) = Q(\bar{t}, j|\hat{s}, d(\hat{s})).$$

The expected infinite-horizon discrete-time discounted reward is

$$v_\alpha^{d^\infty}(\hat{s}) = r_d(\hat{s}) + \sum_{j \in S} \int_0^\infty e^{-\alpha \bar{t}} Q_d(d\bar{t}, j|\hat{s}) v_\alpha^{d^\infty}(j), \qquad (12)$$

where $Q_d(d\bar{t}, j|\hat{s}) = q_d(j|\hat{s})F_d(d\bar{t}|\hat{s})$ is derived from (10).

According to (9), the long-term reward (12) can be simplified as:

$$\begin{aligned} \nu_\alpha^{d\infty}(\hat{s}) &= r_d(\hat{s}) + \sum_{j \in S} \left[\int_0^\infty \tau_d(\hat{s})^{-1} e^{-[\alpha+\tau_d(\hat{s})^{-1}]\bar{t}} d\bar{t} \right] q_d(j|\hat{s}) \nu_\alpha^{d\infty}(j) \\ &= r_d(\hat{s}) + \frac{1}{1+\tau_d(\hat{s})\alpha} \sum_{j \in S} q_d(j|\hat{s}) \nu_\alpha^{d\infty}(j). \end{aligned} \quad (13)$$

To simplify the calculation, we assume that $\tau_d(\hat{s})^{-1}$ is a constant, and $\tau_d(\hat{s})^{-1} = k$ for all $\hat{s} \in S$. Then the equation (13) can be rewritten as:

$$\nu_\alpha^{d\infty}(\hat{s}) = r_d(\hat{s}) + \lambda \sum_{j \in S} q_d(j|\hat{s}) \nu_\alpha^{d\infty}(j), \quad (14)$$

where $\lambda = \frac{k}{k+\alpha}$. Thus the optimal reward has the discrete-time discounted evaluation equation as:

$$\nu(\hat{s}) = \max_{a \in Act_{\hat{s}}} \left\{ r(\hat{s}, a) + \lambda \sum_{j \in S} q(j|\hat{s}, a) \nu(j) \right\}. \quad (15)$$

Since the system cost $g_{<s,e>}$ is a continuous-time Markov decision process with constant transition rate k, it can be uniformized so that the results and algorithms for discrete-time discounted models can be used directly. We define an uniformization of the continuous-time Markov decision process with components denoted by "∼". Let $\tilde{S} = S$, $\tilde{Act}_{\hat{s}} = Act_{\hat{s}}$, $\tilde{Q}_d$ denote the matrix with components $q_d(j|\hat{s})$ for all $\hat{s} \in \tilde{S}$. We use the same assumption given by [7], where

$$[1 - q(\hat{s}|\hat{s}, a)]\tau(\hat{s}, a)^{-1} \leq \tilde{k}. \quad (16)$$

Based on this assumption, we define a constant $\tilde{k} = \lambda_n + \lambda_t + K*\max(\mu_n, \mu_t) < \infty$ satisfying any $\hat{s} \in S$. The uniformization maximum $v(\hat{s})$ of optimal rule d can be obtained as:

$$\nu(\hat{s}) = \max_{a \in \tilde{Act}_{\hat{s}}} \left\{ \tilde{r}(\hat{s}, a) + \lambda \sum_{j \in S} \tilde{q}(j|\hat{s}, a) \nu(j) \right\}. \quad (17)$$

where $\lambda = \frac{\tilde{k}}{\tilde{k}+\alpha}$, $\tilde{r}(\hat{s}, a) \equiv r(\hat{s}, a) \frac{1+\alpha\tau(\hat{s},a)}{(\alpha+\tilde{k})\tau(\hat{s},a)}$, and

$$\tilde{q}(j|\hat{s}, a) = \begin{cases} 1 - \frac{[1-q(\hat{s}|\hat{s},a)]}{\tau(\hat{s},a)\tilde{k}}, & j = \hat{s} \\ \frac{q(j|\hat{s},a)}{\tau(\hat{s},a)\tilde{k}}, & j \neq \hat{s}. \end{cases} \quad (18)$$

Since the state space and action space is limited, then the maximum of equation (17) exists for all $\nu \in V$. In [7], the author proved that if the maximum of (17) is obtained for each $\nu \in V$, then there exists a stationary deterministic optimal policy d^*. Thus, we have

$$d^* \in \arg\max_{d \in D} \left\{ \tilde{r}_d + \lambda \tilde{Q}_d \nu(\hat{s}) \right\}, \quad (19)$$

which means that $(d^*)^\infty$ is optimal. To obtain the maximum $\nu(\hat{s})$ and optimal d^*, we can use Value Iteration Algorithm that is described in [7].

4 Inter-domain Transfer Probability

One of an important QoS metrics of the cloud system is the inter-domain transfer probability for end users. This is because the inter-domain service transfer may cause service disruptions or incur longer service delay. In this section, we discuss and attain the inter-domain probability based on the presented SMDP-based mobile cloud computing model.

From (17), the expected total discounted reward $\nu(\hat{s})$ at state $\hat{s} \in S$ is only related with $\lambda_n, \lambda_t, \mu_n, \mu_t$ and K. For a state of weblet leave, there is no action (i.e., $a = 0$). Therefore, we only need to consider the state with weblet migration arrivals. If $\lambda_n, \lambda_t, \mu_n, \mu_t$ and K are fixed, then $\nu(\hat{s})$ is also fixed at state $<s_n, s_t, A>, A \in \{A_n, A_t\}$. Moreover, the action $a \in \{0, 1\}$ at state $<s_n, s_t, A>, A \in \{A_n, A_t\}$ is fixed, i.e., accept or transfer (i.e., inter-domain service transfer). From the system point of view, the purpose to accept or transfer a weblet migration request is to achieve higher long-term rewards at state $<s_n, s_t, A>, A \in \{A_n, A_t\}$. Let $\pi_{<s_n,s_t,e>}, e \in \{A_n, A_t, F\}$ denote the steady-state probability of state $<s_n, s_t, e>, e \in \{A_n, A_t, F\}$, $\pi_{<s_n,s_t,A>}, A \in \{A_n, A_t\}$ denote arrival steady-state probability of state $<s_n, s_t, A>, A \in \{A_n, A_t\}$. From [8], we can simply use $P_{inter-transfer} = P^n_{inter-transfer} + P^t_{inter-transfer}$ as the inter-domain transfer probability for the entire system, where $P^n_{inter-transfer}$ and $P^t_{inter-transfer}$ are inter-domain transfer probabilities for new weblet migration requests and intra-domain transfer requests, respectively. The entire system inter-domain transfer probability $P_{inter-transfer}$ is a ratio of all inter-domain transferred weblets migration requests to all arrived weblets migration requests, which is defined as:

$$P_{inter-transfer} = \frac{\sum_{s_n=0}^{N} \sum_{s_t=0}^{H} \left((1-a_{<s_n,s_t,A_n>})\pi_{<s_n,s_t,A_n>} + (1-a_{<s_n,s_t,A_t>})\pi_{<s_n,s_t,A_t>}\right)}{\sum_{s_n=0}^{N} \sum_{s_t=0}^{H} \left(\pi_{<s_n,s_t,A_n>} + \pi_{<s_n,s_t,A_t>}\right)},$$

$$0 \leq N+H \leq K, 0 \leq N \leq L \quad (20)$$

where $a_{<s_n,s_t,A_n>} \in \tilde{A}ct_{\hat{s}}$ is the action adopted at state $<s_n, s_t, A_n>$ and $a_{<s_n,s_t,A_t>} \in \tilde{A}ct_{\hat{s}}$ is the action adopted at state $<s_n, s_t, A_t>$.

According to the result of [5], we can derive $\pi_{<s_n,s_t,e>}, e \in \{A_n, A_t, F\}$ as:

$$\pi_{<s_n,s_t,A_n>} = \\ (1 - a_{<s_n,s_t,A_n>})\pi_{<s_n,s_t,A_n>}\frac{\tilde{k}+\lambda_n-\beta}{\tilde{k}} + (1 - a_{<s_n,s_t,A_t>})\pi_{<s_n,s_t,A_t>}\frac{\lambda_n}{\tilde{k}} \\ + a_{<s_n,s_t,A_n>}\pi_{<s_n,s_t,A_n>}\frac{\tilde{k}-\beta-\mu_n}{\tilde{k}} + \pi_{<s_n,s_t,F>}\frac{\lambda_n}{\tilde{k}} \\ + a_{<s_n^{\max},s_t,A_n>}\pi_{<s_n^{\max},s_t,A_n>}\frac{\lambda_n}{\tilde{k}} + a_{<s_n,s_t^{\max},A_t>}\pi_{<s_n,s_t^{\max},A_t>}\frac{\lambda_n}{\tilde{k}},$$

$$\pi_{<s_n,s_t,A_t>} = \\ (1 - a_{<s_n,s_t,A_n>})\pi_{<s_n,s_t,A_n>}\frac{\lambda_t}{\tilde{k}} + (1 - a_{<s_n,s_t,A_t>})\pi_{<s_n,s_t,A_t>}\frac{\tilde{k}+\lambda_t-\beta}{\tilde{k}} \\ + a_{<s_n,s_t,A_t>}\pi_{<s_n,s_t,A_t>}\frac{\tilde{k}-\beta-\mu_t}{\tilde{k}} + \pi_{<s_n,s_t,F>}\frac{\lambda_t}{\tilde{k}} \\ + a_{<s_n^{\max},s_t,A_n>}\pi_{<s_n^{\max},s_t,A_n>}\frac{\lambda_t}{\tilde{k}} + a_{<s_n,s_t^{\max},A_t>}\pi_{<s_n,s_t^{\max},A_t>}\frac{\lambda_t}{\tilde{k}},$$

$$\pi_{<s_n,s_t,F>} = \pi_{<s_n^{\min},s_t,F>}\frac{s_n^{\min}\mu_n}{k} + \pi_{<s_n,s_t^{\min},F>}\frac{s_t^{\min}\mu_t}{k} + \pi_{<s_n,s_t,F>}\frac{\tilde{k}-\beta}{k}$$
$$+ (1 - a_{<s_n^{\min},s_t,A_n>})\pi_{<s_n^{\min},s_t,A_n>}\frac{s_n^{\min}\mu_n}{k}$$
$$+ (1 - a_{<s_n,s_t^{\min},A_n>})\pi_{<s_n,s_t^{\min},A_n>}\frac{s_t^{\min}\mu_t}{k}$$
$$+ (1 - a_{<s_n^{\min},s_t,A_t>})\pi_{<s_n^{\min},s_t,A_t>}\frac{s_n^{\min}\mu_n}{k}$$
$$+ (1 - a_{<s_n,s_t^{\min},A_t>})\pi_{<s_n,s_t^{\min},A_t>}\frac{s_t^{\min}\mu_t}{k}$$
$$+ a_{<s_n,s_t,A_n>}\pi_{<s_n,s_t,A_n>}\frac{s_n^{\min}\mu_n}{k} + a_{<s_n^{\max},s_t^{\min},A_n>}\pi_{<s_n^{\max},s_t^{\min},A_n>}\frac{s_t^{\min}\mu_t}{k}$$
$$+ a_{<s_n,s_t,A_t>}\pi_{<s_n,s_t,A_t>}\frac{s_t^{\min}\mu_t}{k} + a_{<s_n^{\min},s_t^{\max},A_t>}\pi_{<s_n^{\min},s_t^{\max},A_t>}\frac{s_n^{\min}\mu_n}{k},$$
(21)

where $\beta = s_n\mu_n + s_t\mu_t + \lambda_n + \lambda_t$, $0 \leq s_n + s_t \leq K$ and $0 \leq s_n \leq L$. To cover the boundary conditions, we define $s_n^{\min} = \min(s_n + 1, K - s_t, L)$, $s_t^{\min} = \min(s_t + 1, K - s_n)$, $s_n^{\max} = \max(s_n - 1, 0)$ and $s_t^{\max} = \max(s_t - 1, 0)$.

The summation of the steady-state probability for all states is equal to 1, and thus we have:

$$\sum_{s_n=0}^{N}\sum_{s_t=0}^{H} \pi_{<s_n,s_t,e>} = 1, e \in \{A_n, A_t, F\}, \ 0 \leq N + H \leq K, 0 \leq N \leq L.$$
(22)

Based on Equations (21) and (22), the steady-state occurring probability $\pi_{<s_n,s_t,e>}, e \in \{A_n, A_t, F\}$ can be obtained. Thus, the entire system inter-domain transfer probability $P_{inter-transfer}$ can be attained.

5 Performance Evaluation

The inter-domain transfer probabilities of our presented SMDP-based mobile cloud computing model are compared with that computed by using *Guard occupation model* [8]. We conduct a simulation-based study, in which the comparative results are presented in Figure 3. In this simulation, we set the new weblet migration request arrival rate λ_n as 5, the intra-domain transfer weblet migration request arrival rate λ_t as 2, and the leave rates of both new and intra-domain transfer migration requests (μ_n and μ_t) as 4. We set the maximal number of CPUs for the new weblet migration requests $L = \lfloor 0.8K \rfloor$. Thus, the number of reserved CPUs for the weblet intra-domain transfer migration requests is $K - L$. For each value of K, we run the simulation for 5 times.

In Figure 3, we observe that the inter-domain transfer probabilities of both SMDP-based occupation model and Guard occupation model decrease with the increase of the number of CPUs. Additionally, we can see that the inter-domain transfer probability is only related to the total number of CPUs in the cloud when the arrival rate and leave rate of weblets are fixed. This also confirms our discussion about the inter-domain tranfer probability presented in Section 4.

We also observe that if the number of CPUs is smaller than 10 or larger than 22, the differences of the inter-domain transfer probabilities between the SMDP-based occupation model and Guard occupation model is very small. In addition,

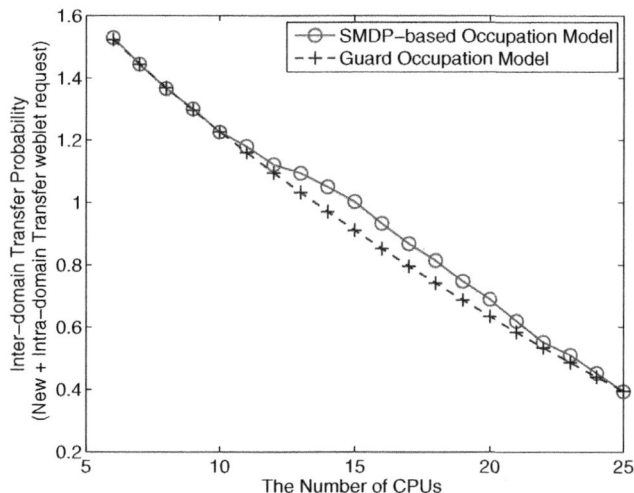

Fig. 3. An example to compare the inter-domain transfer probabilities of using SMDP occupation model and Guard occupation model

if the number of CPUs is between 10 and 22, then, the differences of the inter-domain transfer probabilities are increased. This phenomena can be explained as follows:

- If the number of CPUs is small (i.e., less than 10 in our simulation), then, both SMDP-based occupation model and Guard occupation model cannot accommodate the incoming weblet migration requests for the given simulation setting. As a result, both inter-domain transfer probabilities are high.
- If the number of CPUs is large (i.e., larger than 22 in our simulation), then, SMDP-based occupation model and Guard occupation model have sufficient CPUs to accommodate the coming weblet migration requests for the given simulation setting. Thus, both inter-domain transfer probabilities are low.
- If the number of CPUs is moderate (i.e., between 10 and 22 in our simulation), the inter-domain transfer probability of the SMDP-based occupation model is higher than that computed by using the Guard occupation model. This is because the SMDP-based occupation model focuses more on the maximal system reward that involves the system income and cost, service expenses of MDs, and conservation of energy consumption of MDs. However, the Guard occupation model purely focuses on the reduction of inter-domain transfer rate, which may not be the optimal in terms of system reward.

In general, the increase of the inter-domain transfer probability not only means the decrease of the revenue of a mobile cloud service provider, but also means the disruption of a service. Thus, the system reward should be obtained under a given inter-domain transfer probability to satisfy the desired QoS, i.e., the optimal policy d^* should also consider the restriction enforced by the given inter-domain transfer probability.

If the inter-domain transfer probability is given by P_B, then the system reward (17) can be rewritten as:

$$\nu(\hat{s}) = \max_{P_{inter-transfer} \leq P_B, a \in Act_{\hat{s}}} \left\{ \tilde{r}(\hat{s},a) + \lambda \sum_{j \in S} \tilde{q}(j|\hat{s},a)\nu(j) \right\}. \quad (23)$$

The optimal policy (17) can be rewritten as:

$$d^* \in \arg\max_{P_{inter-transfer} \leq P_B, d \in D} \left\{ \tilde{r}_d + \lambda \tilde{Q}_d \nu(\hat{s}) \right\}. \quad (24)$$

6 Conclusion

In this paper, we present an economic mobile cloud computing model based on Semi-Markov Decision Process. In our approach, both the maximal system reward and expenses of mobile devices are considered. We present the inter-domain transfer probability of the SMDP-based mobile cloud computing model using both theoretical approach and simulation comparative studies. Particularly, we derive both the constraint maximal system reward and the optimal decision policy under a given inter-domain transfer probability. In the future, we will incorporate more system metrics into the constructions of the reward function such as different application tasks or security levels based on multi-threads CPUs. Moreover, we will investigate the optimal CPUs allocation issues using the SMDP-based occupation model.

Acknowledgement. Hongbin Liang's research is supported by the China Scholarship Council and Dijiang Huang's research is supported by the Office of Naval Research's Young Investigator Program.

References

1. Armbrust, M., Fox, A., Griffith, R., Joseph, A., Katz, R., Konwinski, A., Lee, G., Patterson, D., Rabkin, A., Stoica, I., et al.: Above the clouds: A berkeley view of cloud computing. EECS Department, University of California, Berkeley, Tech. Rep. UCB/EECS-2009-28 (2009)
2. Chun, B., Maniatis, P.: Augmented Smartphone Applications Through Clone Cloud Execution. In: Proceedings of USENIX HotOS XII (2009)
3. Huang, D., Zhang, X., Kang, M., Luo, J.: Mobicloud: A secure mobile cloud framework for pervasive mobile computing and communication. In: Proceedings of 5th IEEE International Symposium on Service-Oriented System Engineering (2010)
4. Lyons, K., Pering, T., Rosario, B., Sud, S., Want, R.: Multi-display Composition: Supporting Display Sharing for Collocated Mobile Devices. In: Gross, T., Gulliksen, J., Kotzé, P., Oestreicher, L., Palanque, P., Prates, R.O., Winckler, M. (eds.) INTERACT 2009. LNCS, vol. 5726, pp. 758–771. Springer, Heidelberg (2009)

5. Ni, W., Li, W., Alam, M.: Determination of optimal call admission control policy in wireless networks. IEEE Transactions on Wireless Communications 8(2), 1038–1044 (2009)
6. Pering, T., Want, R., Rosario, B., Sud, S., Lyons, K.: Enabling pervasive collaboration with platform composition. In: Proceedings of Perviasive (2009)
7. Puterman, M.: Markov decision processes: Discrete stochastic dynamic programming. John Wiley & Sons, Inc., New York (2005)
8. Ramjee, R., Towsley, D., Nagarajan, R.: On optimal call admission control in cellular networks. Wireless Networks 3(1), 29–41 (1997)
9. Li, X., Zhang, H., Zhang, Y.: Deploying Mobile Computation in Cloud Service. In: Jaatun, M.G., Zhao, G., Rong, C. (eds.) Cloud Computing. LNCS, vol. 5931, pp. 301–311. Springer, Heidelberg (2009)
10. Zhang, X., Schiffman, J., Gibbs, S., Kunjithapatham, A., Jeong, S.: Securing elastic applications on mobile devices for cloud computing. In: Proceedings of the 2009 ACM Workshop on Cloud Computing Security, pp. 127–134 (2009)

The Smartphone and the Cloud: Power to the User

Roelof Kemp, Nicholas Palmer, Thilo Kielmann, and Henri Bal

Vrije Universiteit, De Boelelaan 1081A, Amsterdam, The Netherlands
{rkemp,palmer,kielmann,bal}@cs.vu.nl

Abstract. In this paper we study how smartphones can benefit from the resources available in clouds. Unfortunately, integrating smartphones with cloud resources is challenging and comes with dangers for the user in terms of loss of control of applications and data as portions move into the cloud. In this paper we outline our work on the Interdroid project, where we are building a framework for smart applications for smartphones which includes components for integration with the cloud.

Keywords: Mobile Computing, Cloud Computing.

1 Introduction

In the last decade we have seen a shift in the use of computers. Our personal computers no longer stand on the desks in our homes, but are located in the pockets of our trousers. We use these pocket devices, called *smartphones*, to access our multimedia data, to communicate with our social networks, to navigate to places of interest, to search for online information, and much more.

These smartphones, by nature, are tightly coupled to a person and therefore are usually within arms reach of their owners. They contain personal information, such as the owners social relations, and can detect the context of their owners using various sensors. Being highly interactive for the user and able to communicate with all sorts of other devices makes the smartphone a compelling platform for many personal applications. Unfortunately, smartphones, being small and battery powered, have limitations in computational power, both in terms of processor speed and memory size, as well as space for data storage.

Next to the shift in our personal use of computers, companies have also changed their way of using computers. Techniques such as web services, virtualization, and grid computing stimulated companies to transform their desktop applications into scalable services running on a flexible number of servers. Big companies realized that they could sell some of their compute capacity when their own demand was low [1]. Startups and companies with an irregular compute demand became their customers and a new paradigm was born: *cloud computing*.

In contrast to smartphones, clouds offer a flexible and almost infinite amount of computational power and data storage, not being hindered by batteries, but connected to the power grid. In addition, clouds are permanently online at a fixed address, while smartphones, due to mobility, may suffer from connectivity problems and changing addresses.

Since the shortcomings of the smartphone are the strong points of the cloud, it seems to be a logical step to enhance smartphone applications with cloud computing to reduce the problems related to limited battery, processor speed, memory size, data storage, and changing addresses. Thus, we see the present and future of cloud computing and mobile computing intertwined: today, already many of the rapidly increasing number of smartphones applications are tightly coupled to commercial cloud services [3, 7]. However, while the use of cloud computing reduces the shortcomings of smartphones, the combination of cloud computing and smartphones also introduces new problems for both developers and users, which we detail in the next section. Furthermore, we describe how we address these problems, in order to avoid that the cloud as a solution is worse than the original problem.

2 Cloud and Phone Integration

In order to give a more intuitive understanding of the benefits of cloud computing as well as how problems with such integration arise, we illustrate a typical integration in the following scenario:

2.1 Navigation on a Smartphone

Tom uses his smartphone for navigation. He installs an application which contains map data and an algorithm to find a route. Calculating a route is compute intensive and therefore takes a long time and consumes much battery power. Furthermore, he notices that his map data is soon getting out of date. Therefore, Tom decides to install a new navigation application that uses the cloud. Now, the route calculation runs very fast in the cloud and Tom gets accurate map data on the fly from the cloud. Tom rarely experiences wrong map information, but when he does, he corrects it using his application, which forwards his corrections to the cloud to be shared with other users. Through the use of cloud computing Tom's user experience improves.

For a holiday, Tom decides to go for a trip to a foreign country. While driving through a rural area, he misses a turn. Normally the navigation application would calculate a new route, however, it cannot since there is no network coverage in this area. He finds his way back to the correct road on his own and then enters the foreign country. He immediately switches off his 3G connection, since roaming costs are too expensive. However, this causes his cloud based navigation application to stop working. After coming back from his holiday, Tom finds out that his client application is no longer compatible with the cloud service. He buys an update but then discovers that the company offering the cloud navigation service changed policy and now charges users to get map updates submitted by other users, including those he submitted in the past. Because of cloud computing, Tom's experience on his holiday trip was worse than if he had used his original local application.

2.2 Integration Problems

As shown in the above scenario, cloud computing can both improve and degrade the smartphone's user experience and introduces new problems related to connectivity, application distribution and data ownership. Nevertheless, we believe that cloud computing, when properly applied, will be beneficial to the smartphone users. In this section we will describe which problems cloud computing introduces and our solutions for these problems.

First of all, we note that by using the cloud, applications are transformed into distributed applications running partly on the smartphone and partly in a remote cloud. As a consequence the different components need to connect to each other and to communicate with each other. However, connectivity is not always available and communication costs energy, time and money. Thus, we believe that every application which is a candidate to be distributed, should remain configurable to also run locally to properly handle those situations where the user cannot or does not want to use the cloud. Then, smartphone users can run compute intensive applications fast and energy efficient when cloud resources are available and affordable *and* use slower and less efficient implementations when offline or effectively offline due to roaming.

Another consequence of partitioning applications into a smartphone client and a cloud service is that where users used to own a complete application, now they only own the smartphone client, whereas the service provider owns the cloud service. Compared to a local application, the user now becomes vulnerable to vendor lock in, version mismatches and, more important, no longer has the same guarantees about the applications service level. The user has to trust the service provider, which may turn out to be untrustworthy by stopping or degrading the service, or increasing the price.

We propose to prevent such problems by bundling client and service code together in the smartphone application. Then, the user is protected against version mismatches since he owns both the client and the service code. Although the service code does not run on the smartphone, it can be stored there and, when needed, be installed on a cloud resource controlled by the user. In the above scenario, Tom would have stored the routing algorithm on his smartphone, but could choose to execute it on his own computer or on rented cloud resources.

For using the data services of the cloud, users should have the ability to select which data will be stored on the cloud, and how this data should be secured, in order to protect their own privacy. Programming models dealing with data exchange with a cloud service should always offer both the means to get data on and off the cloud in combination with local replication, so that the user is free to move from one cloud computing provider to another, without being locked into a service. Furthermore, local replication may be required in order for applications to work properly when the smartphone is not connected to the network. Users might even run cloud services on their own home server in order to maintain the privacy of their data, or reduce the cost associated with the service as provided by the cloud vendor.

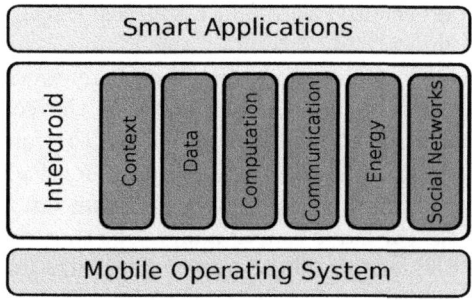

Fig. 1. Abstract overview of the Interdroid project. We strive towards a layer that enables application developers to use social and context information, intensive computing and communication, and data services while keeping energy usage to a minimum.

Finally, we note that not all applications that need to be distributed require a centralized cloud service. Such services are problematic because they introduce a single point of failure in the cloud service. Applications which do not have large computational requirements or data stores associated with them should be implemented in a decentralized way, as we have done in [5], thereby avoiding the centralized point of failure in the cloud. Unfortunately, doing all of these things properly is complicated and so what is needed is a framework which makes developing such applications easier. We detail our work on such a framework in the next section.

3 Interdroid: Smart Applications

We have seen that the shortcomings of smartphones can in principle be improved, by the cloud. We believe that a coherent toolkit that guides developers in applying cloud techniques for mobile applications is of key importance in enabling the synthesis of mobile and cloud platforms.

In our research project, called Interdroid, we aim at providing a broad toolkit for application developers to build really smart applications (see Figure 1). We believe that many applications will run distributed and/or use cloud resources to improve their computational speed, battery life, and data distribution. Therefore, part of the Interdroid project focuses on integrating cloud computing and mobile computing. In the next sections we will detail our ongoing work on the components of Interdroid which involve cloud computing.

3.1 Computation in the Cloud

The most obvious combination of mobile computing and cloud computing is probably moving computations from the mobile device to the cloud, as we are working on in the Cuckoo [4] component. Using Cuckoo, compute intensive parts of applications can be executed on cloud resources, a technique known as *computation offloading* [2].

We implemented our computation offloading framework in such a way that both the code that runs on the smartphone and the code that runs on the cloud are bundled together, so that the user owns the full application, instead of just a client. This protects the user from vendor lock-in and forced paid upgrades. Furthermore, application developers using our framework can write two implementations of the same functionality: one that runs on the cloud when connected to it and one local implementation that runs on the phone when the smartphone is offline or roaming. Finally, we added a runtime decision point in our framework, so that intelligence, based on the current context of the user, can be plugged into the system to avoid offloading costs being larger than its gains.

Computation offloading can speedup smartphone applications significantly, particularly if the compute intensive part can be run parallelized in the cloud as we demonstrated with the eyeDentify [6] object-recognition application. Furthermore, it also enables memory intensive applications on smartphones.

While memory size and processor speed on smartphones are still increasing rapidly, the main shortcoming of smartphones is the relative short battery life. Modern smartphone platforms include many optimizations to improve energy efficiency, but the available sensors, radios, the screen and the use of multiple applications at the same time drains the battery rapidly. Unfortunately, battery densities are not keeping pace with power demands of smartphones. Again, computation offloading can lift the burden on the smartphone by offloading energy hungry parts of an application to the cloud [4,6], provided that the energy saved is greater than the energy spent on communicating with the cloud.

3.2 Communication in the Cloud

There is a particular class of smartphone applications that interact with online information services. Some of these applications monitor information services, such as weather, traffic, stock markets, etc., by periodically polling for new content. Setting the polling frequency involves a trade off between accuracy and high energy usage.

Maintaining high accuracy, but with a low energy usage, can be realized by having the information service push updates, instead of having the application poll for updates. While this approach seems promising, most developers cannot employ it, because they cannot modify the information service.

An application in between the information service and the phone which runs in the cloud can be used to transform a polling smartphone application into one that receives push notifications, without changing the information service. The application in the cloud polls the information resource at a high rate, and upon an update pushes this notification to the smartphone. This way, much of the communication is offloaded to the cloud and a minimal amount of communication, the push messages, are sent to the smartphone.

We are currently implementing a component for Interdroid specifically for this class of applications, which reuses the technique of code bundling as used in

Cuckoo. Instead of providing a standalone service in the cloud with no guarantees for the user, we bundle the service and allow the user to deploy the service on their own cloud resources, including a home server, allowing the user to ensure their quality of service and avoiding problems of vendor lock in. Since monitoring applications always involve communication, the applications on the smartphone will not be able to receive updates while being offline. However, the monitoring application in the cloud is permanently online and can store updates and push them to the phone when it is once again online.

3.3 Data in the Cloud

The cloud not only provides abundant computational power, it also has the opportunity to store large quantities of data, which can be used by smartphone applications. Applications which only store data in the cloud can not work properly when the phone is offline as we saw with the navigation application above. However, applications which only store data on the phone can not take advantage of shared data such as the corrections to map data in our example. What is therefore required is a framework which allows users to replicate portions of the data they are going to use on the smartphone while also sharing that data with the cloud. This raises many issues of data synchronization, replication, and versioning as users share their data with each other and with the cloud.

Thus, one of the components of our Interdroid project is Raven, a framework for synchronizing, versioning and replicating data both between smartphones as well as between smartphones and cloud resources. This framework is designed to make it easy for application designers to add versioning and data sharing features to their application, and returns control over data to the user of the application. Users can replicate locally the data they need when offline and can keep the data even if the cloud service vendor goes offline.

4 Conclusions

In this paper we have discussed the relation between smartphones and computational clouds. There are several possibilities for how smartphones can benefit from the power available in clouds, ranging from enhancements of compute power and battery life through computation and communication offloading to improving data services through synchronization, versioning and replication.

Although the usage of the cloud promises the aforementioned improvements for smartphone users, it also comes with threats, introduced by distributing both applications and data. In particular, the user becomes dependent on the cloud service providers in many cases and applications become unusable when the smartphone is offline. We have discussed our work on Interdroid, which makes it easy for application designers to take advantage of the cloud without causing these problems for the user.

We argue that the integration of cloud computing with mobile computing is the future, but that this must be done in such a way that the power of the cloud enhances the mobile experience, while at the same time, the user maintains control of this power at all times.

References

1. Amazon Elastic Compute Cloud Website, http://aws.amazon.com/ec2
2. Chun, B.-G., Maniatis, P.: Augmented smart phone applications through clone cloud execution. In: Proceedings of the 12th Workshop on Hot Topics in Operating Systems, HotOS XII (2009)
3. Gumiyo, http://aws.amazon.com/solutions/case-studies/gumiyo/
4. Kemp, R., Palmer, N., Kielmann, T., Bal, H.: Cuckoo: a Computation Offloading Framework for Smartphones. In: MobiCASE 2010: Proceedings of The Second International Conference on Mobile Computing, Applications, and Services (2010)
5. Kemp, R., Palmer, N., Kielmann, T., Bal, H.: Opportunistic Communication for Multiplayer Mobile Gaming: Lessons Learned from PhotoShoot. In: MobiOpp 2010: Proceedings of the Second International Workshop on Mobile Opportunistic Networking, pp. 182–184 (2010)
6. Kemp, R., Palmer, N., Kielmann, T., Seinstra, F., Drost, N., Maassen, J., Bal, H.E.: eyeDentify: Multimedia Cyber Foraging from a Smartphone. In: IEEE International Symposium on Multimedia (2009)
7. SnapMyLife, http://aws.amazon.com/solutions/case-studies/snapmylife-interview

Towards Cloud Mobile Hybrid Application Generation Using Semantically Enriched Domain Specific Languages

Ajith Ranabahu, Amit Sheth,
Ashwin Manjunatha, and Krishnaprasad Thirunarayan

Ohio Center of Excellence in Knowledge-Enabled Computing (Kno.e.sis) Center
Wright State University, Dayton, Ohio 45435
{ajith,amit,ashwin,tkprasad}@knoesis.org,
http://knoesis.org/cloud

Abstract. The advancements in computing have resulted in a boom of cheap, ubiquitous, connected mobile devices as well as seemingly unlimited, utility style, pay as you go computing resources, commonly referred to as Cloud computing. Taking advantage of this computing landscape, however, has been hampered by the many heterogeneities that exist in the mobile space as well as the Cloud space.

This research attempts to introduce a disciplined methodology to develop Cloud-mobile hybrid applications by using a Domain Specific Language (DSL) centric approach to *generate* applications. A Cloud-mobile hybrid is an application that is split between a Cloud based back-end and a mobile device based front-end. We present *mobicloud*, our prototype system we built based on a DSL that is capable of developing these hybrid applications. This not only reduces the learning curve but also shields the developers from the native complexities of the target platforms. We also present our vision on propelling this research forward by enriching the DSLs with semantics. The high-level vision is outline in the ambitious Cirrocumulus project, the driving principle being *write once - run on any device*.

1 Introduction

Lately there have been interesting changes at both ends of the *spectrum of computing power*. On one end there has been a boom in mobile computing devices, fueled by fast growing communication networks. On the other end, there has been substantial growth in high-end data centers that offer cheap, on-demand and virtually unlimited computing resources, popularly named *Cloud computing*. In the backdrop of these advances in computing and the growth of data intensive domains such as social networks, a new class of applications have emerged taking advantage of not only the on-demand scalability of computing clouds but also the sophistication of current mobile computing devices.

This class of applications, named *cloud-mobile hybrids*, are characterized by the need for data-intensive computations or extreme scalability in the back-end and mobile device based front-ends. Figure 1 illustrates the structure of a cloud-mobile hybrid application.

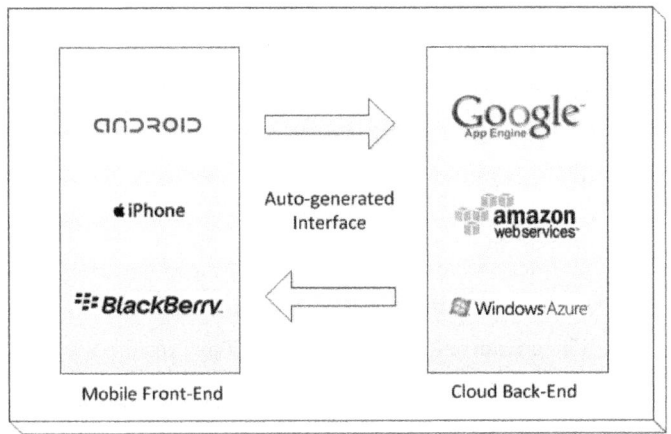

Fig. 1. Structure of a Cloud-Mobile hybrid

An illustrative example of a cloud-mobile hybrid would be an implementation of the *Privacy Score* [11] algorithm. Privacy score is a numerical indicator of the level of private details exposed by an individual in a social network. This score is a relative measure and requires substantial computations in the back-end. These computations can be performed in parallel. Presenting the score to the user is preferred to be via a mobile device, prompted by the increasing use of mobile devices to interact with social networks. Developing such an application, however, is significantly difficult than any other application development effort.

The state-of-the-art in mobile front-ends has changed from mobile-enabled Web sites to platform native applications. These native applications offer a better user experience by tightly integrating with the host platform and taking full advantage of the capabilities of the device. There is no universal development methodology and developers must pick and choose from a multitude of different mobile platforms. Similar choices need to be made in the Cloud space which is fragmented due to vendor specific service interfaces, restricted run-times and many others. Hence, developers have to cope with fragmentation at two different levels and often the efforts are focused on only selected mobile platforms and clouds. Developing portable hybrid applications in an economical and efficient manner is clearly a challenge.

We believe the key in overcoming portability issues is to follow a model-driven development pattern. However, there is no one level of abstraction that can be applied to modeling. Instead, one requires multiple models with varying granularities to cover different aspects of the application. Although semantic modeling is favored at a higher level, developers prefer detailed, concrete syntactic representations such as DSLs. The relationship between the representations of different granularities need to be established, often through explicit annotations. A slicing of the Cloud modeling space and the different types of models required have been discussed in [21] and showing the relevance of semantic models. We discuss four different types of semantics, data, functional, non-functional and system in Section 4.1 where each type addresses a specific aspect in an application. For example data semantics provide platform agnostic data definitions that support data portability.

In this paper we present MobiCloud, our early attempt to introduce a methodology for developing Cloud-mobile applications using DSLs. DSLs offer a midlevel abstraction that is developer friendly but also allows room for high level models to be attached. We also discuss the role of semantic models and the vision outlined by the ambitious Cirrocumulus project[1]. The goal of Cirrocumulus is to enable platform agnostic application development, deployment and management, intended to be achieved using DSLs infused with high-level semantic models as well as semantic-enriched middleware.

2 Motivation and Background

Our motivation for this research primarily comes from the lack of a clear methodology to develop portable applications for Clouds and mobile devices. The recent attention on Cloud-mobile hybrids and the difficulty in developing such applications clearly indicated the void in this space.

Portability issues arise primarily due to the heterogeneity (fragmentation) in both Cloud and mobile platforms. There is ample evidence that such heterogeneities exist and they are indeed the root of many of the issues the industry is facing today.

The Consumer Electronics Show (CES)[2] is the premier showcase of the consumer electronics devices and is indicative of trends in the current and future mobile device markets. During the last CES event, developers openly expressed frustration over a lack of consolidation of mobile platforms [10]. This is one of many complaints about the state of the fragmentation in the mobile space where there is no standardization in how applications are developed or deployed into mobile devices.

Similar fragmentation has happened in the Cloud space with each vendor developing their own paradigm [4]. The Cloud remains a largely non-standard space despite the efforts from National Institute of Standards and Technology (NIST) to standardize it. Recent industry surveys indicate that the practitioners still consider vendor lock-in a serious hindrance to Cloud computing adoption[19].

---

[1] http://knoesis.wright.edu/research/srl/projects/cirrocumulus/
[2] http://www.cesweb.org/

Some experts have also suggested that vendors may purposely promote the Cloud to be a heterogeneous patchwork of frameworks for business reasons [3].

These heterogeneities force developers to be locked-in to a selected set of platforms. Catering for multiple platforms is a time consuming and highly expensive venture only a few would attempt. This research focuses on overcoming the platform lock-in by using an independent language to develop applications.

2.1 Use of DSLs

A DSL is a programming language or executable specification language that offers, through appropriate notations and abstractions, expressive power focused on, and usually restricted to, a particular problem domain [2]. DSL centric approaches have been used in many domains, particularly due to the expressiveness in the domain of interest, runtime efficiency and reliability due to the narrow focus [22]. For example, mathematicians are quite familiar with specialized languages such as MATLAB [8] that provide a convenient way to write matrix oriented programs. Domain of a DSL can be arbitrarily scoped, i.e. a DSL may cater for a generic domain such as Mathematics or be extremely narrow, say configurations for a particular computer game [23].

The emergence of powerful interpreted languages, such as Ruby, have been a key enabler for many modern DSLs. A Ruby based DSL has been successfully used in the IBM Sharable Code (ISC) project [14] to provide programming abstractions for light weight service compositions (a.k.a. mashups).

DSLs are considered as the key component in the software factories approach by Greenfield et al.[7]. Some of these philosophies have played a critical role in adding features to the Microsoft Visual Studio development suite. One of the pertinent arguments Greenfield presents is that many of the goals of Object Oriented Programming (OOP) were impossible to achieve in practice due to the lack of sufficient level of abstraction. A DSL is capable of raising the level of abstraction to achieve convenience in developing and software reuse.

A DSL however is not the silver bullet that provide a universal solution. DSLs by definition, cater to only a specific domain and become inapplicable outside the targeted domain. For example, the IBM Sharable Code DSL is only useful to prepare service compositions. However, given a class of applications, a DSL greatly reduces the effort required to create programs and lowers the barriers to entry.

3 A DSL for Cloud Mobile Hybrids

We now present the prototype DSL focused in this research. Named *MobiCloud* to indicate the presence across mobile and Cloud spaces, the DSL caters for interactive Web applications driven by Create, Retrieve, Update and Delete (CRUD) operations. These applications typically use multiple data structures in a data centric back-end and use a mobile or Web based front-end to manipulate these data structures. The use of Cloud in these applications is primarily for scalability, i.e., the application itself may not require a massive processing capability but is likely to receive a large number of simultaneous requests and hence needs to scale accordingly. Typically, these can be horizontally scaled, i.e. the load can be shared across multiple replicas.

An example of such an application is a *to-do list manager* similar to the very popular task manager application offered by *Remember the Milk*[3]. This application allows users to create *to-do items* using their mobile devices and stores them in a Cloud data store. These reminders can later be retrieved as a list, either on a mobile device or on the Web. Creating an application of this nature from scratch requires developing the following components:

(1) A data storage mechanism tied to the storage technology of choice.
(2) A service layer capable of exposing the operations on the data store.
(3) A service access layer in the targeted front-end capable of accessing the services defined on the server side.
(4) Relevant user front-end components.

Long running software engineering research on design patterns has identified the most appropriate design pattern for this type of applications is the Model-View-Controller (MVC) pattern. Figure 2(a) illustrates the major components present in a MVC based design. Figure 2(b) illustrates the split of these components across the back-end and the front-end.

We designed a DSL that closely resembles the MVC pattern giving separate specifications of each of the major components. This has been a conscious design decision since many developers are already familiar with the MVC pattern thus it would be natural for them to use this DSL.

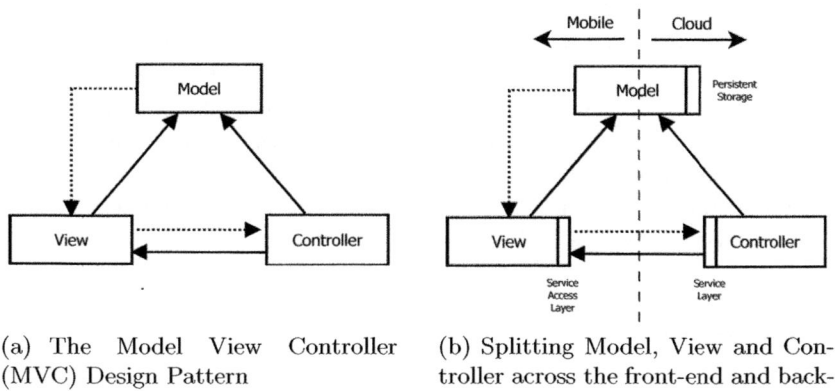

(a) The Model View Controller (MVC) Design Pattern

(b) Splitting Model, View and Controller across the front-end and back-end

Fig. 2. MVC Design Pattern and the its usage for Cloud-Mobile Hybrids

We now present a *hello world* application written using this DSL to exemplify the features of the language. Listing 1.1 depicts the DSL script for this application. The intention of this application is to *illustrate* the main language features.

---

[3] http://www.rememberthemilk.com/

(1) A minimal *model* with only one attribute.
(2) A minimal *controller* with only one action.
(3) A minimal *view* demonstrating a minimal user interface.

This application displays a greeting message on the mobile device by fetching it from the remote, cloud based data storage via a RESTful service interface.

Listing 1.1. The DSL script for the *hello world* application

```
recipe :helloworld do
  metadata :id => 'helloworld-app'
    # models
      model :greeting ,
              {:message => :string}

    #controllers
      controller :sayhello do
        action :retrieve ,:greeting
      end

    # views
      view :show_greeting ,
        {:models =>[:greeting] ,
         :controller => :sayhello ,
         :action => :retrieve}
end
```

We limit the elaboration on our DSL for brevity. Further details of the implementation of this prototype language is available from the MobiCloud technical report [12]. An on-line tool kit and a number of examples, including the complete language specification in BNF are also available [4]. The current system is capable of generating functionally equivalent back-end applications for Google Appengine and Amazon EC2. The front-end capabilities include Android 1.5 and Blackberry platforms.

4 Discussion

4.1 Role of Semantics

Semantic models have been applied in many domains to provide platform-neutral specifications. For example, our faceted classification and search system APIHut uses a taxonomy to organize the functional characteristics of Web APIs [6]. Another domain dependent example is GoodRelations, a standardized vocabulary for product, price, and company data [9].

---
[4] http://knoesis.org/mobicloud

Semantic Web community has been using semantic models to overcome issues of portability and interoperability for years and these are the very issues the Cloud computing community is facing today. A particularly relevant research work was on semantics for Web services where four types of semantics has been identified for a service [20]. Figure 3(a) illustrates an adaptation of the four types of semantics to the Web application domain.

Figure 3(b) illustrate the analysis of the modeling space we presented in [21]. This slicing indicates the applicability of the existing models as well as the voids that are present.

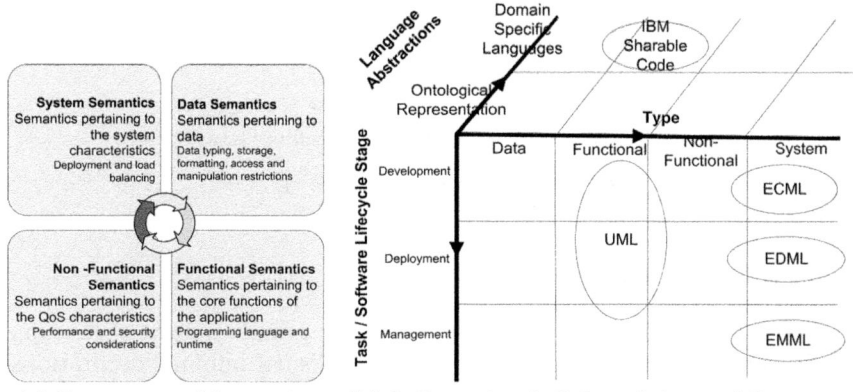

(a) Four types of Semantics   (b) 3 dimensional slicing of the modeling space

Fig. 3. Four types of semantics and its relevance to the modeling space

The key in creating such a breakdown is to direct the model creation and usage towards specific aspects. For example semantic data models (ontologies) can be applied independent of the other aspects of the application. While the direct adaptation of the service oriented semantic categorization may not be the best in the application domain, a similar categorization would be immensely helpful in applying semantics to Cloud-mobile hybrids.

There are three potential uses of semantics in cloud-mobile hybrid application generation.

(1) A key limitation, even with the use of the DSL is reusable data modeling. This is very important in the back-end when the need to migrate applications arise. Existing clouds use a myriad of data models, making the task of migrating across data stores that follow different models a challenge. Model-agnostic semantic data definitions, coupled with the lifting-lowering data migration strategy [16] originally proposed for Web service data mediation is directly applicable here. A semantic data model can be referenced with in the DSL rather than defining one in-line. There are many well established, public semantic data models such as Friend-of-a-friend (FOAF) that can be reused in data definitions.
A mechanism to enable data references is illustrated in Listing 1.2.

(2) Non-functional details of an application are generally interleaved into the logic. However, many of these capabilities can be separated from the functionality and layered on the core functional implementations. Aspect Oriented Programming (AOP) [5] is a relatively new philosophy that advocates a clean separation of cross cutting non-functional concerns. Semantic models can be used to specify these non-functional capabilities and linked to the DSL via annotations.

(3) System details for the application including the deployment parameters and scaling configuration can be expressed via semantic models. In fact such descriptions are being used commercially today. Elastic Computing Modeling Language (ECML), Elastic Deployment Modeling Language (EDML) and Elastic Management Modeling Language (EMML) by Elastra Inc. [1], highlighted in Figure 3(b) is a prime example of a system oriented semantic model. This type of configuration may also be linked to the DSL via annotations.

Listing 1.2. Using a Reference to Define Data Types

model :person, {:**ref** => "foaf:Person"}

4.2 Application UI Features

A potential limitation of the current tool is the generic nature of the applications that are being generated. The generated UI's use minimal decorations and are focused on functionality, rather than visual appeal. Even if the generic UI features can be improved, developers may want to customize their application's visual components. There are two possible solutions:

(1) Use a secondary DSL to define custom UI components and attach them to the views. The XAML [15] UI language is one such well established DSL.
(2) Use the generated projects to bootstrap custom development. This is similar to the model driven development process followed by many major software companies where a high level model, such as UML diagram, is used to bootstrap the development process.

Listing 1.3 is a UI description written in XAML and Listing 1.4 shows how this could be incorporated in to the DSL.

Listing 1.3. An Example XAML template for the Greetings UI

```
<Canvas>
    <Rectangle   Fill="PowderBlue"  />
    <TextBlock
      Foreground="Teal"
      FontFamily="Verdana"
      FontSize="18"
      FontWeight="Bold"
      Text="<%@model.message%>"  />
</Canvas>
```

Listing 1.4. Using a Reference to XAML based UI template

```
view   :show_greeting,
{:models  =>[:greeting],
   :controller  =>  :sayhello,
   :action  =>  :retrieve,
   :uiref  =>  "hello.xaml"}
```

4.3 User Defined Back-end Functions

Custom actions beyond the simple CRUD operations become an absolute necessity when the applications grow in complexity. Similar to the customization of the UI, this DSL may be enhanced to enable plug-in-in actions using user defined functions. These actions may also be written in other DSLs such as PIGLatin [18] scripts. Listing 1.5 shows a possible extension of the DSL to embed a PIGLatin script for a custom back-end function based on Apache Hadoop.

Listing 1.5. Embedding a PIGLatin script in a custom action

```
action  :sort_items,
     :item,{:lang  =>  'PIG'}  do
  %{
        A=load  'items'  using  PigStorage()
               as  (a,  b,  c);
        B=sort  A  by  a;
     }
end
```

4.4 Deployment Complexity

Although the generated applications can be tested on the provided mobile device emulators, deployment to the actual device may require a signing step (using an authenticated key) and optionally an upload to a vendor controlled *app store*. Some of these work flows have been deliberately kept as human centric operations by the vendors. Even if there are Web APIs present, managing keys, security certificates and other deployment operations require the presence of a different layer of automation. Although such facilities are out of scope of this work, adding a middleware layer capable of managing deployments and subsequent management tasks, such as Altocumulus [13], would improve the reach and the usability of the DSL.

5 Vision for the Future

Our vision on the future of the Cloud applications indeed include DSLs as well as semantic-aware middleware layer [17]. The mobicloud DSL is the early attempt to realize this vision as outlined in Figure 4.

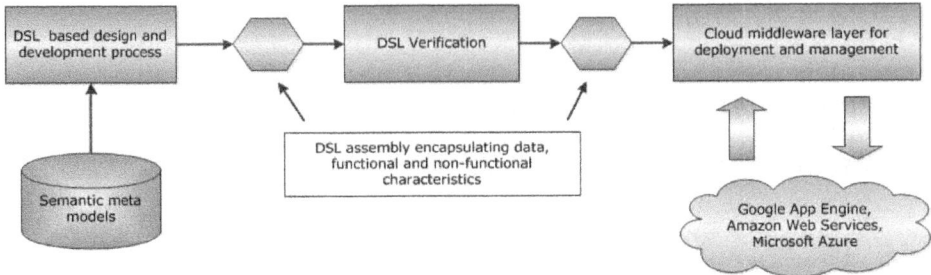

Fig. 4. The High Level Objective of the Cirrocumulus Project

The Cirrocumulus project attempts to provide the following capabilities to support portability and interoperability objectives.

(1) The ability to design and develop (program) with no assumptions about a specific target platform, data model or runtime behavior.
(2) The ability to deploy the artifacts to multiple platforms with no re-architecture or re-programming.
(3) The ability to manage and tune the deployed artifacts with no consideration of *where* they are deployed and migrate them to a different platform if necessary. *Management* refers to the tasks such as taking backups, moving log files etc.

Although these objectives were formed to cater for Cloud portability, they also apply to Cloud-mobile hybrids and to mobile application portability as well. MobiCloud system gave us an opportunity to *test the water* with respect to our development philosophy.

Several insights were gained from the feedback received on the MobiCloud on line tool kit.

(1) Some experienced developers considered the top-down design and development process not flexible enough to create presentable applications. This is indeed the case with high levels of abstractions. However, the default applications with the basic functionality would serve the majority case. Experienced developers can still use the DSL to generate the boiler plate code and continue to customize it as mentioned in Section 4.2.
(2) There was great interest in having a reverse engineering tool to convert an existing application to the DSL. Such a tool in practice would be semi-automated rather than fully automated. A conversion, even with human involvement, would bring value by enabling migrations at a later point and act as an incentive to convert existing programs to the DSL.

6 Conclusion

Our experimental DSL has clearly demonstrated the applicability of DSLs to generate cloud-mobile hybrids, as part of a larger goal of bring portability to

Cloud applications. Although there are many possible improvements, we believe that our philosophy is promising in transforming the Cloud-mobile hybrid application development process. By reusing many existing and well-established semantic technologies, this approach will be able to create, deploy, and manage Cloud-mobile hybrids efficiently and cost-effectively.

References

1. Charlton, S.: Model Driven Design and operations for the Cloud. In: Towards Best Practices in Cloud Computing Workshop, pp. 17–26 (2009), http://bit.ly/cSPAin (last accessed August 27, 2010)
2. van Deursen, A., Klint, P., Visser, J.: Domain-specific languages: an annotated bibliography. SIGPLAN Not. 35(6), 26–36 (2000)
3. Durkee, D.: Why cloud computing will never be free. Communications of the ACM 53(5), 62–69 (2010)
4. Economist Opinion Section: Clash of the Clouds. The Economist (2009), published online at http://bit.ly/cBRAfB (last accessed August 27, 2010)
5. Elrad, T., Filman, R.E., Bader, A.: Aspect-oriented programming: Introduction. Communications of the ACM 44(10), 29–32 (2001)
6. Gomadam, K., Ranabahu, A., Nagarajan, M., Sheth, A.P., Verma, K.: A faceted classification based approach to search and rank web apis. In: IEEE International Conference on Web Services, pp. 177–184 (2008)
7. Greenfield, J., Short, K.: Software Factories: Assembling Applications with Patterns, Models, Frameworks and Tools. In: Companion of the 18th Annual ACM SIGPLAN Conference on Object-Oriented Programming, Systems, Languages, and Applications, pp. 16–27. ACM (2003)
8. Hanselman, D., Littlefield, B.C.: Mastering MATLAB 5: A comprehensive tutorial and reference. Prentice Hall PTR, Upper Saddle River (1997)
9. Hepp, M.: GoodRelations: An Ontology for Describing Products and Services Offers on the Web. In: Gangemi, A., Euzenat, J. (eds.) EKAW 2008. LNCS (LNAI), vol. 5268, pp. 329–346. Springer, Heidelberg (2008)
10. Johnson, A.: Apps call, but will your phone answer?, published online, at http://bit.ly/7OfKeO (last accessed August 27, 2010)
11. Liu, K., Terzi, E.: A Framework for Computing the Privacy Scores of Users in Online Social Networks. In: Proceedings of the 2009 Ninth IEEE International Conference on Data Mining, pp. 288–297. IEEE Computer Society (2009)
12. Manjunatha, A., Ranabahu, A., Sheth, A., Thirunarayan, K.: A Domain Specific Language Based Method to Develop Cloud-Mobile Hybrid Applications. Tech. rep., Kno.e.sis Center, Wright State University (2010), http://knoesis.wright.edu/library/publications/MobiCloud.pdf (last accessed August 27, 2010)
13. Maximilien, E., Ranabahu, A., Engehausen, R., Anderson, L.: Toward cloud-agnostic middlewares. In: Proceeding of the 24th ACM SIGPLAN Conference Companion on Object Oriented Programming Systems Languages and Applications, pp. 619–626. ACM (2009)
14. Maximilien, E.M., Ranabahu, A., Gomadam, K.: An Online Platform for Web APIs and Service Mashups. IEEE Internet Computing 12(5), 32–43 (2008)
15. Microsoft Corporation: Extensible Application Markup Language. Microsoft Developer Network, MSDN (2008)

16. Nagarajan, M., Verma, K., Sheth, A.P., Miller, J., Lathem, J.: Semantic Interoperability of Web Services-Challenges and Experiences. In: IEEE International Conference on Web Services (ICWS), pp. 373–382 (2006)
17. Oberle, D.: Semantic Management of Middleware (Semantic Web and Beyond: Computing for Human Experience). Springer-Verlag New York, Inc., Secaucus (2006)
18. Olston, C., Reed, B., Srivastava, U., Kumar, R., Tomkins, A.: Pig Latin: A not-so-foreign language for data processing. In: Proceedings of the 2008 ACM SIGMOD International Conference on Management of Data, pp. 1099–1110. ACM (2008)
19. Rightscale.com: The Skinny on Cloud Lock-in (2009), published online at http://bit.ly/LZc80 (last accessed August 27, 2010)
20. Sheth, A.: Semantic Web Process Lifecycle: Role of Semantics in Annotation, Discovery, Composition and Orchestration. In: Workshop on E-Services and the Semantic Web (ESSW 2003) in 12th International World Wide Web (WWW) Conference, Budapest, Hungary (2003) (invited presentation)
21. Sheth, A., Ranabahu, A.: Semantic modeling for cloud computing, part 1. IEEE Internet Computing 14, 81–83 (2010)
22. Spinellis, D.: Notable design patterns for domain-specific languages. The Journal of Systems & Software 56(1), 91–99 (2001)
23. Sweeney, T.: Unreal Script Language Reference (1998), http://unreal.epicgames.com/UnrealScript.html (last retrieved August 27, 2010)

Augmenting Pervasive Environments with an XMPP-Based Mobile Cloud Middleware

Dejan Kovachev, Yiwei Cao, and Ralf Klamma

Information Systems & Database Technologies, RWTH Aachen University
Ahornstr. 55, 52056 Aachen, Germany
{kovachev,cao,klamma}@dbis.rwth-aachen.de
http://dbis.rwth-aachen.de

Abstract. Despite the rapid advances in mobile technology, many constraints still prohibit smartphones to run resource-demanding applications in pervasive environments. Emerging cloud computing opens an access to unlimited resources for mobile devices. However, the combination of both technologies to deliver sound mobile cloud applications and services raises new challenges and requirements. Based on a scenario-based requirement analysis and a comprehensive study on existing work for augmenting mobile devices, we propose a XMPP-based mobile cloud computing architecture employing module partitioning and adaptive offloading to nearby computing infrastructure. Research has also been done in the underlying offloading mechanism based on context-aware cost model. Further problems related to this approach are discussed as well, including selection of most optimal offloading plan, application partitioning and issues with XMPP on mobile systems.

Keywords: mobile cloud computing, dynamic software architecture, pervasive computing, augmented execution, XMPP.

1 Introduction

In recent years, smart mobile devices such as iPhones, Android-based smartphones, BlackBerrys, has become highly accessible. They have rich sensing functionality and data exchange connectivity. The applications running on these smart mobile devices seamlessly integrate with realtime data streams and Web 2.0 applications, such as mashups, open collaboration, social networking and mobile commerce [23,14]. Despite the significant improvements of the mobile hardware technologies, mobile devices will continue to be resource-poor, with unstable connectivity, and with less energy. Users cannot expect that the mobile execution platform supports the same functionality as their desktop's or notebook's counterpart. Resource poverty is a major obstacle for many mobile applications [19]. Therefore, computation capacities on mobile devices will compromise the applications.

On the other hand, cloud computing technologies have been emerging recently as a solution to scalable on-demand computing and storage resources that can be accessed via networks. The conceptual model of cloud computing provides a platform for services and applications with high elasticity. Elasticity means that storage and computation resources are put into use according to actual current requirements. These properties would benefit certainly resource-poor mobile devices. The never ending strife for increasing mobile processing power and more data, clouds can be the best possible solution to augment the mobile execution platform.

However, mobile pervasive computing environments are characterized by severe resource constraints and frequent changes in operating conditions, i.e. unstable connectivity, less energy, limited memory and processing power. The existing cloud computing methods available for desktop clients cannot be applied directly. For example, the computation of current mobile applications happens either mostly on the cloud side (e.g. Twitter) or on the device (e.g. a video game). Such static partitioning of mobile applications does not provide optimal user experience in different scenarios, since mobile devices come with different capabilities. Furthermore, accessing distant clouds introduces more latency, monetary costs and limited interactivity.

In this paper, we envision how to deploy mobile applications that are dynamically partitioned between limited mobile devices and the cloud with "unlimited" resources. Furthermore, to avoid the latency of accessing distant cloud centers and improve the user experience for resource-demanding applications, we consider the idea of cloudlets [19], a computing platform exposing its functionality via wireless access to nearby mobile devices. Such a possibility opens the doorways to more powerful interactive mobile applications (see Fig. 1). Moreover, the partitioned modularized applications are deployed on top of an XMPP-based middleware which enables real-time, flexible, scalable and extensible software architecture. XMPP acts as glue joining the heterogeneous parts of the system, providing bi-directional XML streaming with rich publish/subscribe functionality and presence detection.

In the rest of the paper, we elaborate more on the mobile cloud computing augmentation for pervasive environments. Section 2 surveys briefly the related work. Section 3 motivates this approach with scenarios, from which a set of requirements is derived and explained how these requirements can be met with cloud-related methods. In the Section 4, we propose a modular architecture and describe the enabling technology. Furthermore, we underline the importance of a context-ware cost model for dynamic mobile cloud applications. Finally, we discuss some open issues in Section 5 and conclude the paper in Section 6.

2 Related Work

Over the last two decades, there have been several approaches developed on how to augment capabilities of resource-constrained mobile devices. These approaches include software replication, application partitioning and modularization, process and virtual machine (VM) migration. The previous efforts investigate how to increase the processing power, how to increase data storage or memory [4,21],

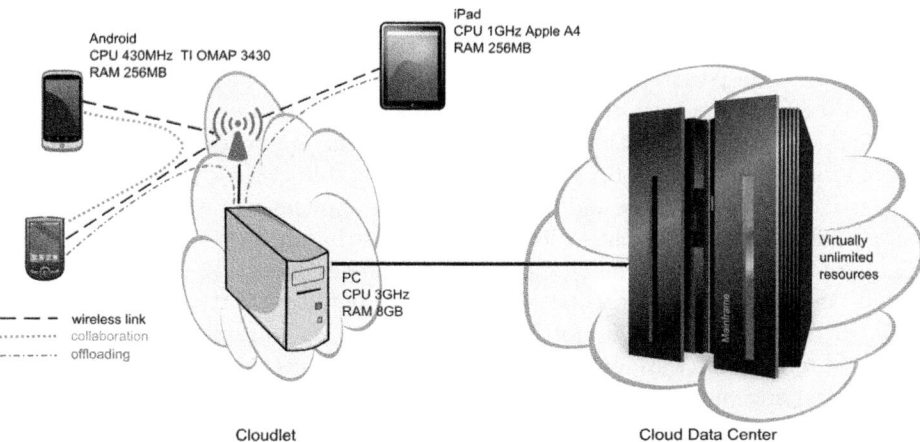

Fig. 1. Mobile devices accessing nearby connected cloudlet and large cloud data centers via Internet

how to balance the performance or QoS with energy consumption, how to ensure application mobility [5], how to secure [7,15] or personalize [20] nearby untrusted computing environments. Our approach draws the best features of previous work, trying to overcome some of their limitations.

The simplest way to augment weak devices such as mobile phones is the application delivery based on the traditional client-server model [12]. However, the client-server model does not consider the changing conditions in pervasive computing environments, causing limited interactivity (thin clients, Web applications) or less portability (fat clients, native mobile applications) [24]. In Spectra [6], programmers define execution plans that run several application partitioning variants which deliver different quality of service. Although Coign [11] is a nice example of automatic partitioning of DCOM applications without source code modification, it outputs client-server applications again statically.

Gu et al. [10] demonstrate an adaptive offloading of certain code classes based on the available resources, class memory footprint and class access frequency. The MAUI system achieves the similar fine grained runtime code offloading to the cloud infrastructure [3]. The primary goal of MAUI is to maximize device battery life with code offload. Developers annotate while programming which methods can be offloaded for remote execution. Since MAUI allows a fine grained offloading mechanism on the level of single methods, the experimental results from MAUI show that the separate method offloading can be contraproductive, i.e. several methods should be combined to achieve benefits. Rellermeyer et al. [16] consider the offloading on complete software modules. Similarly, Zhang et al. [26,25] develop a reference framework for partitioning a single application into elastic components with dynamic configuration of execution. The components, called *weblets*, are platform independent and can be executed transparently on different computing infrastructures. Our approach partitions applications on module level where developers express in the code the mobility of each individual module and its relations with other modules.

Besides code partitioning, the augmented execution can be further achieved with a software replication. Chun and Maniatis [2] propose an architecture that replicates the whole smartphone images and run the application code with few or no modifications in powerful VM replicas. Such architecture, e.g., enables smooth execution of computation and memory-hungry applications, such as virus scanning or file system indexing, without blocking other applications. A similar approach to using virtual machine (VM) technologies executing the computation-intensive software from mobile devices is presented by Satyanarayanan et al. [19]. In this architecture, a mobile user exploits VMs to rapidly instantiate customized service software on a nearby cloudlet and uses the service over WLAN. Rather relying on a distant cloud, the cloudlets eliminate the long latency introduced by wide-area networks for accessing the cloud resources. As a result, the responsiveness and interactivity on the device are enhanced by low-latency, one-hop, high-bandwidth wireless access to the cloudlet. A mobile device delivers a small VMs overlay to the cloudlet infrastructure that already owns the base VM. This approach enables the user to resume a complete personal computing environment including own choices of operating system, applications, settings, and data. We also consider offloading to a nearby computing infrastructure in order to reduce the latency.

3 Scenarios and Requirements

For the purpose of test and evaluation, use case scenarios are designed to illustrate the special usage of the proposed mobile cloud computing framework. The scenarios we consider require more resources than the conventional mobile devices ca provide. We address the issues that may occur related to the scenarios, which helps an in-depth requirement analysis of the framework design.

Scenario 1: Architectural modeling is important in architectural design to see whether the design draft fits to the physical environment. In 1980s, the architect had to build a glass pyramid of the same size in front of the Louvre in order to persuade the Paris citizens to accept his design of the entrance to the national museum. It is almost impossible for the most architects, especially architectural students, to make such a costly experiment. The design models made of plastic foams with a different scale still cannot exactly tell whether a building fits into the real environment. Currently, advanced 3D scanner, 3D printer as well as 3D modeling technologies are able to deploy an augmented model in the physical environment [1]. With advanced cloud services, localization and projection of the model onto the real environment could be realized.

Scenario 2: Virtual archeology deals with those tasks such as 3D reconstruction and context-aware augmented reality [22,9]. These technologies provide researchers appropriate means to explain destroyed structures better or show the original sight. In many cases, research or heritage conservation work is carried out on site at a historical location. The widespread mobile devices help archaeologists during their on-site research. However they are usually expected to access large amounts of heterogeneous multimedia data and run resource-demanding applications, which mobile devices cannot fully deliver. Archaeologists also need to exchange messages and share data collaboratively with the team members.

Based on these and similar use case scenarios, it is obvious that mobile devices like smart phones with an embedded camera and different sensors have great potentials to assist researchers at field work. For example, researchers should able to consume 3D objects from remote data repositories and see virtual objects blended on the video stream from their mobile phone camera. They would be able to see in the ruins virtual historic objects which have been moved to some museums faraway for protection from the severe outdoor environments.

In the reality, problems and challenges are raised from many aspects. A heavy equipment with enough powerful computers lowers the mobility and flexibility of on-site researchers. A large storage capacity for prepared and on-site acquired multimedia data is beyond a mobile device. The advanced 3D technologies cannot be easily used on site, because they require more processing and storage capacity than modern mobile device can provide. Light mobile devices are not able to process that raw data immediately due. A trade-off between reduction of technical infrastructure and provision of high computing capabilities should be made to fulfill the required tasks. Flexibility and mobility feature the applications that need to realize the tasks described in the scenarios.

The tasks in the both scenarios can be well divided into several sub tasks. They address requirements onto the functionality and features of a mobile cloud computing platform as listed in Figure 2.

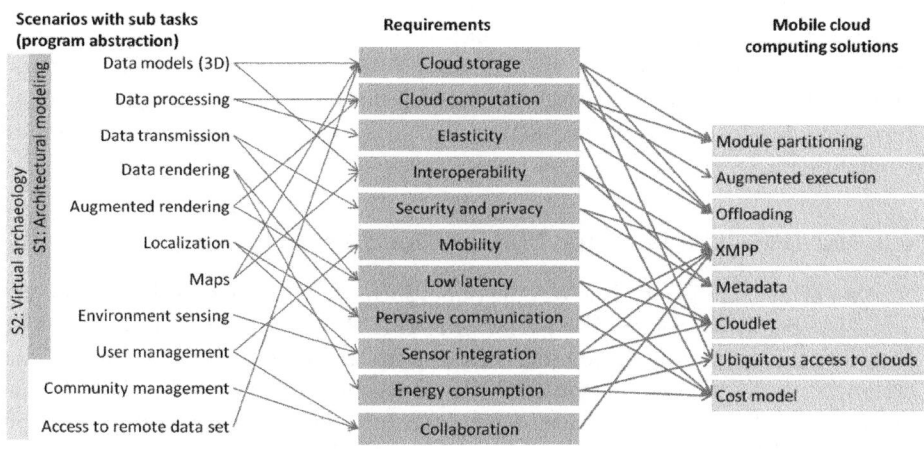

Fig. 2. Scenarios, requirement analysis and the mobile cloud computing solutions

The mobile cloud computing platform should provide a set of solutions to support the communication, collaboration and offloading. The cultural heritage and architecture domains address the scenarios which require an appropriate mobile cloud computing framework.

4 The XMPP-Based Mobile Cloud Middleware

The scenario-based requirement analysis shows that a XMPP based mobile cloud computing solution is promising to meet some requirements simultaneously. A cost model with clear definition and analysis of goals and constraints is the efficient approach to mobile cloud computing. We propose the mobile cloud computing architecture with two crucial aspects: the standardization of XMPP as cloudlet protocol and a context-aware cost model.

4.1 Context-Aware Cost Model

In order to dynamically shift the computation between mobile devices and cloud, applications need to be split in loosely-coupled modules interacting with each other. The modules are dynamically instantiated on and moved between mobile devices and cloud. It depends on several metric parameters modeled in the cost model as depicted in Figure 3. These parameters can include the module execution time, resource consumption, battery level, monetary costs, security, or network bandwidth. The main objective of the cost model is to help make intelligent offloading decisions that fulfill the given goals under certain device constraints based on several input parameters with the minimal overhead.

Several approaches to modeling the parameters have been proposed. Giurgiu et al. [8] model the application modules as resource consumption graph, and try to find a cut in the graph. Every bundle or module composing the application has certain memory consumption, generated input and output traffic, and the code size. The partitioning problem is to find an optimal cut in the graph under device constraints. Meanwhile, end-to-end interaction between phones and servers should be minimized. Gu et al. [10] model the relation between Java classes and use fuzzy logic for decision-making. Zhang et al. [25] use Naïve Bayesian Learning classifiers to find the optimal execution configuration from all possible configurations.

4.2 XMPP: The Glue for Mobile Cloud Services

An essential enabling technology in our approach is the Extensible Messaging and Presence Protocol (XMPP) [18]. The XMPP protocol provides a pure XML foundation for real-time messaging, opening up tremendous possibilities for more advanced real-time applications. XMPP together with its extensions is a powerful protocol for cloud services that demonstrate several advantages beyond traditional HTTP-based Web services (e.g. SOAP and REST):

- decentralized, open and flexible (extensible) communication protocol
- services being discoverable without the need of an external registry
- federation of services enabling easy weaving of cloud services together
- built-in presence functionality providing resource and availability discovery at runtime. XMPP not only allows discovering of services out of the box, but also supports determining their status and availability.

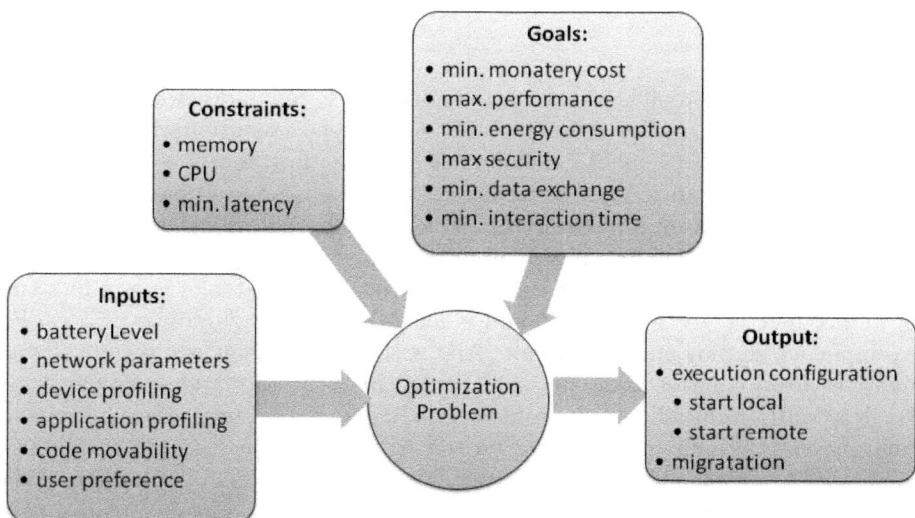

Fig. 3. The cost model with parameters for elastic mobile pervasive cloud applications

- support for real-time data streaming in two directions. Asynchronous invocation eliminates the need for ad-hoc solutions like polling
- interoperability with other protocols and programming language independent
- event notifications
- remote procedure calls (e.g. SOAP over XMPP)
- multimedia session management

Many advantages over existing technologies make XMPP a highly interesting candidate for next generation online services. HTTP was originally designed to accommodate query and retrieval of web pages and does not aim to rather complex communication. The intrinsically synchronous HTTP protocol is unsuitable for time-consuming operations, like computation-demanding database lookups or video processing.

An XMPP network can be seen as a complete XML-based routing framework upon which a messaging middleware can be built. Hence, an XMPP-based middleware can be used to integrate different services into a distributed computing environment. The application modules, external sensors and external services are XMPP entities identified by unique JIDs. They exchange messages through XML stanzas. However, a suitable mechanism to define the interconnections between different entities is needed.

4.3 Modular Architecture of the XMPP-Based Mobile Cloud Middleware

To enable the new mobile cloud application model, the middleware should provide an infrastructure for seamless and transparent execution of elastic applications and offer convenient development support. Furthermore, future mobile applications may utilize and synthesize capabilities from multiple clouds.

We adopt a concept similar to the cloudlets [19] and Slingshot [6] where the computation is offloaded to a nearby more powerful computer called surrogate. A cloudlet is a trusted, resource-rich computer or a cluster of computers that is well connected to the Internet and is available to nearby mobile devices. For example, we employ an access point (AP) with enough resources to run software modules from connected mobile devices. In this way, the latency amongst can be drastically reduced [19]. In contrast to [19] and [6], our approach does not migrate complete virtual machines on the AP. Using VMs for each mobile user connected to the cloudlet will make the cloudlet useless. Therefore, applications coming form different mobile users should share the same cloudlet platform for executing their software modules.

Our architecture relies on the possibility to split applications on loosely-coupled modules that encapsulate a unit of functionality. The developers split the application into loosely-coupled modules. Furthermore, they annotate the modules that can be offloaded and executed remotely. The software development tools determine the dependency between each module which helps further while deciding about module remote execution. The module functionality is exposed through service interface, so modules can easily be instantiated on the mobile device, on the nearby cloudlet, or on a distant cloud infrastructure. The offloading control makes decisions based on a cost model for the optimal execution.

Figure 4 depicts the main architecture of our approach. The XMPP-based mobile cloud middleware runs on mobile devices and cloudlet tier, and connects to other traditional cloud services. The execution manager/validator is responsible to instantiate, migrate and stop the application modules, based on the decisions from the offloading control component. The offloading control component implements the cost model and gives offloading decision for the movable modules based on the sensed context input parameters and designated performance goals . The messaging manager is responsible for synchronizing the different parts of the application. The application itself is split in loosely-coupled modules which can be executed locally on the mobile device or on a nearby cloudlet that hosts the pervasive cloud middleware. There exist some parts of the application which cannot be offloaded such as UI or accessing device hardware functionality. These parts constitute the application core running on the mobile device. The application metadata describes the dependencies and interconnection between modules, and their functionality. The module offloading proxy bridges the module invocations on separate tiers, so that the applications run as one logical entity.

To use XMPP as a communication protocol has several benefits. First, the XMPP publish/subscribe functionality enables easy realtime integration of heterogeneous entities, e.g. modules, services, and sensor streams, which greatly

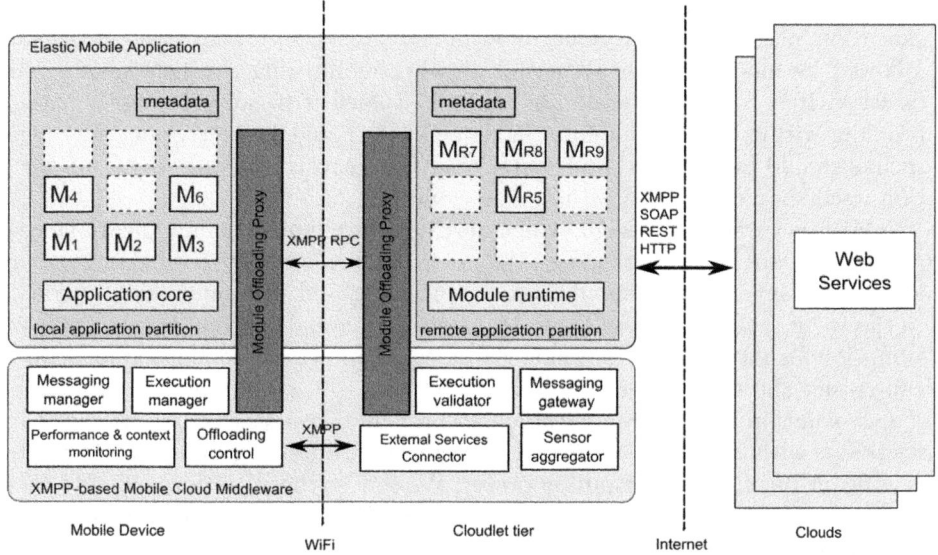

Fig. 4. Modular architecture of the XMPP-based Mobile Cloud Middleware

extends the possibilities of the mobile pervasive cloud applications. This is important in our case since the modules are dynamically started or stopped on different tiers. The XMPP presence awareness simplifies greatly the discovery of new resources and their availability. Second, XMPP enables better interoperability since XMPP uses XML streams for communication. The XML-based communication makes the protocol easily extendable, e.g., in multimedia session management where the signaling is handled by XML messages and media data is sent via other protocols such as Real-Time Transport Protocol (RTP) [17]. Third, XMPP inherently supports bi-directional communication, providing means for collaborative applications. Fourth, the federation model of XMPP enables interconnection with other servers, making the cloudlet a gateway to external cloud services.

5 Discussion

Since the execution of application modules shifts dynamically between mobile devices and cloud, the validity of execution results needs to be checked. The underlying middleware needs to transfer the state of migrating modules and related data. Moreover, requirements for real-time execution integration depend on the application itself. For example, in the previous use scenarios the mobile application would require fast updates from the rendering modules that run on the cloudlet tier. In contrast, an collaborative data sharing between mobile users on site would require more modest communication rate.

Due to the nature of pervasive environments, selection of the most optimal execution plan cannot be done deterministically. Continuous context sensing followed by module, application and network monitoring can help making the decision. It is resource-consuming for mobile devices to solve the optimization problem within itself. Therefore, lightweight, fast and predictive decision algorithms should be executed on the device. Alternatively, the optimization execution itself should be offloaded into the cloudlets.

XMPP is verbose by design, which may influence mobile communications negatively. Although verbosity makes the protocol open and easy for debugging and learning, it introduces also additional processing and communication overhead to the weak mobile devices. That means additional battery and bandwidth consumption for mobile devices. Using the stream compression extension for XMPP can reduce the size of messages.

Our solution is mostly applicable to smart mobile devices with Internet connectivity, with handheld form and a large, high quality graphics display, and significant but limited computing power. We choose these platforms due to their pervasive availability and large sensing capabilities. Their hardware improves constantly, but will still be resource-poor. Therefore, the mobile cloud middleware takes advantage of nearby computers, i.e. cloudlet, serving as a computation and collaboration platform with lower latency. However, the middleware can be extended easily to operate on the other cloud infrastructures. Furthermore, we give developers the control to split the application instead of automatic partitioning as introduced in [11].

6 Conclusions

Mobile cloud computing can augment the limited capabilities of mobile devices via access to elastic cloud computing resources. Based on two scenarios in the domains of cultural heritage and architecture, requirements on mobile cloud computing are derived. Challenges and problems raise related to large-scale data sets, limitations of mobile devices, required nearby equipment, and changing requirements of pervasive environments etc. Starting from the requirements, we propose a mobile cloud computing architecture to provide solution in the domains. Mobile devices are regulated to offload resource-demanding applications onto cloudlets in the pervasive environment aiming to lower the latency, and thus improve user experience. Access to remote cloud services are also supported. The XMPP protocol shows a promising component to meet the platform requirements. Moreover, a cost model is specified to model the factors that influence the performance and cost of this mobile cloud computing model. This cost model is still given at abstract level. Further plans of setting up this mobile cloud computing platform will be carried out with our on-going research on the underlying cost model. Additionally, our goal also is to enbable interoperability between the cloudlet and our existing cloud infrastructure (cf. [13]).

Acknowledgment. This work was supported in part by NRW State within the B-IT Research School and the Excellence Initiative of German National

Science Foundation (DFG) within the research cluster UMIC. We thank our colleague Dominik Renzel, and Georgios Toubekis (RWTH Aachen Center for Documentation and Conservation) for fruitful discussions with them concerning technical issues, and scenarios in the domains of architecture and archeology.

References

1. Broll, W., Lindt, I., Ohlenburg, J., Wittkämper, M., Yuan, C., Novotny, T., Gen Schieck, A.F., Mottram, C., Strothmann, A.: Arthur: A Collaborative Augmented Environment for Architectural Design and Urban Planning. Journal of Virtual Reality and Broadcasting 1(1) (December 2004); urn:nbn:de:0009-6-348
2. Chun, B.-G., Maniatis, P.: Augmented Smartphone Applications Through Clone Cloud Execution. In: Proceedings of the 12th Workshop on Hot Topics in Operating Systems (HotOS XII), Monte Verita, Switzerland. USENIX (2009)
3. Cuervo, E., Balasubramanian, A., Cho, D.-K., Wolman, A., Saroiu, S., Chandra, R., Bahl, P.: MAUI: Making Smartphones Last Longer with Code Offload. In: Proceedings of the 8th International Conference on Mobile Systems, Applications, and Services (ACM MobiSys 2010), San Francisco, CA, USA, pp. 49–62. ACM (2010)
4. Das, S., Agrawal, D., Abbadi, A.E.: ElasTraS: An Elastic Transactional Data Store in the Cloud. In: Proceedings of the 1st USENIX Workshop on Hot Topics in Cloud Computing (HotCloud 2009), San Diego, CA, USA. USENIX Association (2009)
5. David, F.M., Donkervoet, B., Carlyle, J.C., Chan, E.M., Pasquale, F.: Supporting Adaptive Application Mobility. In: Meersman, R., Tari, Z. (eds.) OTM-WS 2007, Part II. LNCS, vol. 4806, pp. 896–905. Springer, Heidelberg (2007)
6. Flinn, J., Narayanan, D., Satyanarayanan, M.: Self-tuned remote execution for pervasive computing. In: Proceedings Eighth Workshop on Hot Topics in Operating Systems (HotOS), Schloss Elamu, Germany, pp. 61–66. IEEE (2001)
7. Garriss, S., Cáceres, R., Berger, S., Sailer, R., van Doorn, L., Zhang, X.: Trustworthy and Personalized Computing on Public Kiosks. In: Proceeding of the 6th International Conference on Mobile Systems, Applications, and Services (MobiSys 2008), Breckenridge, CO, USA, pp. 199–210. ACM (2008)
8. Giurgiu, I., Riva, O., Juric, D., Krivulev, I., Alonso, G.: Calling the Cloud: Enabling Mobile Phones as Interfaces to Cloud Applications. In: Bacon, J.M., Cooper, B.F. (eds.) Middleware 2009. LNCS, vol. 5896, pp. 83–102. Springer, Heidelberg (2009)
9. Gleue, T., Dähne, P.: Design and implementation of a mobile device for outdoor augmented reality in the archeoguide project. In: VAST 2001: Proceedings of the 2001 Conference on Virtual Reality, Archeology, and Cultural Heritage, pp. 161–168. ACM, New York (2001)
10. Gu, X., Nahrstedt, K., Messer, A., Greenberg, I., Milojicic, D.: Adaptive Offloading Inference for Delivering Applications in Pervasive Computing Environments. In: Proceedings of the First IEEE International Conference on Pervasive Computing and Communications (PerCom 2003), Dallas-Fort Worth, TX, USA, pp. 107–114. IEEE (2003)
11. Hunt, G.C., Scott, M.L.: The Coign Automatic Distributed Partitioning System. In: Proceeedings of the Third Symposium on Operating System Design and Implementation (OSDI 1999), New Orleans, LA, USA, pp. 187–200. USENIX Association (1999)

12. Jing, J., Helal, A.S., Elmagarmid, A.: Client-server Computing in Mobile Environments. ACM Computing Surveys (CSUR) 31(2), 117–157 (1999)
13. Kovachev, D., Klamma, R.: A Cloud Multimedia Platform. In: Proceedings of the 11th International Workshop of the Multimedia Metadata Community on Interoperable Social Multimedia Applications (WISMA 2010), Barcelona, Spain, pp. 61–64. CEUR (2010)
14. Kovachev, D., Renzel, D., Klamma, R., Cao, Y.: Mobile Community Cloud Computing: Emerges and Evolves. In: Proceedings of the First International Workshop on Mobile Cloud Computing (MCC), Kansas City, MO, USA. IEEE (2010)
15. Oberheide, J., Veeraraghavan, K., Cooke, E., Flinn, J., Jahanian, F.: Virtualized In-Cloud Security Services for Mobile Devices Categories and Subject Descriptors. In: Proceedings of the First Workshop on Virtualization in Mobile Computing, Breckenridge, CO, USA, pp. 31–35. ACM (2008)
16. Rellermeyer, J.S., Duller, M., Alonso, G.: Engineering the Cloud from Software Modules. In: Proceedings of the Workshop on Software Engineering Challenges in Cloud Computing (ICSE-Cloud, in Conjunction with ICSE 2009), Vancouver, Canada, pp. 32–37. IEEE (2009)
17. Saint-Andre, P.: Jingle: Jabber Does Multimedia. Multimedia 14(1), 90–94 (2007)
18. Saint-Andre, P., Smith, K., Tronçon, R.: XMPP: The Definitive Guide. O'Reilly (2009)
19. Satyanarayanan, M., Bahl, P., Cáceres, R., Davies, N.: The Case for VM-Based Cloudlets in Mobile Computing. IEEE Pervasive Computing 8(4), 14–23 (2009)
20. Satyanarayanan, M., Kozuch, M.A., Helfrich, C.J., Hallaron, D.R.O.: Towards Seamless Mobility on Pervasive Hardware. Pervasive and Mobile Computing 1(2), 157–189 (2005)
21. Tolvanen, J., Suihko, T., Lipasti, J., Asokan, N.: Remote Storage for Mobile Devices. In: Proceedings of the 1st International Conference on Communication Systems Software & Middleware, New Delhi, India, pp. 1–9. IEEE (2006)
22. Vlahakis, V., Ioannidis, M., Karigiannis, J., Tsotros, M., Gounaris, M., Stricker, D., Gleue, T., Daehne, P., Almeida, L.: Archeoguide: an augmented reality guide for archaeological sites. IEEE Computer Graphics and Applications 22(5), 52–60 (2002)
23. Wright, A.: Get Smart. Communications of the ACM 52(1), 15–16 (2009)
24. Wu, H., Hamdi, L., Mahe, N.: TANGO: A Flexible Mobility-Enabled Architecture for Online and Offline Mobile Enterprise Applications. In: Proceedings of 11th International Conference on Mobile Data Mangement (MDM 2010), Kanas City, MO, USA, pp. 230–238. IEEE (2010)
25. Zhang, X., Jeong, S., Kunjithapatham, A., Gibbs, S.: Towards an Elastic Application Model for Augmenting Computing Capabilities of Mobile Platforms. In: The Third International ICST Conference on MOBILe Wireless MiddleWARE, Operating Systems, and Applications, Chicago, IL, USA (2010)
26. Zhang, X., Schiffman, J., Gibbs, S., Kunjithapatham, A., Jeong, S.: Securing Elastic Applications on Mobile Devices for Cloud Computing. In: CCSW 2009: Proceedings of the 2009 ACM Workshop on Cloud Computing Security, Chicago, IL, USA, pp. 127–134. ACM (November 2009)

Elastic HTML5: Workload Offloading Using Cloud-Based Web Workers and Storages for Mobile Devices

Xinwen Zhang, Won Jeon, Simon Gibbs, and Anugeetha Kunjithapatham

Computer Science Laboratory, Samsung Information Systems America
75 W. Plumeria Drive, San Jose, CA 95134

Abstract. In this position paper, we propose the concept of *Elastic HTML5*, which enables web applications to offload workload using cloud-based web workers and cloud-based storage for mobile devices. Elastic HTML5 is a collection of software components and functions in for a web runtime agent (e.g., web browser); this includes components and methods to create and manage web workers in the cloud so as to augment the computation functionality of a browser-based application running on a device. Elastic HTML5 also includes the components and methods to create and manage elastic storage between the main thread of a web application and its web workers. With these functions, a web application can run in elastic manner such that, whenever necessary, the mobile device can obtain resources from the cloud, including computation and storage, and when necessary, it can run offline at the device side completely.

1 Introduction

HTML5 is being proposed by W3C as the next generation of HTML. Although the specification has been under development since 2004, many features of HTML5 are supported by the latest builds of browsers including Firefox, Chrome, Safari and Opera. Microsoft is also starting to support HTML5 within Internet Explorer. Web workers and local storage are important concepts and features in HTML5. A web worker is a script that runs in the background within the browser. For example, in the Chrome browser, a web worker is an independent background renderer process. A web application (e.g., an HTML page) can communicate with a web worker with onMessage() and postMessage() methods. A web page usually can only talk to web workers from the same origin. Each web application or web worker can have persistent local storage (such as SQL database). Compared with traditional cookies, local storage has the benefits of larger data size, supporting variant data types, and first class object names to access data with JavaScript.

Mobile devices such as cellular phones, in general, are constrained platforms with limited computational power and storage, support for a fixed set of codecs and data formats, and limited capability to access and process web services. To address these restrictions for computing-, networking- and storage-intensive web applications, we propose the concept of *Elastic HTML5*, which enables an elastic version of the web worker model in HTML5 and allows workers to run on remote cores. Elastic storage between a device and cloud dynamically extends the capability of a web application and synchronizes data between the web application's main thread and its web workers.

Our elastic scheme enables transparent distribution of the computation and storage of a single web application into multiple locations including client browser environments and remote cloud nodes. The main function to support this feature is a pair of proxy web workers on client and cloud sides to relay and route messages between other web workers. Rather than a simple and rigid solution where nearly all processing and storage is done either on the cloud or on the device, we want to be able to migrate functionality between the device and cloud, as and when it is required. This ability allows the device to adapt to different workloads, performance goals, and network latencies. For example, an application could run locally when the device workload is light, but as the workload increases, more and more of its computation can be shifted off the device to the cloud.

Some of the advantages of elastic HTML5 include:

- It can transparently augment mobile browser capability with cloud computing and storage resources.
- It needs no change to the existing web programming model (i.e., HTML, CSS, and JavaScript) and facilitates cross-platform development.
- It introduces no new programming language to application developers.
- It uses a simple extension for integration with web applications, via web worker and storage APIs.
- It requires a simple extension and integration with mobile platforms which support HTML5 web worker and storage specifications.

The rest of this paper is organized as follows. Section 2 covers some background and related work about HTML5 and computation offloading. We highlight some features and functions of elastic HTML5 in Section 3, including launching web workers in the cloud and runtime message flow. We discuss some future research directions towards practical elastic HTML5 at the end of this paper.

2 Background

HTML5. The W3C HTML5 working draft specifies the APIs of HTML5 applications. Web workers and local storage are important features of HTML5. A web worker can be forked from the main browsing context in the browser runtime and can communicate via `postMessage()` and `onMessage()`. An HTML5 application can use local storage of a client, such as SQL database, to store application-specific data. So far, there has been no prior consideration of running web workers and storage in the cloud for the client-side browsing sessions.

Remote Execution Systems. Cyber foraging [7,8,16] is a common approach explored by many to augment the capability of resource-constrained mobile devices. The basic idea is to dynamically discover and make use of nearby resources, called *surrogates*, to offload the execution of an application or parts of an application running on a mobile device. Adaptive Offloading [10], Coign [11], and R-OSGi [14] leverage programming language and application runtime middleware to transform applications into distributed systems. However these remote execution mechanisms and architectures are not for the browser environment and many of them are tightly coupled with specific programming languages such as Java. In addition, most of them require a new application model, which target splitting legacy applications to run in distributed manner.

CloneCloud [9] takes the approach of cloning the entire user's mobile device environment on a remote server(s). Applications can then be quickly restarted on or migrated to the remote machine when the user's machine is running low on resources. Whereas CloudCloud needs to change existing applications to make them partial on device and partial on cloud, our scheme does not need to change the application model used by web applications. It uses existing interfaces for communication between JavaScript contexts and works transparently with HTML5 applications. More specifically, existing HTML5 applications can use our techniques without any modification.

Virtual Machine Migration. Virtual machine migration [12,17] and VM-based cloudlet [15] are complementary approaches to enable users to seamlessly access their applications and data across multiple and heterogeneous devices in general. It also enables users to instantly continue/restore an application on a different device, when their current machine is running low on resources. The fundamental design objective of our elastic web application model is to remove the constraints of specific mobile platforms by providing a distributed framework that extends the device into the cloud. The salient feature of the elastic model is that it can offer a range of elasticity patterns between resource-constrained devices and Internet-based clouds [19]. Each pattern in turn can be realized by several execution configurations. A comprehensive cost model can be used to dynamically adjust execution configurations thus optimizing application performance in terms of power savings, monetary savings, or throughput.

Browsing Through Remote Rendering. Flashproxy [13] supports active web content on mobile devices using a proxy to splice active content out of web pages and replace it with an Ajax-based remote display component. The spliced active content executes within a remote sandbox on the proxy. However, Flashproxy is specific for the Flash plugin, and it has to change Flash bytecode to access a proxy. Similarly, Opera Mini is a JavaME application that displays web pages that are reformatted and compressed by Opera servers [4]. Deepfish is Microsoft's experiment in remote rendering for mobile browsing [2].

Compared with our scheme, Flashproxy has only one proxy server and does not offer an elastic storage mechanism. Also, Opera Mini and Deepfish work for web page format/rendering only, so they do not distribute computation or storage of a single web application into many locations.

3 Elastic HTML5

3.1 HTML5 Web Worker

The W3C WHATWG specifies HTML5 web worker [5]. Web workers allow JavaScript to run in parallel on a web page, without blocking the user interfaces. Typically, a web worker is an independent browser execution context (i.e., thread or process) from any web page. It runs in background and talks with the web page that creates it via dedicated message channels. It can talk with web servers via `XMLHttpRequest` (for HTML4) and `WebSocket` (for HTML5) methods. Furthermore, it can store data persistently in local database. A web worker can be either dedicated, in that it works for one web page, or shared by multiple pages, e.g., to transfer data or messages between them. Also, it can create sub-workers, delegate tasks, aggregate results, and can be a shared library for other web workers.

We have tested the support of web worker functionalities in various mainstream browsers. As a web worker runs in an independent browser execution context, it cannot access global variables and DOM (Data Object Model) from the main thread of a web page. Also, as accessing to session and local storage are through global variables in the current HTML5 specification, a web worker cannot access a web page's session and local storages. However, according to the specification, a web worker can access local database storage, which can be independent to each web worker, or can be shared with the main thread of the application, although so far based on our experiment, none of the current mainstream browsers support this function. Therefore, this position paper focuses on the elasticity of web worker and database-based storage. However, if other forms of local storage are supported for web worker in the future, it can be applicable to them.

3.2 Elastic Web Worker

The basic idea for elastic web application with HTML5 is to run web worker in a cloud platform, e.g., private cloud such as home and enterprise, or public cloud such as Amazon EC2 [1] and Google AppEngine [3]. Figure 1 shows the overview of our elastic web worker. To transparently support this elasticity, the following requirements need to be supported for web worker:

- A web worker should communicate with a web page's main thread with existing APIs (i.e., message-based communication channels), no matter it is running on local browser environment or in cloud environment.
- Web workers can communicate with each other with existing APIs no matter they are located together in one location or in separate locations.
- A web worker should be able to access database storage with existing APIs, no matter the database is in device side or in cloud. That is, the database should appear "locally" no matter where the web worker is running.
- When running in cloud, a web worker should be able to access external web servers and comply with the same security policies as it runs in the client-side browser environment.

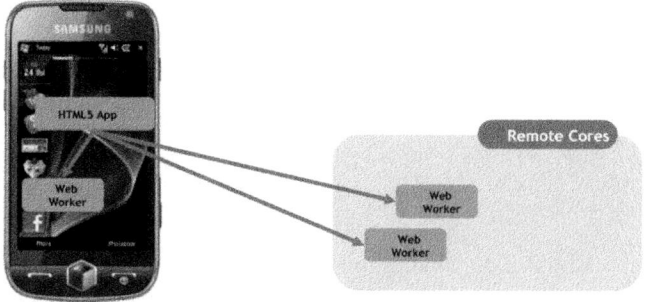

Fig. 1. Concept of Elastic Web Worker

3.3 Launching Web Worker

Figure 2 shows the infrastructure to launch a web worker in elastic HTML5. The high-level work flow is following:

1. Whenever the main thread of an HTML5 application (e.g., a loaded web page) invokes a web worker, it sends the request to a device-side elastic web worker manager (EWWM), which decides where to launch the web worker - in the same browser environment of the device, or on cloud. EWWM provides the same web worker message APIs to JavaScript such that there is no need to change any existing HTML5 application.
2. EWWM checks security policies on launching the web worker, such as the same-origin policy (SOP), according to the HTML5 specifications. A web worker is invoked by EWWM only when it passes the security checks.
3. EWWM maintains a table with each loaded web page, which includes the information of all the names of active web workers (accessed by the main thread in the web page) and others including URLs if they are running in the cloud, and the message ports if they are running locally.
4. EWWM decides where to launch the target web worker. When the worker is launched locally, EWWM invokes the web worker with parameters from the first step, and updates the table with the related information of the launched web worker. The decision can be made by considering some cost objectives [19].
5. If EWWM decides to launch the target web worker in the cloud, it files the request to cloud web worker service (CWWS), along with the parameters and code of the web worker. CWWS in turn arranges necessary resources such as a specific CPU core, storage, etc. and launches the web worker with the transferred parameters.
6. CWWS returns the end point (URL) of the web worker to EWWM, which in turn updates its active web worker table with the related information.
7. After these steps, the main thread and invoked web workers can talk to each other, using existing message-based web worker APIs.

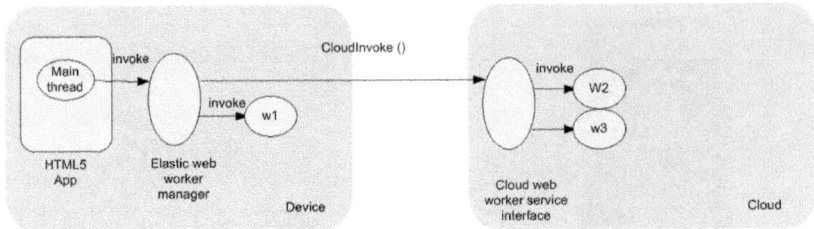

Fig. 2. Launching Web Worker in Elastic HTML5

In order to save the cost of transferring the code from device to cloud, EWWM can send the web worker invocation request with parameter of the web worker's absolute URL. Then, CWWS fetches the web worker code and invokes it in cloud. Again, the security policies are enforced by EWWM and CWWS such that web workers will be executed with the same origin as the device-side web application, although its code can be retrieved from a different origin.

In real implementation, EWWM can be a browser extension, a plugin, or a usual web worker. Also, CWWS can be web service interface, or simply a cloud-based web worker.

3.4 Runtime Message Flow

Figure 3 shows the runtime interaction among a web application and elastic web workers. Two proxy web workers, residing on device and cloud side and representing the role of EWWM and CWWS, respectively, route and relay messages between original web workers and web application main thread. All communication between web workers are through postMessage() and onMessage() interfaces specified by HTML5. The communication between two proxy web workers are bidirectional, e.g., via WebSocket APIs [6], a dedicated TCP connection, or a long lived HTTP connection, which is not visible to the main thread and other web workers. The proxy web workers act as *distributed elasticity layer* for web applications.

A cloud web worker can access external web servers (e.g., via Ajax) directly, according to the same security policies in the client-side browser environment. Optionally, a cloud web worker can only access external web servers via the device-side proxy web worker, which can enforce security policies in more efficient and consistent way. To support this, the Ajax APIs should be wrapped or rewritten to embody this indirect calling. With this design, each web worker and the main thread have exactly the same interfaces as specified by HTML5 or its future versions. All communications and location-awareness are handled by the proxy web workers transparently.

A web worker can be migrated from device to the cloud or vice versa. As each web worker communicates with others and external web servers with the

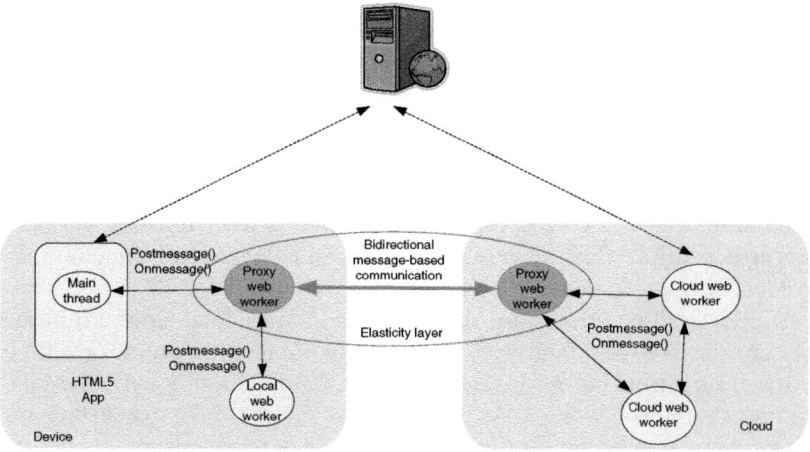

Fig. 3. Message Flow of Elastic Web Worker

same interfaces as specified by HTML5, a web worker can seamlessly run after migration. The decision for the migration can also be based on some objectives.

3.5 Elastic Storage

Each web worker can have its local database storage. To maintain transparency, a web worker can access database no matter where it is running, thus management and synchronization of local storage between device and the cloud has to be supported. Figure 4 shows the method to enable this idea. By this description, we assume that there is only one active copy of a data item in a database - either the data is in cloud-side or in device-side.

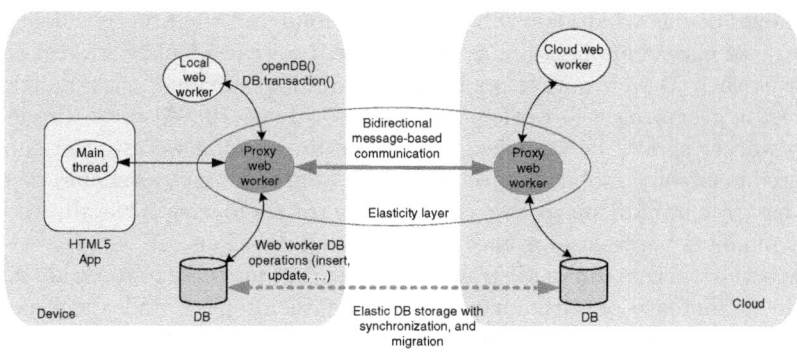

Fig. 4. Database Storage and Web Worker in Elastic HTML5

Similar to web worker APIs, the elasticity layer provides the same interfaces specified by HTML5 to local database storage. when a web worker invokes APIs to access database such as `openDB()`, the proxy web worker looks up at both device and cloud sides to check if the target database exists. If not, the database is created locally (local to the invoking web worker) and database transactions are performed as usual, e.g., with `DBtransaction()` method. Otherwise, the elastic layer routes the message to the corresponding database either in device or cloud side, and the corresponding proxy web worker finishes the operations and returns results. Thus, although a database is located in the cloud, it works like a local storage to device web workers.

To support offline operation, the device-side and cloud-side database should be synchronized or even migrated. This can be done either online or offline. Also, a database synchronization or migration can be done along with the invocation or migration of a web worker. As alternatives, the database synchronization and migration can be done by elasticity layer or other mechanisms.

According to different operations from different web workers, the databases at device and at cloud can be split from a single database, e.g., each has different tables, or each has different attributes of the same table. Database splitting and union operations are possible along with invocation and migration of web workers.

4 Concluding Remarks and Future Work

While we believe elastic HTML5 bridges the web-based programming model for mobile devices and powerful computing resources from cloud in a friendly manner, there are several challenges to be resolved before it can be widely useful for general web applications. First, the network latency is a major issue for many highly user interactive web applications. To minimize latency, web workers having tight dependency on updating UI parts of an application should run at the device side. In general, there can be a break-even point by considering the benefit of offloading web workers to cloud to have fast computation and the cost of network latency, which needs comprehensive study in our future work.

With distributed web workers and storage, running an elastic HTML5 application should decide what logic will run on the cloud and what on the client. As the topology of elastic applications is more varied, we can identify several common patterns, such as web worker pools and shadowing on cloud, task splitting web workers, aggregating web workers, and many others [19]. Also, to decide where a web worker should be launched, and when an running web worker should be migrated between a mobile device and cloud, some cost factors should be considered, e.g., for minimizing power consumption of the device. Overall, we believe a cost model is necessary for elastic HTML5 applications.

Since web workers run in different locations, it is desirable to replicate database storage to increase performance, but then data integrity and synchronization become issues. Further, code and application state computation migration is a traditional problem in many systems [8,18]. It is a challenging task to support runtime web worker migration thus enhance mobile user experience, but at the same time achieve the transparency and seamlessness. Furthermore, integrity and data security of web workers running on cloud are potential problems. We have designed a lightweight protocol to distribute shared secrets and session keys between distributed application components running on mobile devices and cloud nodes for mutual authentication purposes [20]. However, how to build strong trust between web worker runtime environments in the cloud is an open problem that we will explore.

References

1. Amazon Elastic Compute Cloud (Amazon EC2), http://aws.amazon.com/ec2/
2. Deepfish, http://en.wikipedia.org/wiki/Microsoft_Live_Labs_Deepfish
3. Google AppEngine, http://code.google.com/appengine/
4. Opera Mini, http://www.opera.com/mobile/
5. Web Workers, Draft Recommendation (August 23, 2010),
   http://www.whatwg.org/specs/web-workers/current-work/
6. WebSocket API, Editor's Draft (August 10, 2010),
   http://dev.w3.org/html5/websockets/
7. Balan, R., Flinn, J., Satyanarayanan, M., Sinnamohideen, S., Yang, H.: The case for cyber foraging. In: Proc. of ACM SIGOPS European Workshop (2002)
8. Balan, R., Satyanarayanan, M., Park, S., Okoshi, T.: Tactics-based remote execution for mobile computing. In: Proc. of MobiSys (2003)
9. Chun, B.-G., Maniatis, P.: Augmented smartphone applications through clone cloud execution. In: Proc. of USENIX HotOS XII (2009)
10. Gu, X., Messer, A., Greenberg, I., Milojicic, D., Nahrstedt, K.: Adaptive offloading for pervasive computing. IEEE Pervasive Computing 3(3) (2004)
11. Hunt, G.C., Scott, M.L.: The Coign automatic distributed partitioning system. In: Proc. of OSDI (1999)
12. Kozuch, M., Satyanarayanan, M.: Internet suspend/resume. In: Proc. of IEEE WMCSA (2002)
13. Moshchuk, A., Gribble, S.D., Levy, H.M.: Flashproxy: transparently enabling rich web content via remote execution. In: Proc. of MobiSys (2008)
14. Rellermeyer, J.S., Alonso, G., Roscoe, T.: R-OSGi: Distributed Applications Through Software Modularization. In: Cerqueira, R., Pasquale, F. (eds.) Middleware 2007. LNCS, vol. 4834, pp. 1–20. Springer, Heidelberg (2007)
15. Satyanarayanan, M., Bahl, P., Caceres, R., Davies, N.: The case for VM-based Cloudlets in mobile computing. IEEE Pervasive Computing 8(4) (2009)
16. Sousa, J., Garlan, D.: An architectural framework for user mobility in ubiquitous computing environments. In: Proc. of IEEE/IFIP Working Conference on Software Architecture (2002)
17. Travostino, F.: Seamless live migration of virtual machines over the man/lan. In: Proc. of SC (2006)
18. Xian, C., Lu, Y.H., Li, Z.: Adaptive computation offloading for energy conservation on battery-powered systems. In: Proc. of ICPADS (2007)
19. Zhang, X., Jeong, S., Kunjithapatham, A., Gibbs, S.: Towards an Elastic Application Model for Augmenting Computing Capabilities of Mobile Platforms. In: Cai, Y., Magedanz, T., Li, M., Xia, J., Giannelli, C. (eds.) Mobilware 2010. LNICST, vol. 48, pp. 161–174. Springer, Heidelberg (2010)
20. Zhang, X., Schiffman, J., Gibbs, S., Kunjithapatham, A., Jeong, S.: Securing elastic applications on mobile devices for cloud computing. In: Proc. of ACM Cloud Computing Security Workshop (2009)

Cost-Sensitive Detection of Malicious Applications in Mobile Devices

Yael Weiss[1,2], Yuval Fledel[1,2], Yuval Elovici[1,2], and Lior Rokach[1,2]

[1] Department of Information Systems Engineering,
Ben-Gurion University of the Negev, Be'er Sheva 84105, Israel
[2] Duetsche Telekom Laboratories at Ben-Gurion University, Israel
{wiessy,fledely,elovici,liorrk}@bgu.ac.il

Abstract. Mobile phones have become a primary communication device nowadays. In order to maintain proper functionality, various existing security solutions are being integrated into mobile devices. Some of the more sophisticated solutions, such as host-based intrusion detection systems (HIDS) are based on continuously monitoring many parameters in the device such as CPU and memory consumption. Since the continuous monitoring of many parameters consumes considerable computational resources it is necessary to reduce consumption in order to efficiently use HIDS. One way to achieve this is to collect less parameters by means of cost-sensitive feature selection techniques. In this study, we evaluate ProCASH, a new cost-sensitive feature selection algorithm which considers resources consumption, misclassification costs and feature grouping. ProCASH was evaluated on an Android-based mobile device. The data mining task was to distinguish between benign and malicious applications. The evaluation demonstrated the effectiveness of ProCASH compared to other cost sensitive algorithms.

Keywords: Intrusion Detection, Mobile Devices, Malware, Security, Android, sCost sensitive feature selection.

1 Introduction

Smart mobile phones have become a primary communication device for many individuals. In 2008, the converged mobile device segment outpaced the rest of the industry growing 22.5% compared to 2007[1]. Integrating the traditional functionality of mobile phones with special computer-enabled features not previously associated with telephones, smart phones require various security solutions in order to maintain their proper functionality and to protect against malicious behavior. Many of these solutions have migrated from desktop computers where they were initially introduced.

Some of the more sophisticated solutions, such as host-based intrusion detection systems (HIDS), continuously monitor many parameters in the device. Data mining techniques are then applied to the collected data in order to detect abnormal states. As opposed to desktop devices which have evolved over the years into robust instruments with massive resources, mobile devices are constraint-based devices since they are limited, mainly in battery power but in memory and CPU as well. Therefore, since the continuous monitoring of many parameters consumes considerable computational resources it is necessary to reduce consumption in order to efficiently use HIDS on mobile devices. One way to reduce the power consumption is to monitor those features whose acquisition requires less power while maintaining the data mining process performance. To implement this task it is necessary to make a smart decision on which variables to monitor.

Another important aspect involved in determining which subset of features to monitor is feature grouping which arises occasionally when feature costs vary with the choice of a prior feature. For instance, let us assume that the raw data about two feature values lies on the same sector on a rotating disk. In order to monitor the first feature value we need to pay two cost units. The first cost unit is for uploading the page from the hard disk to the main memory; the second supplementary cost unit is paid in order to read the raw data from the main memory and then to perform the required calculations to induce the feature value from the raw data. Then, in order to monitor the second feature, we only need to pay the second supplementary cost unit since the page is already in main memory. Hence, information about the feature grouping must be tailored into the feature selection process.

In addition to the resource costs, when deciding which features subset to monitor, it is also necessary to consider misclassification costs. In some applications the cost of false positive (FP) and false negative (FN) costs bear a different penalty. For example, if the false positive (FP) cost is substantially higher than the false negative (FN) cost, we would prefer to sample features that would reduce the false positive (FP) errors over features that reduce the false negative errors.

In this study, we evaluate ProCASH, a new cost-sensitive feature selection algorithm sensitive to the resources, and misclassification costs as well as feature grouping. ProCASH was evaluated on an Android-based mobile device and the data mining task focused on smart phone security. Extensive experimentation s demonstrated the effectiveness of ProCASH compared to other algorithms.

The rest of the paper is structured as follows. In section 2 we present related work; in section 3 we introduce the ProCASH cost sensitive feature selection algorithm; in section 4 we describe the datasets we used for the evaluation and the evaluation results; section 5 presents a summary and concluding remarks.

2 Related Work

Cost-sensitive learning is an essential task in several real-world applications. Turney et al.[2] presented a taxonomy of the main types of costs involved in inductive concept learning. Two costs, misclassification cost and test cost (which is equivalent to the resource consumption which is consumed when a feature is being monitored) are particularly relevant to this paper.

Turney[3] was the first to consider both test and misclassification costs. Turney's approach presents the inexpensive classification with an expensive test (ICET) system. ICET uses a method that employs a hybrid approach, combining a greedy search heuristic (decision tree) with a genetic algorithm for building a decision tree that minimizes the total cost objective which is composed of test and misclassification costs. Furthermore, ICET considers feature grouping. Although the ICET is robust, it is very time consuming as well. Chai et al. [4] offered csNB, a new cost-sensitive classifier based on a naïve Bayes algorithm. Several works, such as [8][10] use a cost-sensitive decision tree for the classification task. Ling et al.[5] proposed a decision tree algorithm which applies a new splitting criterion, minimal total cost, to training data instead of the well known minimum entropy measurement. In another paper, Sheng et al. [6] offered a framework where a decision tree is built for each new test case. Ling et al. [7]subsequently updated their strategy for building cost-sensitive decision trees by incorporating possible discounts when obtaining the values of a group of attributes with missing values in the tree building algorithm. Sheng et al. [9] suggested a hybrid cost-sensitive decision tree, DTNB that reduces the minimum total cost by integrating the advantages of a cost-sensitive decision tree and those of the cost-sensitive naïve Bayes. While it uses the cost-sensitive decision tree in order to decide which tests to choose, for the classification task it uses the cost-sensitive naïve Bayes. Freitas et al. [10] suggest a new splitting criterion in building decision trees that considers different aspects of test costs. They examine several cost-scale factors that regulate the influence of tests costs as it can make trees more sensitive to tests costs. They also suggest how to embed the risk cost for performing the test in the new cost-sensitive splitting criterion. The risk cost captures the change in the quality of life due to performing these tests on the patient. However, no experiments were carried out in regard to risk cost.

3 ProCASH- A Cost-Sensitive Algorithm

In this paper we introduce ProCASH, a cost-sensitive algorithm which takes as its starting point a similar preprocessing step as the CASH algorithm. However, unlike the CASH algorithm, ProCASH does not assume that all cost types have the same cost units; While CASH evaluation measurement is the summation result of the different costs, ProCASH evaluation metric is, the achieved average misclassification cost given a maximal budget of resource cost. Furthermore, in contrast to CASH which used the genetic algorithm as its search algorithm, ProCASH uses a new search algorithm.

CASH [11] is a cost-sensitive feature selection method which uses a new fitness function based on comparing histograms. This algorithm follows the filter approach. The CASH algorithm takes into account resource costs as well as feature grouping and misclassification costs. The

CASH algorithm consists of four main steps: preprocessing; creating an initial population of individuals; computing the fitness of each individual; and applying a genetic algorithm to the initial population. CASH assumes that all types of costs are given in the same scale and therefore, it uses as an evaluation metric the average total cost which is composed of the summation of the average misclassification and resource cost.

CASH's preprocessing step is composed of four sub-steps. In the first sub-step, CASH computes the average a priori cost which indicates when a features subset should not be obtained. That is to say, if the average a priori cost, computed according to the distribution of classes in the training set, is lower than the average misclassification cost achieved by the features subset, CASH will not choose this subset. Then, in the second sub-step CASH computes histograms for each class value of each feature in the training dataset. In the third sub-step, CASH computes for each feature how it classifies the records in the training dataset. This computation is based on a cost- sensitive majority rule that classifies all the records in a certain bin to the class which minimizes the misclassification cost. Finally, in the fourth sub-step, CASH calculates for each feature the misclassification cost ratio it assigns to each record in the training dataset. The motivation for calculating the misclassification cost ratio of a certain feature is to supply the algorithm with the knowledge of whether or not the decision is sufficiently distinctive. That is to say, based on the distribution of the classes in a certain bin of a feature, the CASH algorithm tries to estimate what is the likelihood that the algorithm's classification was correct.

As opposed to CASH, ProCASH adds an alteration to the preprocessing step by changing the way that the classification decision of each feature to each record is carried out. For each feature and record, ProCASH checks if the misclassification cost ratio is larger than a certain threshold. If so, ProCASH add the record to a list of records that have not been correctly classified by that feature. Furthermore, in contrast to CASH which used the genetic algorithm as its search algorithm, ProCASH uses a new search algorithm. In the beginning of the search, ProCASH's search method first looks for an initial subset with which it starts the search by performing the two following steps. Firstly, it gathers each feature in the training set that: (1) has classified correctly more than a predefined threshold of the training set's records; (2) whose feature resource cost is no higher than the resource constraint; and (3) whose average misclassification cost of all the records in the dataset is lower than the average a priori cost Secondly, ProCASH builds the initial selected features subset by iteratively selecting features with the minimal average misclassification cost from the features that were gathered in the previous step. ProCASH continues to add features to the initial subset until the point is reached where the addition of an extra feature causes the initial chosen subset's resource cost to exceed the resource constraint.

Then, after the initial feature subset necessary to start the search has been chosen, ProCASH computes a list of all the records that have not been correctly classified by at least one of the chosen subset features. ProCASH then starts the search with the initial chosen subset by iteratively adding to the chosen subset one feature which holds the minimum misclassification cost. ProCASH continues to add features to the chosen subset as long as: (1) the number of records that have not been correctly classified by at least one feature in the chosen subset is larger than five percent of the records in the training set and (2) the chosen subset's features resource cost does not exceed the resource constraint. Finally, ProCASH returns as output the chosen subset of features.

4 Experiments

In this section we present and analyze empirical results obtained from evaluating a cost-sensitive malware detection framework designed for Android devices [12]. Our goal was to explore malware detection when using ProCASH in comparison to several other cost-sensitive algorithms, all constrained by a specific CPU consumption budget cost.

ProCASH was implemented in Java. In order to evaluate the performance of ProCASH, we compared it to four algorithms: csDT_csf1 classifier[10], csDT_csf2 classifier[10], GA+META+CsId3[3] and GA+META_ICF[3]. The csDT_csf is a classifier which follows the embedded approach. csDT_csf employs a cost-sensitive decision tree to obtain a setting for the cost-scale factor (csf), that adjusts the strength of the bias towards lower cost attributes. The GA+META_GA+META_ICF is a wrapper algorithm which uses GA+META_ICF [13] as its fitness function and the genetic search as its search algorithm. The GA+META_CsId3 is a wrapper algorithm which uses the same heuristic function as in CSID3[14][15] algorithm and the genetic search as its search algorithm. For each feature selection algorithm (ProCASH, GA+META+ICF and GA+META+CSID3) based on the training set, the algorithm selects a feature subset. Then, features that were not selected were eliminated from the corresponding training and testing set. Afterward, a decision tree was induced on each of the training sets and its performance was evaluated on the corresponding test set. We used the J48 classifier, a Java implementation in WEKA[16] data mining applications of the C4.5 decision tree algorithm that Quinlan[17] introduced. In order to make our classifier cost-sensitive to misclassification costs, we used a meta-learner, implemented in Weka MetaCostClassifier[18].

The rest of this section is composed of the following subsections: In subsection 4.1 we describe the datasets we used; in subsection 4.2 we show the metrics and the statistical tests for measuring the performance of the algorithms and compare them; in subsection 4.3 we describe the experiment plan; in subsection 4.4 we describe the evaluation results followed by a discussion in subsection 4.5.

4.1 Datasets

There are several types of threats targeting mobile malware. In our research we focus on attacks against the phones themselves and not the service provider's infrastructure [34]. Four classes of attacks on a mobile device were identified [6]: unsolicited information, theft-of-service, information theft and denial-of-service (DoS). Since Android is a new platform and there are yet no known instances of Android malware, we developed four applications that perform denial-of-service and information theft.

The first malware we developed was the "Tip Calculator", a calculator which unobtrusively performs a DoS attack. When a user clicks the "calculate" button to calculate the tip, the application starts a background service that waits for a period of time and then launches several hundreds CPU-consuming threads. The attack almost absolutely paralyzes the device. The system becomes very unresponsive and the only effective choice is to shutdown the device (which also takes some time). An interesting observation is that the Android system often kills a CPU-consuming service but always keeps on re-launching it a few seconds later.

Also, we developed three malicious information theft applications. The first malware includes a set of two Android applications exploiting the Shared User ID feature. In Android, each application requests a set of permissions which is granted at installation time. The Shared User ID feature enables multiple applications to share their permission sets, provided they are all signed with the same key and explicitly request the sharing. It is noteworthy that the sharing is done behind the scenes without informing the user or asking for approval, resulting in implicit granting of permissions. The first Android application is Schedule SMS, a truly benign application that enables one to send delayed SMS messages to people from a contact list for which the application requests necessary permissions. The second application, Lunar Lander, is a seemingly benign game that requests no permissions.

Once both applications are installed and the Lunar Lander obtains the capability to read the contacts and send SMS messages, it exhibits a Trojan-like behavior, leaking all of the contact names and phone numbers through SMS messages to a pre-defined number. This resembles RedBrowser - a Trojan masquerading as a browser that infects mobile phones running J2ME by obtaining and exploiting SMS permissions.

The second information theft application masquerades as a Snake game and misuses the device's camera to spy on the unsuspecting user. The Snake game requests Internet permission for uploading top scores and while depicting the game on the screen, the application is unobtrusively taking pictures and sending them to a remote server.

The third and last malicious information theft application we developed, HTTP Upload, also steals information from the device. It exploits the fact that access to the SD-card does not require any permission. Therefore, all applications can read and write to/from the SD-card. The application requires only Internet permission and in the background it accesses the SD card, steals its contents and sends it through the Internet to a predefined address.

The small number of malicious applications left us with a class imbalance problem. The class imbalance problem occurs when one of the classes is represented by a very small number of cases compared to the other classes. This problem has been recognized as a crucial problem in machine learning and data mining since it causes serious negative effects on the performance of standard learning methods (which assume a balanced distribution of the classes). Several solutions to the imbalance problem have been proposed: assigning distinct costs to the classification errors; internally biasing the discrimination-based process; re-sampling the original data set (either by over-sampling the minority class and/or under-sampling the majority class) until the classes are approximately equally represented, or by duplicate vectors from the minority class. We decided to cope with this problem by under-sampling the benign class, i.e., using only part of the benign applications that were generated for the first set of experiments and duplicating the vectors of malicious applications.

The four malicious applications were installed on two Android devices. A monitoring application, which continuously sampled various features on the device, was installed and activated on the devices. The conditions were regulated and measurements were logged on the SD-card. Three of the four malicious applications (Tip Calculator, Snake, and HTTP Uploader) were used for 10 minutes by each user, while in the background the application collected feature vectors every 2 seconds. The Lunar Lander game was not used for 10 minutes. The only malicious functionality of the Lunar Lander game was to send SMSs to a predefined destination. Therefore, when the Lunar Lander game was the only application used on the device, we expected that all of the vectors would be identical. Hence, we decided to sample this application once during the short period of the attack and the sampled. These feature vectors were then duplicated and aligned with the number of vectors from the rest of the applications.

In addition to malicious applications, 20 benign tool applications from the Android framework and Android market were used. All the benign tool applications were verified to be virus-free before installation by manually exploring the permissions that the applications required and by using a static analysis of .dex files. Each of the two Android devices had one user who used each of the 20 applications for 10 minutes; in the background, the monitor application collected new feature vectors every 2 seconds. All the vectors were labeled with their true class: 'tool' or 'malicious'.

Then, for each device, five dataset were generated. To create the 5 datasets for each device, we divided randomly the 20 tool application that were used for the first sub experiment to 5 groups of size 4, while none of the tools overlapped across the different groups. For each device, the feature vectors of each group were added to a different dataset out of the 5 datasets of the device. The feature vectors for the malicious application were collected and added to each one of the 10 datasets (5 dataset for each user of the two users).The reason for choosing only 4 tool applications is to guarantee that the different classes are equally represented in each dataset. After the division we had 10 groups, 5 for each device.

Table 1 indicates the CPU power consumption costs, with and without group discount, of each feature when it is being monitored by the monitoring application. Costs were estimated using the CPU profiler which can be found in Android's SDK

The first column represents the different features that the agent extracted from the device. The extracted features are clustered into two primary categories: Application Framework and Linux Kernel. Features belonging to groups such as Messaging, Phone Calls and Applications belong to the Application Framework category and were extracted through APIs provided by the framework. Features from such groups as Keyboard, Touch Screen, Scheduling and Memory belong to the Linux Kernel category. Some Linux Kernel features can be extracted directly from Java. For example, Memory and Scheduling parameters can be accessed directly through a special filesystem-based interface of the Kernel and so can be extracted through the usage of ordinary Java I/O classes.

A total of 88 features were collected for each monitored application. The second column presents the feature group assignment indication. There are 15 unique groups. Tests carried out on a group are discounted in terms of CPU power consumption cost. The most common reason that features were in the same group is that they are read from the same file-like source. Therefore, the entire file must read even if only a single feature is needed. However, once we paid the reading cost for the first feature, we no longer need to pay the reading cost for a second feature on the same file. As an example, all the features in the group Binder were written on the same file and the agent only needed to open the file, which has a common CPU power consumption cost (18.25%) that is shared for all the features in that group. Then, in order to process each one of the features in that file (10 features in total), we only had to pay for each feature an additional CPU power consumption cost of 0.2. The other 14 groups are as follows: Sysfs, Scheduler_Statistics, Load_Average, Virtual_Machine_Statistics, Keyboard_Dynamics, System_configuration, Pressure_Dynamic, Process_Statistcs, Phone_Call, SMS, Logcat, Keyboard, Network and Misc.

Table 1. Feature CPU consumption cost before and after discount and feature grouping

Feature	Group	Before discount cost	After discount cost
Local_RX_Packets	Network	13.2	0.2
Local_RX_Bytes	Network	13.2	0.2
Local_RX_Packets	Network	13.2	0.2
Local_TX_Bytes	Network	13.2	0.2
WiFi_TX_Packets	Network	13.2	0.2
WiFi_TX_Bytes	Network	13.2	0.2
BC_Transaction	Binder	18.45	0.2
BC_Reply	Binder	18.45	0.2
BC_Acquire	Binder	18.45	0.2
BC_Release	Binder	18.45	0.2
Binder_Active_Nodes	Binder	18.45	0.2
Binder_Total_Nodes	Binder	18.45	0.2
Binder_Ref_Active	Binder	18.45	0.2
Binder_Ref_Total	Binder	18.45	0.2
Binder_Death_Active	Binder	18.45	0.2
Binder_Death_Total	Binder	18.45	0.2
Binder_Transaction_Active	Binder	18.45	0.2
Binder_Transaction_Total	Binder	18.45	0.2
Binder_Trns_Complete_Active	Binder	18.45	0.2
Binder_Trns_Complete_Total	Binder	18.45	0.2
Free_Pages	Virtual_Machine_Statistics	11.35	0.2
Inactive_Pages	Virtual_Machine_Statistics	11.35	0.2
Active_Pages	Virtual_Machine_Statistics	11.35	0.2
Anonymous_Pages	Virtual_Machine_Statistics	11.35	0.2
Mapped_Pages	Virtual_Machine_Statistics	11.35	0.2
File_Pages	Virtual_Machine_Statistics	11.35	0.2
Dirty_Pages	Virtual_Machine_Statistics	11.35	0.2
Writeback_Pages	Virtual_Machine_Statistics	11.35	0.2
DMA_Allocations	Virtual_Machine_Statistics	11.35	0.2
Page_Frees	Virtual_Machine_Statistics	11.35	0.2
Page_Activations	Virtual_Machine_Statistics	11.35	0.2
Page_Deactivations	Virtual_Machine_Statistics	11.35	0.2
Minor_Page_Faults	Virtual_Machine_Statistics	11.35	0.2
Batery_Voltage	Sysfs	1	0.95
Battery_Current	Sysfs	1	0.95
Battery_Temp	Sysfs	1	0.95
Button_Backlight	Sysfs	1	0.95
Keyboard_Backlight	Sysfs	1	0.95
LCD_Backlight	Sysfs	1	0.95

Table 1. (*continued*)

Battery_Level_Change	Sysfs	1	0.95
Avg_Key_Flight_Time	Keyboard_Dynamics	0.56	0.2
Del_Key_Use_Rate	Keyboard_Dynamics	0.56	0.2
Avg_Trans_To_U	Keyboard_Dynamics	0.56	0.2
Avg_Trans_L_To_R	Keyboard_Dynamics	0.56	0.2
Avg_Trans_R_To_L	Keyboard_Dynamics	0.56	0.2
Avg_Key_Dwell_Time	Keyboard_Dynamics	0.56	0.2
Yield_Calls	Scheduler_Statistics	4.5	0.2
Schedule_Calls	Scheduler_Statistics	4.5	0.2
Schedule_Idle	Scheduler_Statistics	4.5	0.2
Running_Jiffies	Scheduler_Statistics	4.5	0.2
Waiting_Jiffies	Scheduler_Statistics	4.5	0.2
Load_Avg_1_min	Load_Average	2.55	0.2
Load_Avg_5_mins	Load_Average	2.55	0.2
Load_Avg_15_mins	Load_Average	2.55	0.2
Runnable_Entities	Load_Average	2.55	0.2
Total_Entities	Load_Average	2.55	0.2
CPU_Usage	Process_Statistics	13.9	0.2
Running_Processes	Process_Statistics	13.9	0.2
Context_Switches	Process_Statistics	13.9	0.2
Processes_Created	Process_Statistics	13.9	0.2
Outgoing_SMS	Log_Cat	1.25	0.2
Garbage_Collections	Log_Cat	1.25	0.2
Camera	Log_Cat	1.25	0.2
Orientation_Changing	System_Configuration	0.75	0.75
Keyboard_Opening	Keyboard	0.75	0.2
Keyboard_Closing	Keyboard	0.75	0.2
Incoming_SMS	SMS	0.65	0.2
Package_Changing	Misc	0.65	0.2
Package_Restarting	Misc	0.65	0.2
Incoming_Calls	Phone_Calls	2.55	0.2
Outgoing_Calls	Phone_Calls	2.55	0.2
Missed_Calls	Phone_Calls	2.55	0.2
Avg_Touch_Pressure	Pressure_Dynamics	0.25	0.25
Avg_Touch_Area	Pressure_Dynamics	0.25	0.25

4.2 Experiment Plan: Detecting Android Malware in a Cost Sensitive Fashion

In this study, our two types of costs were not measured by the same unit cost. One cost, the CPU power consumption, was measured on a scale of the CPU percentage utilized when monitoring a certain feature from the device. The other cost, misclassification, was measured in dollar ($) terms. Therefore, when given a maximal budget of CPU power consumption cost, our evaluation metric was the achieved average misclassification cost. The lower the evaluation measure value was, the better the algorithm performed.

We used hypothesis tests in order to examine if ProCASH's average misclassification costs and execution times were statistically significant lower than the other algorithms'. We performed an Adjusted Friedman cost hypothesis test [19] with a 5% significance level. If the null hypothesis was rejected, we then conducted a Bonferroni-Dunn post Hoc[19] with a 5% significance level.

The purpose of the experiment was to evaluate the ability of the proposed methods to distinguish between malicious and benign applications given a specific CPU processing power budget. We used 10 datasets extracted from two different devices, evaluated five cost sensitive methods, four misclassification cost matrices and nine CPU processing power budgets. The division of the dataset into training and testing sets was performed in such a way that the benign and malicious applications in the training and testing set were different. For each one of the 10 datasets (generated by selecting 5 times 4 different benign applications for each device), each time a different malicious application and different benign application were not included in the training set but were included in the testing set.

4.3 Experiment Results

Table 2 presents the average misclassification cost obtained in all of the runs. The first column represents the CPU consumption power budget cost. The second column represents the algorithms that were compared. Then, each of the following columns showed the average misclassification cost on the different cost matrices. As can be seen from Table 2, the average misclassification cost of the ProCASH algorithm tends to be better than that of all the other algorithms. ProCASH outperforms csDT_csf2, GA+META_ICF and GA+META_CSID3 algorithms in all of the cost matrices and under all the CPU consumption budgets. Furthermore, ProCASH algorithm outperforms csDT_csf1 in almost all of the cases except three. Tables 3 and 4 present the results of adjusted Friedman and Bonferroni-Dunn statistical hypothesis tests respectively. From Table 3 we can see that the null-hypothesis, that all classifiers perform the same, was rejected using the adjusted Friedman test on all of the misclassification cost matrices. The adjusted Friedman test was conducted with a confidence level of 95%. Table 4 indicates that the ProCASH algorithm significantly outperforms all of the four algorithms at confidence levels of 95% in all of the 4 misclassification cost matrices. Additionally, Figures 1 and 2 represent the performance of each algorithm in each of the CPU budget costs, given a misclassification matrix cost when the FP cost and FN cost are equal to $10 and when the FP cost is equal to $10 and the FN cost is equal to $5 respectively. We can see in Figures 1 and 2 that the ProCASH algorithm outperformed the rest of the algorithms with significant differences in the majority of the datasets. Moreover, we can see that ProCASH performance improved as the budget cost increased, until a certain budget was reached where ProCASH performance stayed the same. This implies that at a certain point, ProCASH decided that sampling more features would cause overfitting to the training set and would not improve the classification performance on the testing set.

Table 2. Comparing cost sensitive algorithms: summary of experimental results

Budget cost	Algorithms	Misclassification cost matrices (FN cost_FP cost)			
		FN=$10 FP=$10	FN=$8 FP=$10	FN=$7.5 FP=$10	FN=$5 FP=$10
1	ProCASH	**$1.72**	**$1.66**	**$1.49**	**$1.50**
1	csDT_csf1	$4.27	$3.95	$2.96	$1.86
1	csDT_csf2	$4.67	$4.21	$3.34	$2.56
1	GA+META_ICF	$3.35	$3.18	$2.85	$1.87
1	GA+META_CSID3	$3.35	$3.16	$2.77	$2.26
5	ProCASH	$1.81	**$1.68**	$1.77	**$1.49**
5	csDT_csf1	**$1.61**	$1.87	**$1.73**	$2.10
5	csDT_csf2	$2.40	$2.46	$2.05	$1.86
5	GA+META_ICF	$2.24	$2.42	$2.34	$2.66
5	GA+META_CSID3	$2.70	$2.24	$2.50	$2.58
10	ProCASH	$1.81	**$1.68**	**$1.71**	**$1.49**
10	csDT_csf1	**$1.61**	$1.87	$1.74	$2.10
10	csDT_csf2	$2.41	$2.47	$2.06	$1.85
10	GA+META_ICF	$2.27	$2.00	$1.78	$2.63
10	GA+META_CSID3	$2.66	$2.54	$2.23	$2.46
15	ProCASH	**$1.21**	**$1.04**	**$0.99**	**$0.94**
15	csDT_csf1	$1.61	$1.87	$1.74	$2.10
15	csDT_csf2	$2.41	$2.47	$2.06	$1.85
15	GA+META_ICF	$2.95	$2.74	$2.55	$2.58
15	GA+META_CSID3	$2.44	$2.12	$2.35	$1.75
20	ProCASH	**$1.21**	**$1.04**	**$0.99**	**$0.94**
20	csDT_csf1	$1.61	$1.87	$1.74	$2.10
20	csDT_csf2	$2.41	$2.47	$2.06	$1.85
20	GA+META_ICF	$2.49	$2.36	$2.39	$2.63
20	GA+META_CSID3	$2.51	$2.99	$3.00	$2.96
25	ProCASH	**$1.21**	**$1.04**	**$0.99**	**$0.94**
25	csDT_csf1	$1.61	$1.87	$1.74	$2.10
25	csDT_csf2	$2.41	$2.47	$2.06	$1.85
25	GA+META_ICF	$2.13	$1.84	$1.79	$2.26
25	GA+META_CSID3	$2.59	$2.71	$2.52	$2.69
30	ProCASH	**$1.21**	**$1.04**	**$0.99**	**$0.94**
30	csDT_csf1	$1.61	$1.87	$1.74	$2.10
30	csDT_csf2	$2.41	$2.47	$2.06	$1.85
30	GA+META_ICF	$3.12	$3.07	$2.64	$3.65
30	GA+META_CSID3	$2.25	$2.06	$1.79	$3.00
35	ProCASH	**$1.21**	**$1.04**	**$0.99**	**$0.94**
35	csDT_csf1	$1.61	$1.87	$1.74	$2.10
35	csDT_csf2	$2.41	2$.47	$2.06	$1.85

Table 2. (*continued*)

35	GA+META_ICF	$2.64	$2.37	$2.32	$2.95
35	GA+META_CSID3	$2.48	$2.66	$2.03	$2.98
40	ProCASH	**$1.21**	**$1.04**	**$0.99**	**$0.94**
40	csDT_csf1	$1.61	$1.87	$1.74	$2.10
40	csDT_csf2	$2.41	$2.47	$2.06	$1.85
40	GA+META_ICF	$2.25	$1.82	$1.94	$2.12
40	GA+META_CSID3	$3.52	$3.19	$2.65	$2.83

Table 2. adjusted Friedman tests results for each one of the misclassification cost matrices

F(4,8) with Critical Value of 2.69				
Misclassification cost (FP-FN)	FN=$10 FP=$10	FN=$8 FP=$10	FN=$7.5 FP=$10	FN=$5 FP=$10
Friedman statistic value	13.6	17.92	18.55	21.72

Table 3. The superscript "+" indicates that the average total cost of ProCASH was significantly higher than the corresponding algorithm at a confidence level of 95%

Average Misclassification cost	FN=$10 FP=$10	FN=$8 FP=$10	FN=$7.5 FP=$10	FN=$5 FP=$10
csDT_csf1	+0.12	+1.61	+2.41	+2.25
csDT_csf2	+0.17	+0.38	+0.28	+0.36
GA+META_ICF	+0.12	+0.32	+3	+2.77
GA+META_CSIC3	+0.25	+0.17	+0.38	+0.38

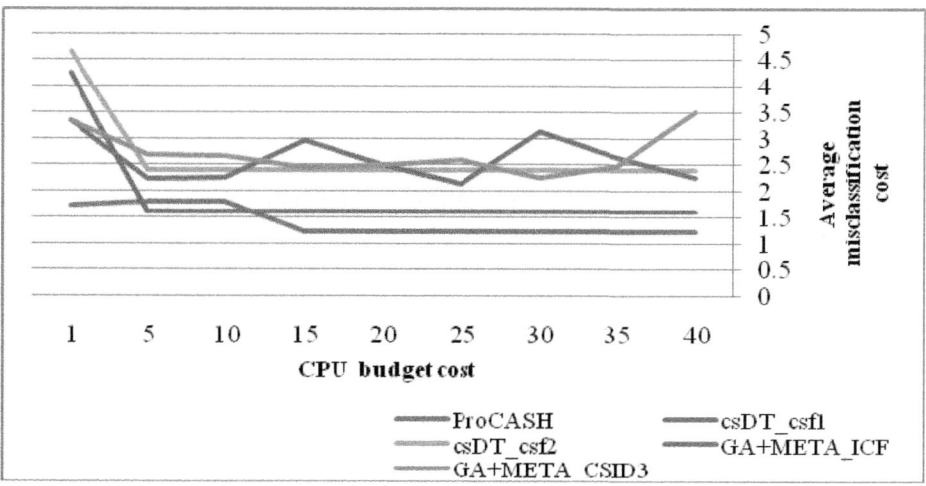

Fig. 1. The average misclassification cost that was obtained for each algorithm in each CPU consumption budget constraint, given a misclassification cost matrices of FP cost=10 and FN cost =$10

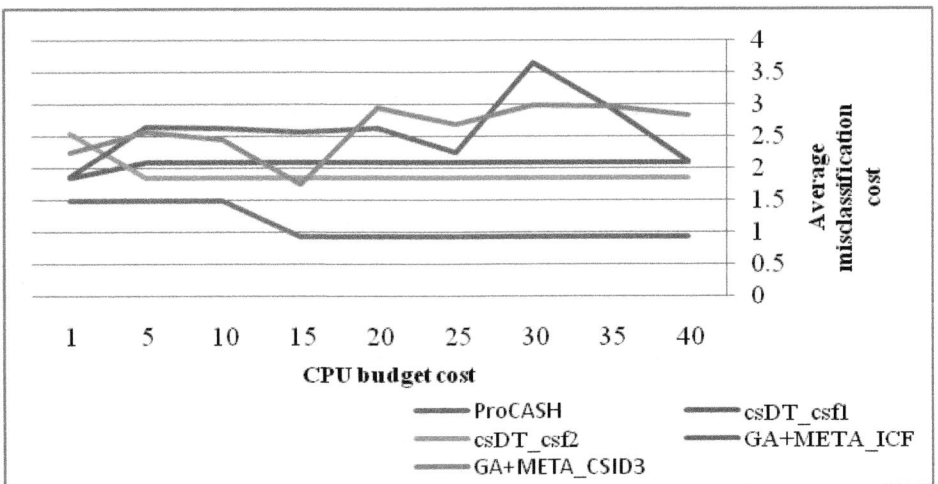

Fig. 2. The average misclassification cost that was obtained for each algorithm in each CPU consumption budget constraint, given a misclassification cost matrices of FP cost=FN cost =$5

4.4 Discussion

The advantages of the new ProCASH algorithm, as the experimental study indicates, can be summarized as follows:

1. When compared to state-of-the-art, cost sensitive algorithms, ProCASH performed better in distinguishing between malicious and benign applications on Android mobile devices under a wide range of CPU resource budget constraints and misclassification cost matrices.

2. Since ProCASH follows the filter approach, it can be used in conjunction with any classification algorithm and not only decision tree classifiers. Potentially, there might be domains in which using other classifiers will dramatically reduce the average misclassification cost.

5 Experiments

Host-based intrusion detection systems (HIDS) are based on continuously monitoring many parameters in the device such as CPU and memory consumption. Then, data mining techniques are applied to the collected data in order to detect abnormal states. Since the continuous monitoring of many parameters consumes considerable computational resources it is necessary to reduce consumption in order to used HIDS on mobile devices.

One way to achieve this is to collect less parameters by means of cost-sensitive feature selection techniques. One way to reduce computational resource consumption is to collect fewer parameters by means of cost-sensitive feature selection techniques. In this study, we evaluated ProCASH, a new cost-sensitive feature selection algorithm sensitive to resource and misclassification costs and feature grouping. ProCASH was evaluated on an Android mobile device. The data mining task was to distinguish between benign and malicious applications. The results indicate that ProCASH outperforms other cost-sensitive algorithms in terms of average misclassification costs in all of the CPU resource budget constraints.

References

1. http://ems007.iconnect007.net/pages/zone.cgi?a=48033
2. Turney, P.D.: Types of Cost in Inductive Concept Learning. In: Proc. Workshop Cost-Sensitive Learning, 17th Int'l Conf. Machine Learning, pp. 15–21 (2000)
3. Turney, P.D.: Cost-Sensitive Classification: Empirical Evaluation of a Hybrid Genetic Decision Tree Induction Algorithm. J. Artificial Intelligence Research 2, 369–409 (1995)
4. Chai, X., Deng, L., Yang, Q., Ling, C.X.: Test-Cost Sensitive Naive Bayes Classification. In: Proc. 4th Int. Conf. Data Mining, pp. 51–58 (2004)
5. Ling, C.X., Yang, Q., Wang, J., Zhang, S.: Decision Trees with Minimal Costs. In: Proc. 21st Int. Conf. Machine Learning, p. 69 (2004)
6. Sheng, S., Ling, C.X., Yang, Q.: Simple Test Strategies for Cost-Sensitive Decision Trees. In: Proc. 16th European Conf. Machine Learning, pp. 365–376 (2005)
7. Ling, C.X., Sheng, V.S., Yang, Q.: Test Strategies for Cost-Sensitive Decision Trees. IEEE Transactions on Knowledge and Data Engineering 18(8), 1055–1067 (2006)
8. Sheng, V.S., Ling, C.X., Ni, A., Zhang, S.: Cost-Sensitive Test Strategies. In: Proc. 21st Nat'l Conf. Artificial Intelligence (2006)
9. Sheng, S., Ling, C.X.: Hybrid Cost-Sensitive Decision Tree. In: Jorge, A.M., Torgo, L., Brazdil, P.B., Camacho, R., Gama, J. (eds.) PKDD 2005. LNCS (LNAI), vol. 3721, pp. 274–284. Springer, Heidelberg (2005)
10. Freitas, A., Costa-Pereira, A., Brazdil, P.B.: Cost-Sensitive Decision Trees Applied to Medical Data. In: Song, I.-Y., Eder, J., Nguyen, T.M. (eds.) DaWaK 2007. LNCS, vol. 4654, pp. 303–312. Springer, Heidelberg (2007)
11. Weiss, Y., Elovici, Y., Rokach, L.: The CASH Algorithm-Cost-Sensitive Attribute Selection using Histograms. Lecture Notes in information system engineering. Ben-Gurion University (2010)
12. Shabtai, A., Weiss, Y., Kanonov, U., Elovici, Y., Glezer, C.: "Andromaly": An Anomaly Detection Framework for Android Devices. Lecture Notes in information system engineering Ben-Gurion University (2009)
13. Núñez, M.: The use of background knowledge in decision tree induction. Machine Learning 6, 231–250 (1991)
14. Tan, M., Schlimmer, J.: Cost-sensitive concept learning of sensor use in approach and recognition. In: Proceedings of the Sixth International Workshop on Machine Learning, ML 1989, pp. 392–395 (1989)
15. Tan, M.: Cost-sensitive learning of classification knowledge and its applications in robotics. Machine Learning 13, 7–33 (1993)
16. Frank, E., Hall, M.A., Holmes, G., Kirkby, R., Pfahringer, B., Witten, I.H.: Weka: A machine learning workbench for data mining. In: Maimon, O.Z., Rokach, L. (eds.) Data Mining and Knowledge Discovery Handbook. Springer, Heidelberg (2005)
17. Quinlan, J.: C4.5: Programs for machine learning. Morgan Kaufmann, San Francisco (1993)
18. Domingos, P.: MetaCost: A general method for making classifiers cost-sensitive. In: Proceedings of the Fifth International Conference on Knowledge Discovery and Data Mining, pp. 155–164 (1999)
19. Botha, R.A., Furnell, S.M., Clarke, N.L.: From desktop to mobile: Examining the security experience. Computer & Security 28, 130–137 (2009)
20. Demšar, J.: Statistical comparison of classifiers over multiple data sets. Journal of Machine Learning Research 7, 1–30 (2006)

On-line Signature Verification on a Mobile Platform

Nesma Houmani, Sonia Garcia-Salicetti, Bernadette Dorizzi,
and Mounim El-Yacoubi

Institut Telecom; Telecom SudParis; Intermedia Team,
9 rue Charles Fourier, 91011 Evry, France
{nesma.houmani,sonia.salicetti,bernadette.dorizzi,
mounim.el_yacoubi}@it-sudparis.eu

Abstract. This paper concerns the implementation of our online signature verification system on a mobile device. Verification involves confirming or denying a person's claimed identity. Our system is based on a Hidden Markov Model and outputs two complementary scores: the first one is related to the likelihood given by the HMM of the claimed identity; the second one is related to the segmentation given by such an HMM on the input signature. A claimed identity is confirmed when the arithmetic mean of the two scores obtained on such an input signature is higher than a threshold. Also, a personal normalization of the local parameters of the signature is carried out to make the system robust to changes of platforms. A patent was submitted with special emphasis on the latter claim. This system is implemented on a mobile platform PDA Qtek 2020 ARM 400 MHz. An acquisition interface is developed allowing an enrollment step of a person by acquisition of 5 of his/her signatures, and a verification step of a given signature of a registered person. Enrolment speed depends on the complexity of the signature, while verification is performed in real time. Performance assessment of our system, carried out on two databases acquired on a PDA, shows a degradation of system performance on mobile platform compared to a fixed platform. In order to improve the performance in the case of mobility, we propose a strategy for enhancing the quality of the reference signatures at the enrolment phase.

Keywords: Online signature verification, PDA device, Mobile conditions, Hidden Markov Models.

1 Introduction

Online signature differs from offline (static one, which is affixed to the bottom of a letter or a check) in that it is acquired on a special terminal (digitizing tablet or PDA) or using a special pen (Anoto) that records the sequence of points plotted during the dynamic signing process. In this way, we get directly a temporal signal (see Figure 1), namely the sequential coordinates of plotted points, unlike the static signature that corresponds to an image after digitization. Several studies (including ours) have shown that using online signature allows obtaining better recognition performance than using off-line signature, particularly because online signature can encode the dynamics of the signature, which is less variable for a given person than the shape alone, and also because the gesture of signing is more difficult to falsify than the signature image.

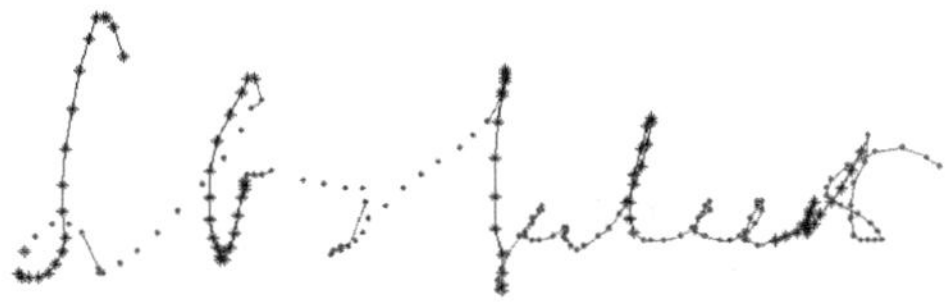

Fig. 1. Example of an online signature and the corresponding point's sequence

Signature is an interesting biometric modality for automatic person authentication due to the fact that it is highly accepted by the population, the latter having been using it for ages on paper documents. Moreover online signature can be recorded without the need of any extra sensor on all mobile devices that possess a tactile screen. Such devices are nowadays largely deployed all over the population for various usages. In general, their protection against theft is rather limited as person authentication is ensured thanks to a code pin, which can even be optional. This is not sufficient for applications such as mobile banking, financial transactions, access to a medical database, contract signing through internet, etc. In these situations, a stronger user authentication on the mobile device is needed before allowing the transaction. Biometric modalities can be used to this end and some smart phones already use fingerprints (with a special acquisition device) [1].

The object of the MOBIO project is the study of the feasibility of face and voice in this mobility context [2]. Online signature is another possibility, which has been envisaged by Martinez-Diaz et al. [3] who nevertheless stress the loss of performance encountered in the mobility context as compared to acquisition on a fixed digitizing tablet.

The aim of this paper is to describe our signature verification system (called TSP) as well as its implementation on a mobile device. We will also present different comparative performance evaluations on a digitizing tablet (fixed platform) and on a PDA (mobile platform). Finally, we propose some possible ways to improve the performance in the case of mobility.

2   The TSP Signature Verification System

Our system, described in details in [4], is based on several important characteristics:

- An extraction of 25 relevant parameters at each point of the signature representing the dynamic information (ex: speed, acceleration) and the local spatial context.
- Modeling the signatures of a person by a Hidden Markov Model that allows taking into account the intra-class variability between the different signatures of the same person.

- The fusion of two complementary scores from this model, corresponding to two levels of signature description: the likelihood score computed locally on each point of the trajectory and the segmentation score computed on portions of the signature.
- Personal normalization of the parameters of the signature in order to make the system robust to different acquisition sensors (digitizing tablet, pen, PDA). A patent [5] was submitted with special emphasis on the latter claim.

This system has been tested on many databases of a hundred people, gained over a digitizing tablet. Note that in each database, imitations of the signature of each person of more or less good quality are available (skilled forgeries). This makes it difficult for the system to both recognize the signatures from the same individual and discriminate the authentic ones from imitations. Tests are also performed with random forgeries (genuine signatures of other persons in the database).

In [4], an average Equal Error Rate (EER, corresponding to as many false rejections as false acceptances) around 3.5% is obtained when the system is tested with skilled forgeries. With these figures, our system is one of the best from the state of the art, and is proposed as a reference system by the BioSecure Association [6].

3 Implementation on a PDA

This system has been implemented on a mobile platform PDA Qtek 2020 ARM 400 MHz in 2006 in the framework of the SecurePhone European project. An acquisition interface has been developed, which allows two types of processing: an enrollment step of an unknown person by acquisition of 5 of his/her signatures, which are stored in the PDA as references of the person; and a verification step of a given signature of a registered person.

4 Performance Evaluation

The system has been evaluated on two different databases acquired on a PDA, namely PDA-64 containing data of 64 people, and BioSecure DS3-210 [6] containing data of 210 people.

The experiments show a degradation of performance when using a mobile digital device instead of a fixed digitizing tablet. Indeed, as mentioned above, an average EER of 3.5% is obtained on fixed platforms with skilled forgeries, while on PDA-64, the EER is of 16.02% (Figure 2), and of 9.95% on DS3-210 (in case of no time variability on Figure 3). This is due to the fact that signatures of a given person are altered on mobile platforms as the writer signs while standing and holding the PDA on his/her hand. This makes signatures in general less complex and more variable, leading to a decrease of performance, which is particularly visible when the reference and test signatures are time-spaced (see Figure 3, in the case of time variability with an EER of 12.32% for skilled forgeries).

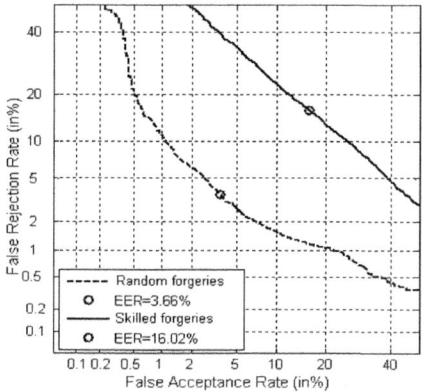

Fig. 2. System performance on a home-made "PDA-64" database for random and skilled forgeries

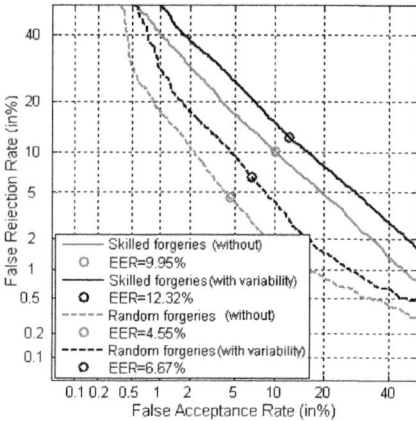

Fig. 3. System performance on "BioSecure DS3-210" [6] database acquired on 2 sessions. Performance is shown for random and skilled forgeries, with and without time variability (intra and inter sessions).

5 Conclusion

In order to improve the performance in mobile conditions, we propose a strategy for enhancing the quality of the signature references. Indeed we have shown in [7] that an entropy quality measure can be used for selecting good quality reference signatures at the enrollment phase. Moreover, our results in [7, 8] show that using "good" quality signatures improves significantly the verification results. We also suggest proceeding at regular time steps when re-enrolling the persons in order to face the time variability of online signatures.

References

1. Pocovnicu, A.: Biometric Security for Cell Phones. Informatica Economica 13(1) (2009)
2. MOBIO (Mobile Biometry) European project, http://www.mobioproject.org/
3. Martinez-Diaz, M., Fierrez, J., Galbally, J., Ortega-Garcia, J.: Towards mobile authentication using dynamic signature verification: Useful features and performance evaluation. In: ICPR (2008)
4. Ly Van, B., Garcia-Salicetti, S., Dorizzi, B.: On using the Viterbi Path along with HMM Likelihood Information for On-line Signature Verification. IEEE Transactions on Systems, Man and Cybernetics, Part B, Special Issue on Recent Advances in Biometric Systems 37(5), 1237–1247 (2007)
5. Ly Van, B., Garcia-Salicetti, S.: French Patent n°FR0553552 "Vérification de signature"
6. http://www.biosecure.info
7. Garcia-Salicetti, S., Houmani, N., Dorizzi, B.: A Novel Criterion for Writer Enrolment based on a Time-Normalized Signature Sample Entropy Measure. EURASIP Journal on Advances in Signal Processing (2009)
8. Houmani, N., Garcia-Salicetti, S., Dorizzi, B.: On Assessing the Robustness of Pen Coordinates, Pen Pressure and Pen inclination to Time Variability with Personal Entropy. In: Proc. IEEE Second International Conference on Biometrics: Theory, Applications and Systems (BTAS 2009), Washington (2009)

Google Android: An Updated Security Review

Yuval Fledel, Asaf Shabtai, Dennis Potashnik, and Yuval Elovici

Deutshce Telekom Laboratories at Ben-Gurion University,
Beer-Shava, Israel
{fledely,shabtaia,dennisp,elovici}@bgu.ac.il

Abstract. Among the most significant smartphone operating systems that have arisen recently is Google's Android framework. Google's Android is a software framework for mobile communication devices. The Android framework includes an operating system, middleware and a set of key applications. Designed as open, programmable, networked devices, Android is vulnerable to various types of threats. This paper provides a security assessment of the Android framework and the security mechanisms incorporated into it. In addition, a review of recent academic and commercial solutions in the area of smartphone security in general and Android in particular is presented.

Keywords: Mobile devices, Google, Android, Security.

1 Introduction

Among the most significant smartphone operating systems that have arisen recently is Google's Android framework. Designed as open, programmable, networked devices, Android is vulnerable to various types of attacks that can make the phone partially or fully unusable, cause unwanted SMS/MMS billing, expose private information, or infect every name in a owner's phonebook [1].

Smartphone market share in the US has increased from 11 percent of all cellular phone subscribers in 2008 to 17 percent in 2009, and it is expected to increase significantly over the next few years, almost fivefold by 2013 [2]. The Android framework has gained much interest by both the developers' community and smartphone users in a relatively short period of time. In fact, according to [3] Android is the fourth most popular smartphone in the US as of February 2010. Smartphones based on the Google Android operating system are expected to increase 10 percent during 2010 [4]. Consequently, smartphones are likely to become a fertile ground for various types of threats. Another major factor attracting hackers is that smartphones are often carried for business purposes and are likely to have sensitive and valuable information. They also provide remote access to a company's most sensitive data, which can lead to data leakage if their phones are hacked into.

The increasing number of attacks on mobile platforms along with the increasing usage has led many security vendors and researchers to propose a variety of security solutions for mobile platforms. As a case in point, Symbian and Google have designed their operating systems to enable applications to run only in specialized sandboxes, minimizing the capability of malware to spread [5]. A robust application signing and certification mechanism was integrated into Symbian's operating system and was proven highly effective in reducing malware attacks. The risks to Android are nevertheless significant, mainly because it's an open source and open platform software stack operated in a heterogenic mobile environment. On one hand, it allows introducing new applications and services very quickly. On the other hand, it raises security issues that the academic community and security vendors attempt to address. This paper reviews and assesses the security mechanisms incorporated into the Android framework. Additionally, a list of security mechanisms which can be incorporated to harden the Android is presented.

2 The Android Framework

The Android[1] software stack is built on the *Linux kernel*, which is used for its device drivers, memory management, process management, and networking. The next level up contains the *Android native libraries*. These libraries are written in C/C++ and are used by various system components in the upper layers. Incorporating these libraries in Android applications is achieved through Java interfaces or native compiled code. The next level is the Android *runtime*, comprising of the *Dalvik Virtual Machine* and the *core libraries*. Dalvik runs .dex (Dalvik executable) files that are designed to be more compact and memory-efficient than Java .class files. The core libraries are written in Java and provide a substantial subset of the Java 5 SE packages as well as some Android-specific libraries. The *Application Framework* layer, written fully in Java, includes Google-supplied tools as well as proprietary extensions or services. The topmost *Application* layer provides applications such as phone, web-browser and email client.

Each application in Android is packaged in an .apk (Android package) archive for installation. The .apk is similar to a standard Java jar file in that it holds all code and non-code resources (e.g., images, manifest) for the application. The applications are written in Java based on the APIs provided by the Android SDK. An Android package is basically a collection of components: *Activities, Services, Broadcast Receivers* and *Content Providers*. Components in one .apk are isolated from components in another .apk and can only communicate with each other and share data through means provided by the system. Each .apk is associated with a primary process in which all of the application's components are executed. Enck et al. [6] and Burns [7] provide an overview of the main components of an Android application and guidelines for using the Android-specific mechanisms correctly in order to protect the application.

Android is a multi-process system, where each application (and parts of the system) runs in its own process. For the most part, security between applications and the system is enforced at the process level through standard Linux facilities, such as POSIX user IDs and group IDs assigned to applications. Files in Android (both application- and system-files) are subject to the Linux file permission mechanism. In

---

[1] www.android.com

addition, access-control is provided through the application level permission mechanism that enforces restrictions on the specific operations that a particular application can perform. Signing applications is another significant security feature in which all of the application's files are signed along with their meta-data in the .apk. A review of Android's inherent security mechanisms is provided in [8].

3 Android Security Analysis

In this section we describe findings from our assessment of various security aspects of the Android framework. The results presented in this section were validated on T-Mobile G1 and HTC Desire devices.

3.1 Analysis of the Android Framework's Cornerstone Layers

This subsection presents the outcome of our analysis of Android's lower layers. We adopted a security-oriented code-review approach to identify potential vulnerabilities. We focused on special locations that might be problematic such as interfaces, structures and configurations. The source code that was reviewed is from the Android open source project repository.

Linux Kernel

In 2009, 111 CVE (Common Vulnerabilities and Exposures) entries were logged for the Linux kernel (i.e., two entries a week). Drivers and vendor-specific additions are the main locations for these vulnerabilities. Android contains a considerable amount of vendor-specific code. Therefore, *Android code should be submitted to mainline*. Code that is integrated into the mainline passes several phases of checks and validations that are likely to identify and remove bugs; some may have security implications. However, Android has diverged from the mainline kernel and Android vendors have so far published the code only after shipping a product on the market.

Android modifications include *hardcode POSIX user ids (uids) and group ids (gids)* in the kernel code. These modifications contradict the basic design decisions in Linux. However, it increases security since the system services are not required to run with root privileges. Two examples of such modifications are the "paranoid network" that limits network access based on gids, and the Binder which accepts the first process that uses it as its master, but only if it has the "system" group.

The *Linux kernel is highly configurable*. On a typical Android device, common Linux options are disabled in order to reduce memory consumption. Less code also means a smaller attack surface. On the other hand, it also means omitting "security enabler" modules. A few examples from HTC Desire that have security implications are: disabling the auditing support and BSD task accounting (less input is available to intrusion detection systems); disabling SYN cookie support (if enabled, it can reduce the chance of SYN flood attack); and disabling security modules (e.g., SELinux). The following modules are enabled: PPTP, L2TP and IPSec-based VPN connections (supported on Android from release 1.6); CFS scheduler group scheduling; disk encryption; and NetFilter (provides firewall capabilities). Enabling these modules in the kernel configuration requires a trivial amount of effort and also consumes minor memory space. However, providing the means for using these modules is not trivial and requires additional user-space components that will provide the API for these modules in order to avoid the need for root access.

Android employs Linux's *Completely Fair Scheduler (CFS)* that ensures that an equal share of CPU is distributed among all processes. In addition, specific processes can be granted a larger share, but are still prevented from monopolizing the CPU. To test this fairness mechanism we have created a simple application that starts 100 threads that loop doing nothing in particular. Running this application on a T-Mobile G1 device resulted in the entire device being frozen.

Linux also supports *storage quota*, but Android does not enable it. This means that an application can create files of any size, on both the internal flash and the SD card. Files can be created outside of the private application folder and therefore do not counted in the application's size figure.

Finally, we tested the applicability of existing *Linux root-kits and key-loggers* that usually require root level access in order to install. We successfully compiled and activated the rathole and Linux rootkit V (lrk5) root-kits, and the vlogger key-logger on Android [9]. The running root-kits on Android can remotely explore running processes, hide or kill specific processes, prevent hidden processes from being stopped, enable packet sniffing and provide several methods to communicate with the root-kit such as encrypted SSH backdoor and remote shell. Using the key-logger we managed to log keystrokes on the devices keyboard [9]. Detailed description of similar experiments is provided in [3].

System Libraries

Android makes use of many native libraries. These libraries are intended to be used by native processes, other native libraries or by Dalvik through Java Native Interface (JNI). JNI is normally used for: (1) providing low-level functionalities; (2) implementing computationally intensive calculations; (3) hiding code and licensing issues; and (4) leveraging existing libraries.

The native libraries are written in C/C++, which is not type safe. Thus, native libraries have a higher chance of bugs than Java code. Since JNI loads native libraries into the memory space of a Dalvik process, bugs in the native library may crash the Dalvik process, corrupt its memory or cause arbitrary code execution. For that reason, system libraries are a target when searching for vulnerabilities. Example vulnerabilities in Android native libraries are CVE-2009-0606, and CVE-2009-0608. Usually, such vulnerabilities stem from using outdated vulnerable versions of ported libraries and not keeping pace with upstream bug fixes.

Dalvik Runtime

Dalvik is a Java Virtual Machine (VM) based on Apache Harmony which was extensively modified and adapted for environments with low memory. Securing Dalvik is crucial since vulnerability in the VM affects all applications. Dalvik provides the possibility of executing native code through JNI without requesting permission for it. Employing native code, however, removes the layer of defense provided by the VM.

A potential weak spot is the .dex file loading code which is required to deal with .dex files from unreliable sources. The verification process of .dex files that is performed during the installation of the application is also applied whenever the .dex files are loaded to memory. By inspection of the Dalvik code we conclude that sanity checking is implemented in the initial loading code. However, we identified unchecked pointer operations and as a result, we were able create a malformed .dex file that during installation caused the Package Installer to crash resulting in a phantom application that cannot be uninstalled because the installer claimed it is already installed, nor can another package with the same package name be installed if it has a different signature.

Pure "desktop" Java malware spreads by injecting code into other class files without harming the valid structure of the victim class file and its verifiability. Java malware such as StrangeBrew and BeanHive search for writable class files in the user's working directory and then modify them to start execution from the viral segment. From our experiments with the StrangeBrew malware we conclude Java malware are not applicable to the Android framework for two reasons. First, they infect class file formats and must be adjusted to support injecting malicious code into .dex files. Second, in Android, applications do not have write privileges to any .apk files. Moreover, since it is not possible to list a folder of another application for its files, any effective search for files to infect is not feasible [9].

Forcing Windows OSs (XP and Vista) to automatically run a malicious Windows executable from a T-Mobile G1 device (located on the SD-card) by using an Autorun.inf file was also tested and found not feasible [9].

3.2 Application-Level Permissions

The application-level permission mechanism is responsible for securing APIs provided by the system and other applications [8]. Whereas some of the core permissions are reserved for Google applications, a large variety of Normal and Dangerous protection level permissions are still available for non-Google applications. As a result, abuse of such permissions is inevitable. As an example, an Internet access with the ability to read various contents stored on the device (e.g., files on the SD card, SMS messages or GPS tracking) can be used to acquire confidential information or to spy on the user without his/her knowledge. Other examples include: Denial of Service (DoS) attacks (e.g., denying the ability to place phone calls or draining the battery) and the abuse of paid services (e.g., phone calls, SMS/MMS messages, and chargeable network traffic). As a case in point, SMobile analysis of 68% of the applications available on the Android market indicates that 20% of the applications request permissions to access private or sensitive information that can be used for malicious attacks. Small portion of the applications have the ability to brick the device, read or use the authentication credentials from another service or application, or send unknown premium SMS messages without the user's authorization [9]. A recent example is the AndroidOS.FakePlayer.a Trojan horse that masquerades as a media player but in the background sends SMS messages to premium rate numbers.

Another source of difficulties arises from the shared user-ID feature. When an application, declaring a shared user-ID, is installed, all of the granted permissions are ascribed to the shared user-ID. At runtime, each of the applications sharing that user-ID will be granted by a combined set of permissions. A simple attack scenario based on exploiting the user-ID feature would probably take place as follows. The user installs two applications sharing a user-ID. The first requests access to the Internet while the second wishes to access the contact list. As soon as both applications are installed, each application is capable of both reading the contacts and sending them through the Internet. The user, however, is unaware of the collaboration between the completely unrelated applications. A major design flaw is that the user does not have the ability to only partially grant the requested permissions.

3.3 Installing Applications

The Package Manager (the service responsible for the installation process) validates the correctness of the .apk during installation. Validation includes: verification of the digital signature; confirmation of legitimacy of shared user-ID and permission requests; and the validation/verification of the included .dex file. The package installation API is guarded by the INSTALL_PACKAGES permission which is of Signature protection level and defined in the core Android package. Thus, malicious applications cannot install applications on their own. The Package Installer, a legitimate application-level wrapper for the Package Manager is included as one of the core applications that are provided with the Android-operated device.

There are three main methods for installing .apk files. The first, which is intended mainly for developers, is using the Android Debug Bridge (a command-line tool that is supplied along with the SDK). The installation command is issued from a PC while the Android device is connected via USB connection. The actual installation is done directly by the Package Manager without any user interaction and therefore has the ability to propagate worms from PCs to devices silently. The two remaining installation options are intended for the device owners. The first one is installing via Android Market. Since the Market application is signed by Google, it can interact directly with the Package Manager. The last installation method is based on installing applications from the SD card using 3rd party applications that enable the user to search for .apk files on the SD card and to initiate the installation process which is carried out by the Package Installer.

When installing an existing application, the installation will be allowed only when the signatures of the existing application and the new application match. The signature-matching safeguards against malicious applications that attempt to gain access to private data through substitution of the original package.

3.4 Web-Browser

Web-browsing exposes Android users to common attacks such as: Cross-Site Scripting (XSS); URL encoding attacks; social engineering; and malicious scripts. WebKit, Android's open-source Web engine, has a history of vulnerabilities. Previous attacks on the Web browser include a buffer overflow in an outdated native library, and an explicit XSS vulnerability. Both attacks enabled the attacker to run any malicious code on the device with all the abilities and privileges assigned to the Web browser application. Since the browser runs with its own POSIX user-id, an attack is limited to the browser, leaving other phone functions (e.g., dialing or messaging) unharmed. The browser is also limited by the application-level permissions it has been granted at installation (Internet access, ability to acquire wake locks, location-based APIs, network-related information retrieval, and writing to the SD-card). Nevertheless, in a successful attack, the attacker could gain information stored by the browser such as cookies, passwords, favorites and form-field values. Having access to all of the browser's private data, the attacker could corrupt it in order to prevent correct operation in the future. The browser provides several security-related configurable options. These include remembering form data and passwords; accepting cookies; displaying security warnings; loading images; enabling JavaScript; blocking pop-ups; and setting (or disabling) the homepage.

3.5 Connectivity and Communication

Multiple communication transports (Bluetooth, Wi-Fi, cellular, cable) provide many options for malware to infiltrate a device. Some malware can propagate through more than one transport. For example, Lasco is a malware which spreads via the Bluetooth on Symbian devices [11]. In addition, it also infects all Symbian Installation Source (SIS)-files using social engineering.

Bluetooth on Android supports pairing and audio headsets. Additional functions can be enabled using 3rd party applications (e.g., Object Push). For the pairing process, Android allows itself to become discoverable, but only for a short duration of two minutes. In addition, the owner needs to accept the connection. This significantly decreases the likelihood of being detected by attackers).

The two USB sub-protocols which are supported by Android are: mass-storage device and the Android Debugger Bridge (adb). By default, adb is disabled, and the device is mounted as mass-storage device where only the device's SD card are exposed. When USB debugging is enabled, the device can be managed with the adb tool which is provided in the Android SDK. This tool makes it possible to push and pull files to and from the device, install .apk files, redirect TCP and UDP packets, etc.

One of the Android-specific Linux kernel changes is the "Paranoid-Network". Usually, on Linux systems, any user-space process can open network connections at will. On Android, a user-space application must receive the INTERNET permission in order to make any kind of network connection. The enforcement of the INTERNET permission is done at the kernel level and therefore even native applications are subject to this setting. The Paranoid Network setting works by hard-coding several

POSIX group IDs in the kernel. An application must be a member of the relevant group before it is allowed to create sockets.

3.6 Conclusions

From our analysis of the Android framework we identified two main threat classes which should be countered by employing proper security solutions/capabilities. First, whenever discovering a bug or vulnerability in one of the core components (such as a native library or a kernel component), an attacker might be able to run malicious code in a highly privileged mode and even gain full control over the device. This threat is amplified due to the fact that Android's code is publicly available; some system processes run with root privileges; and no fine-grained access control mechanism exists for system processes. Second, the application-level permission mechanism is not sufficient and installation of an application that maliciously uses permissions granted by the unaware user is a scenario which is likely to occur. The framework also provides the adb install feature that makes it possible to install applications and to grant permission to an application without any user interaction. In addition, a user cannot approve a sub-set of requested permissions (it is "all-or-none") and cannot verify that an application uses its granted permissions only for benign purposes. Moreover, the shared user-ID mechanism allows sharing permissions between applications without a user's awareness or the need for explicit approval.

These threat classes may results in compromising the availability, confidentiality and/or integrity of private content that is stored on the device (e.g., pictures, contacts, emails, documents), applications and services (e.g., phone, messaging, emailing, Internet) and resources (e.g., battery power, communication, memory and CPU).

4 Applicable Security Mechanisms for Android

In order to further harden an Android device and mitigate the identified high-risk threats, additional safeguards may be employed. Some of these mechanisms were tested and evaluated in our mobile security laboratory. Several security companies, such as SMobile, Mocana, McAfee and DroidSecurity are already providing security solutions for Android. Additionally, several security mechanisms have been proposed and evaluated by the academic community. In the following paragraphs we provide a description of several security mechanisms that can be adapted to harden the Android.

Anti-malware

To identify and remove malware, anti-malware software examines files, email attachments, memory, system configuration, MMS, Bluetooth objects, etc. It usually identifies known malware based on a signature repository. As mentioned earlier, several commercial solutions are available for Android which also provides an anti-malware component. There are also open-source anti-virus and rootkit detectors that can be ported to Android such as the ClamAV [12]. Anti-malware is a well-known solution and is extensively used in other platforms. Signature-based solutions provide low false-positives, but will only detect known malware and require continuous updating of the signature repository. At this time, the anti-malware solution does not seem to be effective for mobile devices.

Firewalls

A firewall running on Android can prevent remote network attacks. It is a well known and highly effective solution; however, it will not protect against attacks via webbrowser, SMS/MMS, email or Bluetooth and will not provide phone call filtering.

We have implemented a preliminary Firewall for Android which is based on NetFilter. NetFilter is a Linux kernel subsystem that provides firewalling capabilities (e.g., packet filtering and connection tracking capabilities). NetFilter is enabled on Android devices including T-Mobile G1 and HTC Desire, thus only a control application is needed. However, in order to update the firewall policy, the control application should run with root privileges.

In the basic firewall, that we activate on Android rules are very simple and provide the ability to block communication to/from specific IP addresses and ports. The more suitable firewall policy for Android is one that allows defining rules at the application level. In such a way the policy will define for each application who can access it and where it can send information to. We can also make sure that port scanning is not preformed from the device by a malicious application. However, firewalling at the application level is hard to achieve since, as mentioned before, any application that is granted with INTERNET permission can open a socket at its will.

Intrusion Detection System

Host-based intrusion detection system (HIDS) monitors the device, applications or user's behavior to detect/prevent abnormal or known malicious behavior. Anomalybased IDS can detect unusual phone call/SMS activity, denial of service attacks, and protect the information on the device in case of theft or loss. While it may detect new and isolated attacks, it will probably suffer from high rate of false positives.

Most academic initiatives to enhance protection of mobile devices have employed host-based intrusion detection systems comprising an agent collecting various features from the device and then applying various machine learning algorithms to classify the behavior of the system as benign or malicious or to detect anomalies [9]. In our Android security research we developed and evaluated the "Andromaly", which is an experimental anomaly-based IDS for Android [13][14]. Andromaly employs various methods, such as anomaly detection and temporal reasoning, to facilitate detection of maliciously behaving applications. An IDS such as the Andromaly can be used for reporting suspicious behavior of applications to Google via the Android Market.

Access Control

Android incorporates several access control mechanisms. While these mechanisms are enforced on the application level or only on files, Linux can provide other tools that are directly enforced by the kernel. As a case in point, we tested the SecurityEnhanced Linux (SELinux) on an Android G1 device [15]. SELinux allows restricting of any process in the system, including root-owned, and by that limiting access of processes and users to resources and/or services, thus limiting the potential damage from malicious or exploited applications. Its decisions are based on an access control policy, which should be deployed together with the base system. Our experimentation with SELinux on Android has shown that it consumes very few resources and incurs a very low overhead [15].

Android provides simple authentication functionality based on a screen lock pattern mechanism. Such mechanism can be extended so that the device can be locked remotely (when the device is lost or stolen), or by protecting sensitive information stored on the device, or on the SD card using password-based encryption.

In the same context, Ni et al. [16] present the DiffUser framework that provides role-based access control mechanism for smartphone users. DiffUser was implemented and evaluated on Android. Each user can be assigned with different rights. For example, only an administrator can install/uninstall applications; the guest user can only use the phone application.

Protecting Android Permissions

During the installation of application on Android, the user may view a list of required permissions, and may decline installation based on this list. However, there is no way for the user to allow only a subset of the required permissions. Nauman and Khan [17] added an advanced feature to the Package Installer enabling the user to decline certain requested permissions but still permit installing of the application. Such a change would be highly beneficial to security aware users. This solution would protect from granting unneeded permissions that could be maliciously used. However, applications granted with a partial set of permissions may crash if the developer did not anticipate and provide a solution for such a situation (i.e., handle cases in which partial permissions were given). This solution can be enhanced for corporate users to provide the option for hardening Android devices by limiting permissions granting based on a predefined policy.

Additional efforts for enhancing Android security at the application level permissions are presented by the Kirin system [18] and Secure Application INTeraction (SAINT) [19]. These two systems presented an installer and security framework that realize an overlay layer on top of Android's standard application permission mechanism. This layer allows applications to exert fine-grained control over the assignment of permissions through explicit policies.

Spam-Filter

A spam filter blocks unwanted MMS, SMS, emails, and calls from an unreliable origin. In the mobile phone arena, spam filters are implemented using the white/black listing approach, with caller ID and words/phrases dictionary being used as the source for allowing/blocking a call or a message. Products for spam filtering on Android are already available. eMail spam filtering can be provided by either the email server (e.g., gmail account) or by Android client side email application.

Application Certification

Android uses certificates in a limited way in order to ensure package integrity and that two or more packages are from the same origin. Applications that define their own permissions may choose to grant such permissions only to packages sign by the same author. There is no support for root Certificate Authorities (CAs) or for certificate chains in Android. In order to employ the application trust mechanism, Android needs to be modified to support trust levels of applications, associating CA certificates to the trust level, as well as verifying certificate chains. This mechanism is highly effective in detecting malicious applications before they are installed on a device. However, this solution is highly expensive in terms of implementation and maintenance.

Certification process was implemented by other mobile operating systems (e.g., Symbian, iPhone and Blackberry). Although certification has been proven very effective, it is not error prone and malicious applications can still unintentionally be approved and signed. In addition we can assume that users will continue to download and install "unapproved" applications that are available from free websites and prefer them over trusted applications that need to be paid for. Furthermore, Android is grounded in an open source approach, while the certification framework contradicts this approach; thus researchers should look for alternatives to capture application semantics without relying on manual code inspection.

Automated Code Analysis and Verification

Android .apk files encapsulate valuable information that can help in understanding an application's behavior. This information includes requested permissions, framework methods called by the application, framework classes used by the application, User Interface widgets and more. We took that avenue by exploring the use of machine learning classifiers on static features extracted from Android's application files [19]. In this approach, the application file is represented by static features extracted from the file and the classifiers are then applied to learn patterns in the code in order to classify new files. Schmidt et al. [12] evaluated a framework for static function call analysis and performed a statistical analysis on function calls used by native applications. Chaudhuri [21] presented a formal language for describing Android applications and data flow among application's components. This formal language can be used for statically analyzing Android applications and data flow between applications and comparing those with security specifications defined in the application's manifest. This provides the ground for security decisions such as is the application safe and does it do what it claimed to do. Therefore it can provide the means for a developer to certify is application, and for the user to verify the proof of the certification before installation.

Such an approach is closely coupled with certification and can provide an automated alternative as a part of the certification process; developers can certify their applications, and users can verify the proof of the certification before installation. Such a method can also be used for rapid examination of Android packages and informing Google team, via the Android Market of suspicious applications.

Data Leakage Prevention (DLP)

DLP is a relatively new field in computer security used mainly by enterprises. DLP mechanisms prevent sensitive/private content from leaking out. Identification of such content is done by applying various content and context inspection mechanisms (e.g., predefined keywords and patterns/regular expressions, fingerprinting of sensitive content and statistical algorithms). These mechanisms haven't been integrated yet into smart mobile platforms despite the fact that these devices can store content that should be protected (e.g., location, documents, contacts, calendar etc.)

We have investigated the implementation of DLP solution on Android. Its main requirement is the possibility to monitor and block outbound communication over the network (Wi-Fi/3G), SMS, or MMS. We analyzed several ways to hook into the data flow. Due to security considerations, such changes require deep OS integration, and can't be supported by just adding a regular (add-on) Android application.

Network data flow can be interrupted in the following locations: Java API, C API (i.e., native libraries) and kernel. *Altering the Java API* is simple but can be easily bypassed by calling the native libraries. *Hooks in native libraries*, such as libc, requires relatively low maintenance yet is can still be bypassed by using statically-compiled binaries. *Kernel hooking* is very difficult to bypass, but requires high maintenance.

Interruption of SMS messages can be done in the following locations: SMS application, application framework, serial line, rild daemon and kernel. *Altering the SMS application* requires low effort but can be easily bypassed by installing a 3rd party SMS application. *Modifying the application framework* is easy to implement and also difficult to bypass. *The serial line* used by rild can be rerouted to a monitoring daemon as demonstrated by Mulliner and Miller [22]. *Altering the rild daemon* and *the kernel* is also possible, however, it is has no additional benefits over the previous methods.

MMS is simply an SMS message that is marked as non-textual, and contains a URL. Therefore monitoring MMS is similar to the combination of file uploads (i.e., network monitoring), and SMS monitoring.

An Additional DLP feature, anti-theft, is provided for smartphones by several security vendors. This feature provides remote control capability over the device in case it gets lost or stolen. This module enables to locate the device, block it and wipe its data remotely.

5 Summary and Conclusions

In this paper we analyzed security issues pertaining to Google's Android in order to identify potential security flaws that should be mitigated using security solutions for Android devices. The risk arising from these vulnerabilities is amplified by the fact that as a smartphone, Android devices are expected to handle personal data and provide PC-compliant functionalities, thereby exposing the user to all the attacks that threaten users of personal computers.

We reviewed the security-related mechanisms that are inherently integrated in the Android framework and surveyed additional security mechanisms that can be applied on Android-based handsets. Several of these mechanisms were tested and evaluated in our laboratory. A security suite for mobile devices, especially open-source and open platform such as the Android, should include a collection of tools, optionally operating in collaboration.

Our review indicates that the defensive shell around Android was designed with extensive care since the security mechanisms embedded in Android address a broad range of security threats. However, despite these Android-integrated measures we conclude that it is highly important to incorporate a mechanism that can prevent or contain potential damage deriving from an attack on the Linux kernel layer such as the SELinux access control mechanism. Also, better protection should be added for hardening the Android permission mechanism and protecting owner's private data by modifying the permission mechanism, using a firewall, Intrusion Detection System, automated static analysis, encryption and a DLP mechanism.

Finally, remote management mechanisms can be used to consolidate several other security mechanisms while providing the ability to remotely control, configure and manage the device (e.g., setting network parameters or firewall policy, pushing security updates, tracking the device location, uninstalling/installing applications, bricking the device and deleting or encrypting data). Context-aware capabilities can also be added to dynamically allow and restrict access to resources (documents, emails) and services (camera, Internet, phone, messaging) based on a predefined policy and on the instantaneous context of the device.

References

1. Piercy, C.: Embedded devices next on the virus target list. Electronic Systems and Software 2(6), 42–43 (2005)
2. Frost, Sullivan: World mobile anti-malware products markets. Frost and Sullivan Report # M154-74 (2007)
3. Papathanasiou, C., Percoco, N.J.: This is not the droid you're looking for. In: DEF CON 18 (2010)
4. Pelino, M.: Predictions 2010: Enterprise Mobility Accelerates Again. Forrester (2009)
5. Lawton, G.: Is It Finally Time to Worry about Mobile Malware? Computer 41(5), 12–14 (2008)
6. Enck, W., Ongtang, M., McDaniel, P.: Understanding Android Security. IEEE Security and Privacy 7(1), 50–57 (2009)
7. Burns, J.: Developing Secure Mobile Applications for Android. Technical Report, iSEC (2008)
8. Shabtai, A., Fledel, Y., Kanonov, U., Elovici, Y., Dolev, S., Glezer, C.: Google Android: A Comprehensive Security Assessment. IEEE Security and Privacy 8(2), 5–44 (2010)
9. Shabtai, A., Fledel, Y., Kanonov, U., Elovici, Y., Dolev, S.: Google Android: A State-of-the-Art Review of Security Mechanisms. CoRR abs/0912.5101 (2009)

10. Vennon, T., Stroop, D.: Threat Analysis of Android Market (2010), http://threatcenter.smobilesystems.com/wp-content/uploads/2010/06/Android-Market-Threat-Analysis-6-22-10-v1.pdf
11. Emm, D.: Mobile Malware – New Avenues. Network Security 2006(11), 4–6 (2006)
12. Schmidt, A.D., et al.: Enhancing Security of Linux-based Android Devices. In: 15th International Linux Kongress, Germany (2008)
13. Shabtai, A., Kanonov, U., Elovici, Y.: Intrusion Detection on Mobile Devices Using the Knowledge Based Temporal-Abstraction Method. Journal of Systems and Software 83(8), 1524–1537 (2010)
14. Shabtai, A., Elovici, Y.: Applying Behavioral Detection on Android-Based Devices. In: Cai, Y., Magedanz, T., Li, M., Xia, J., Giannelli, C. (eds.) Mobilware 2010. Lecture Notes of the Institute for Computer Sciences, Social Informatics and Telecommunications Engineering, vol. 48, pp. 235–249. Springer, Heidelberg (2010)
15. Shabtai, A., Fledel, Y., Elovici, Y.: Securing Android-Powered Mobile Devices Using SELinux. IEEE Security and Privacy 8(3), 36–44 (2010)
16. Ni, X., Yang, Z., Bai, X., Champion, A.C., Xuan, D.: DiffUser: Differentiated User Access Control on Smartphones. In: Proceedings of the 5th IEEE International Workshop on Wireless and Sensor Networks Security (2009)
17. Nauman, M., Khan, S.: Design and Implementation of a Fine-grained Resource Usage Model for the Android Platform. To appear in International Arab Journal of Information Technology (2010)
18. Enck, W., Ongtang, M., McDaniel, P.: On lightweight mobile phone application certification. In: Proceedings of Computer and Communications Security Conference, pp. 235–245 (2009)
19. Ongtang, M., McLaughlin, S., Enck, W., McDaniel, P.: Semantically Rich Application-Centric Security in Android. In: Proceedings of the 25th Annual Computer Security Applications Conference, Honolulu, Hawaii (2009)
20. Shabtai, A., Fledel, Y., Elovici, Y.: Automated Static Code Analysis for Classifying Android Applications Using Machine Learning. In: International Conference on Computational Intelligence and Security, Nanning, China (2010)
21. Chaudhuri, A.: Language-Based Security on Android. In: Proceesings of the ACM Workshop on Programming Languages and Analysis for Security, pp. 1–7 (2009)
22. Mulliner, C., Miller, C.: Fuzzing the Phone in your Phone, Black Hat USA (2009)

SAVED: Secure Android Value addED services

Antonio Grillo, Alessandro Lentini, Vittorio Ottaviani,
Giuseppe F. Italiano, and Fabrizio Battisti

Department of Computer Science, Systems and Production
University of Rome "Tor Vergata",
Via del Politecnico 1, 00133 Rome, Italy
{grillo,lentini,ottaviani,italiano}@disp.uniroma2.com

Abstract. The availability of free Software Development Kits for recent mobile device platforms challenges many developers in realizing applications for the growing Smartphone market. In many cases such applications may interoperate in their working environment using mechanisms similar to the inter-process communication (IPC) and made available by the mobile operating system. Unfortunately, mobile devices lack in flexible solutions for making these communications secure. In this paper we propose a framework to secure the message exchange with the services installed on Google Android mobile devices. VASs realized by different providers are discovered, used and composed by an Application Frame designed for realizing complex goals. We implemented a prototype of our proposed framework on a real device and we performed extensive testing to measure the overhead introduced by the cryptographic operations required to protect the inter process communication.

Keywords: IPC security, Value Added Services, digital certificate, service interoperability.

1 Introduction

Today's smartphones are widespread mobile devices that combine advanced features in managing personal, phone and business data. One of the most interesting features of smartphones lies perhaps in the possibility to install third party applications; as a consequence, one can use his/her own smartphone truly as a PC: accessing social networks, paying bills, checking bank accounts, etc... Smartphone applications are commonly installed and stored in memory, and in modern devices all the application's data are kept safe from the OS by using a sandbox approach. Such approach prevents other applications to access unauthorized data insulating each application from the others [4], [5], [6]. However, users and manufacturers may suffer from mobile malware infections, call-ID spoofing attacks, spam, and problems with third-party applications or accidents that can cause malfunction of network capacity, disclosure of sensitive business data and more in general privacy problems such as loss of personal data.

In this paper we propose a new framework based on Google Android Operating System (OS) for the realization of several value added services (VAS). We call this framework SAVED (Secure Android Value addED services). SAVED enables secure communication between services and applications using such services via Inter Process Communication (IPC)/Remote Procedure Call (RPC). Each VAS is realized through an Android Service. The access to such a service requires the execution of an authentication and authorization phase among the involved parties. Once this initial phase is completed, the application sets up a secure communication with the service using a symmetric encryption scheme.

2   State of the Art

Android is a multi-process system, in which each application (and parts of the system) runs in its own process. Most security between applications and the system is enforced at the process level through standard Linux facilities, such as user and group IDs that are assigned to applications [1]. The Android system requires that all installed applications be digitally signed with a certificate whose private key is held by the application's developer. The Android system uses the certificate as a means of identifying the author of an application and establishing trust relationships between applications. The Android approach grants security of application's data, and prevents access to all services developed by others. Every service publishes in its personal manifest file the permissions required to use the service. One of the permission settings in the manifest file is *Protection level*. The *Protection level* field configures the security policies required by the service; if the level is set to *signature* the service will communicate only with these applications with which it shares the same developer certificate.

The main advantage of the approach followed in the Android design is that developers have to focus their attention only on the application, while the OS grants that all the applications that are not allowed to access the services are prevented from doing so. This simplification comes at an extra cost: only developers sharing certificates and private keys can use services already developed in new applications. This is a huge limitation compared to the growing size of the mall market and the number of organizations and developers enrolled in publishing applications and services. According to a very recent survey on the smartphone OS market published by Gartner [2], both Android and Apple were the only two OSs vendors among the top five to increase their market share: in particular, Android moved to the fourth position displacing Microsoft Windows Mobile for the first time and the Android market has grown 4.4 times in size, going from 10,000 to 20,000 applications in the first four months of the 2010.

The approach of Android prevents third parties to start using the framework's VAS. Developers can use each others' services sharing certificates and credentials: in this case, the applications can interact but the security of the whole framework is granted from a single digital signature; if the developer's digital signature is stolen a hacker could sign his/her own applications, thus getting complete access to all data of the framework.

Our approach wants to promote the framework scalability and grant secure access to services developed by other users without the need to share private data. We propose to insert a new layer that handles security of inter-process communications; in such layer, trustability is granted directly by the security policy of the framework, and each application can require access and publish services interacting with the framework like in a PKI environment. Thanks to SAVED framework it is possible to face different kinds of threats:

- *Service Spoofing:* the application refers to a service by simply using an interface that establishes the name, the package and the methods signatures; if the original service is replaced on the mobile device, applications that exploit that service are unaware of the substitution.
- *Memory Dump:* starting from Android 1.5, a new API has been introduced to generate a memory dump programmatically. The static method *dumpHprofData(String fileName)* of the *Debug* class generate a dump file that can be converted with the *hprof-conv* tool of the Android SDK and, subsequently, analyzed with different memory analysis tools (e.g., Eclipse MAT, JProfile, etc.). If a fake application execute the dump periodically and export the dump data using a connection (e.g., HTTP connection), it is possible to steal the data exchanged among applications and services.

3 The Framework

SAVED (Secure Android Value addED services) is a framework that grants secure communication between services without requiring private data sharing. Our intent is to improve interoperability between applications and services facing the limits of the Android's native approach. The purpose of SAVED is to allow applications to use services developed by others, to add new VAS to the framework or even to create new applications using already existing VAS. All the interactions performed using the proposed frameworks will be performed in a secure way. SAVED adds supplementary security at the process communication level: each application is accredited to the framework which grants privileges to access in a secure way shared services and facilities. Single process security provided using sandboxes with the Android approach is also preserved in SAVED. In our framework we defined two main entities:

- **Application,** which provides graphical user interface, and all the logics implementing the task to be realized. Applications are implemented extending the *Android.Activity* class.
- **Value Added Service (VAS),** which provides to the applications developed using the framework all the certified services. VASs are implemented as remote services extending the *Android.Service* class. The ProxyCA and the ProxyTSA are two special VAS in the framework; these VASs allow the communication with a Certification Authority and a Timestamping Authority, respectively.

In order to realize Applications participating to the framework, developers have to extend specific interfaces and include particular resource packages. When a new VAS is realized, it is required to export its class package. Such class packages will be imported from the Applications that will use the services provided by the VAS. The packages imported will be used to perform inter-process communication. Including such packages and extending the interfaces will provide the supplementary security layer that will grant a secure communication between entities and prevent the access to the services to those applications that are not allowed.

Moreover, we tried to address some best practices to create components participating to the framework enforcing the required security needs. Some examples follow:

- *Activation code:* when the Application/VAS is installed on the device an unlock code should be required to the user; the Application/VAS will remain locked (preventing all interactions) until the user will insert the proper activation code of every entity;
- *Use of standard certificate*: each component should have a proper X509 digital certificate signed from a valid Certification Authority (CA), such certificate will be saved in a keystore inside the component memory area; the component will be responsible to take care of managing correctly the keystore itself to grant a secure saving of the other's certificates;
- *Model View Control Pattern*: VAS and Applications will take care of implementing independently graphical user interfaces to be shown to the end user;
- *Mutual Authentication:* each entity needs to implement a mechanism to grant mutual authentication. The mutual authentication should be ensured by mutually exchanging and verifying the digital certificates. Using a handshake schema (e.g., TLS handshake) the involved entities exchange their digital certificates, check the certificates validity through the ProxyCA, and mutually authenticate themselves (see Fig. 1).

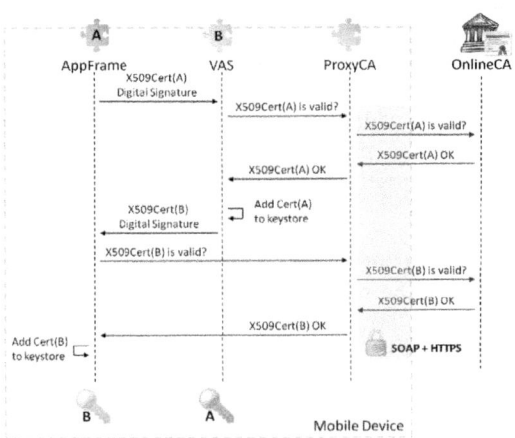

Fig. 1. Mutual Authentication phase

- *Session Authentication:* once the entities are mutually authenticated, a session key (i.e., SK) is shared. According to our approach the SK is generated by both the Application and the VAS using parameter defined by the two parties (i.e., CTRL_A and CTRL_B). Adopting a key agreement protocol (e.g., Diffie-Hellman protocol) the involved entities agree on secret SK that will be used to encrypt subsequent communications (see Fig. 2).

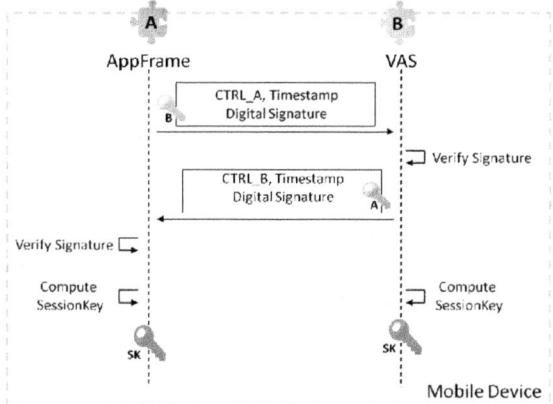

Fig. 2. Session Authentication phase

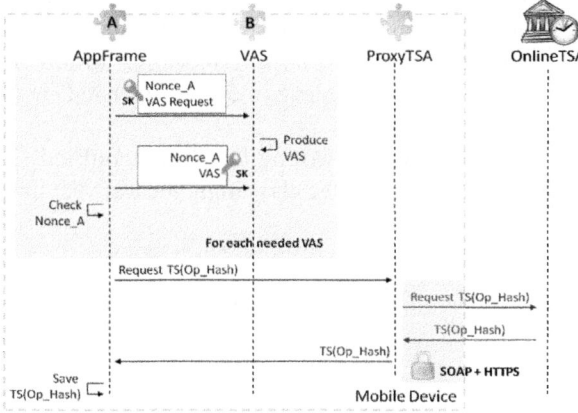

Fig. 3. Session Encryption phase

- *Session Encryption:* Every VAS allows access to its functionalities only to "trusted" Applications; trusted Applications have performed successfully the Mutually Authentication and the Session Authentication phases. In order to enforce the uniqueness of each interaction with VAS a random value (i.e. Nonce_A) is used; the confidentiality is granted by encrypting the exchanged data with the SK.

The Application composes the results of different VAS in order to realize a complex goal. At the end of this phase, the Application interacts with a timestamping authority through the ProxyTSA in order to securely keep track of the creation time of the realized goal (see Fig. 3). The sensitive data of the operation are summarized applying an Hash function (i.e., Op_Hash) and these data are sent to the Timestamping service.

Mutual Authentication, Session Authentication and Session Encryption represent the secure core of SAVED framework and should be carefully performed in order to join the framework.

4 The Framework Implementation

We developed a prototype of the SAVED framework on an Android 1.5 platform. The main features of the proposed framework are encapsulated into the jar files that contains two kind of files (i.e., .aidl, .Stub) for the inter process communication.

AIDL (Android Interface Definition Language) is an IDL [3], [4] with which it is possible to generate automatically the source code that allows two Android applications to exchange information using IPC. AIDL/IPC interface based mechanism is similar to Common Object Model (COM) or Common Object Request Broker Architecture (CORBA). In order to implement an AIDL/IPC service it is required to perform some steps:

- Create an .aidl file to define the interface (*YourInterface*.aidl). The interface defines the access methods and the fields available to a client.
- Add the .aidl file to the makefile and implement the methods of the interface creating a class that extends the *YourInterface.Stub* (.Stub file is automatically generated by the tool) and implements methods declared in the .aidl description file.
- Publish the interface to clients rewriting the Service.onBind (Intent) method; this method will return an instance of the class implementing the interface.

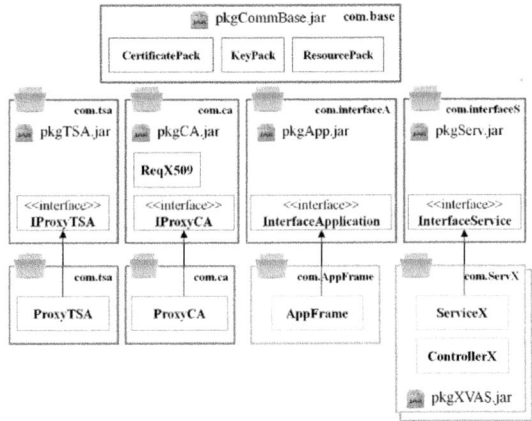

Fig. 4. SAVED framework main packages

This IPC mechanism needs a way to share complex information, such as non-primitive types, between two entities. In order to achieve this goal Android provides Parcelable class able to serialize and deserialize complex types.

Fig. 4 simplifies the package diagram of SAVED. The picture shows on top the following core .jar files:

- *pkgApp.jar* contains the interface *InterfaceApplication* that must be implemented by every class that want to participate SAVED as an Application;
- *pkgServ.jar* contains the interface *InterfaceService* that needs to be implemented by every class that want to be a VAS in the framework;
- *pkgCA.jar* carries the *IProxyCA.aidl* with his relative .Stub file; these files allow the communications between the entities of SAVED and the ProxyCA. Moreover, the jar file contains the parcelable class ReqX509 that is mandatory for the communication;
- *pkgTSA.jar* packages the *IProxyTime.aidl* with his relative .Stub file to grant communication with the Proxy TSA;
- *pkgCommBase.jar* contains the three base parcelable files that grant the communication between the Application and the VASs, namely *CertificatePack.java*, *KeyPack.java* and *ResourcePack.java*.

In order to grant to an Application to contact and receive services from all the VAS inside the framework, and so assemble the services offered from the VAS to create complex applications, it is required to install the ProxyTSA and the ProxyCA Android packages (apk); these entities are shown in the lower left half of Fig. 4.

ProxyCA is one of the underlying VASs that exist in the framework. All entities must submit to the ProxyCA the digital certificates they receive from their communication partners. The service contacts a web service that works as an online Certification Authority, inserts the certificate in a XML file and through a secure HTTP connection (i.e., HTTPS) asks for the certificate verification. The web service checks the certificate validity and answers with an XML response.

ProxyTSA is another basic VAS of SAVED. As the ProxyCA the ProxyTSA takes in account the communication with an external partner, the timestamping web service. All the communications between the proxy and the timestamping web service are managed through XML messages on HTTPS.

A Building a Value Added Service

- Create a new Android project with a class that extends the native Service class;
- Import in the project:
- pkgServ.jar,
- pkgCommBase.jar,
- pkgCA.jar;
- The main class of the project must implement *InterfaceService* interface class and consequently all his methods;

- Create the graphical user interface;
- Create the IServiceX.aidl in the project as described previously;
- Create and export pkgXVAS.jar containing IServiceX.aidl and the corresponding .Stub file generated automatically;
- Service class must implement, all the standard methods of the Android native Service class, and the .aidl interface with all the methods defined through the description language;
- Release the service as an .apk file for the installation on the device.

B Building an Application

- Create a new Android project which contains a class that extends the native Activity Android class;
- Import in the project:
- pkgApp.jar,
- pkgCommBase.jar,
- pkgCA.jar,
- pkgTS.jar;
- The main class of the project must implement the *InterfaceApplication* interface with all his methods;
- Create a graphical user interface to allow the user to interact with the Application;
- Import from each VAS you want to use in the Application the corresponding jar file (i.e., pkgXVAS.jar)
- Use each service in a proper way, taking care of managing and releasing correctly the connection with the involved VAS. Note that early versions of Android platform serialize the access to the services.
- Release the Application as an .apk file for the installation on the device.

Assume we are in a scenario where we have one Application and one VAS, each one with its own digital certificate signed by different CAs. Note that in this scenario, none of the entities "knows" the public key or the certificate of the counterpart. If the two entities wish to cooperate, they need to authenticate each other. After contacting the ProxyCA to verify the communication partner trustability (cfr. the Mutual Authentication phase), an asymmetric cryptography session to exchange the session key can be started (cfr. the Session Authentication phase). Finally, the session between the involved parties is encrypted using symmetric cryptography (cfr. the Session Encryption phase). The need to switch from asymmetric to symmetric cryptography is due to the performance overhead of asymmetric cryptography: indeed, the switch from asymmetric to symmetric cryptography improves the performances of the whole framework reducing the effort due to encryption/decryption operations.

5 Use Cases

The framework described in the previous sections, can be used to develop complex Applications or VAS interfacing basic services in a secure, certified and non-repudiable way. In this section, we will detail two use cases.

A Payment VAS

An explicative use case, which requires security and non-repudiability of the operations can be the access to money transfer services and management of the related information.

A payment gateway (e.g., PayPal^{tm} [8], or Google checkout [10]) eases the transfer of information between payment portal and frontend processor. A new trend (e.g., PayPal Mobile Checkout [9]) for payment gateway is to provide its service to the growing population of mobile users.

A developer who has implemented an application that require any form of money transfer (e.g., buying e-books, music, games, tickets, ... or booking a service) needs to interface his/her application with all available payment gateways. A payment gateway aims to reach as many developers as possible in a simple and secure way, preserving its distinctive user interface. Using SAVED, developers can interface their e-commerce Application with the VASs exposed by different payment gateways (see Fig. 5).

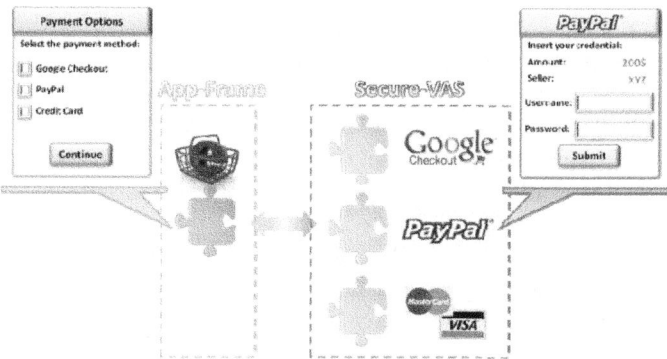

Fig. 5. The Payment VAS Use Case realized in the SAVED framework with three payment opportunities (i.e., Google Checkout, PayPal and Credit Card)

After inserting the payment gateway VAS in our framework, all the communications between SAVED Applications and VAS will be performed in a secure way. The VAS using the framework will transparently provide API ready to use in the Application.

B Event Certification

Producing digital evidence through a report can demonstrate other advantages of using Application and VASs belonging to the SAVED framework. To preserve in time and space a digital document, some accessory information are required to keep in time the authenticity of the digital document and the proof of the evidenced fact. The user probably could produce, as evidence, a document enriched by: a picture of the place where the fact happened or of the fact itself; some georeferenced information; a timestamp; a voice note; and a text note characterizing the digital document.

A modern mobile device, such as a Smartphone, is the most suitable tool to reach the use case's goal as it is equipped with all the hardware capabilities to perform the following services:

- GPS Position; that allows the user to get his/her current location on earth, the two GPS coordinates (latitude and longitude) will be used from the Application to be attached to the digital document; Android provides two classes to interact with the GPS device and access such coordinates:
  - LocationManager which handles the access to the location service of the system. The location service provides periodically updates of the current location of the device on the earth and alerts the user when he/she is in the proximity of some location on earth.
  - LocationListener is used to manage the alerts and the changes of latitude and longitude coming from the LocationManager class. The operating system calls the LocationListener methods automatically.
- Pick a photo; Android provides some classes and methods to access the camera installed in the device, to present a preview of the picture to the user, to get the picture clicking a button, to save the picture in the device filesystem and to handle the camera settings. The classes used to perform all these operations are:
  - SurfaceHolder.Callback, represents the user interface to show the preview of the picture the user is going to take.
  - Camera, which handles all the camera actions such as connect/disconnect, handle the settings and getting data to be converted in a manageable format.
- Getting an audio record; is the service providing to the user the possibility to get an audio record, using the device microphone; the class used to get access to audio/video recorder is the MediaRecorder class.
- Getting a text note; such functionality has not been implemented as a VAS, but like a small extension created using an EditBox.

The Application realized for this use case simplifies in a GUI the interaction with the VASs (see Fig. 6) and generates a data package containing a picture, a GPS location, a text note and an audio message. Such data package will be processed and sent to the TSAProxy that will certify the time in which the digital document has been created as it is connected with a Time Stamping Authority.

The VASs implemented to test the framework can be used in a plethora of different contexts; a lot of applications currently on the Android Market use one of the securized services in a non secure way.

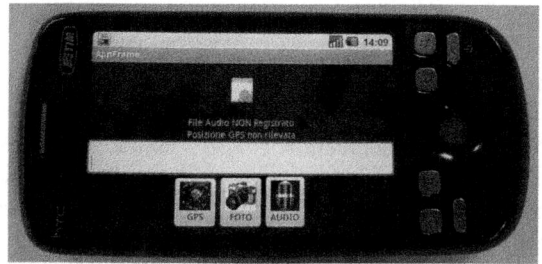

Fig. 6. Screenshot of the Application that realizes the Event Certification use case

6 On a Real Device

The framework has been tested on an Android HTC Magic device. The device was equipped with Android 1.5 OS, 3.2 M-pixel camera, Integrated GPS Antenna, Wi-Fi: IEEE 802.11 b/g. Using Android ADB tool different .apk, created using Eclipse IDE, have been installed on the HTC Magic. The testing phase has highlighted a slower response of the Applications due to security operations, inter-process communications via AIDL interfaces and parcelable classes. We executed some performance tests using our prototype. We aimed at measuring the time computational overhead introduced by the use of SAVED, and thus we measured the time needed to execute security functions. In particular, we have considered the overhead related to each one of the phases described in Section 3.

Table 1. Time overhead for the framework phases

Phase		Time (ms)
1.	Mutual Authentication*	1197
1.	Mutual Authentication	446
2.	Session Authentication	257
3.	Session Encryption	795
Total Framework Overhead		*1498*

In Table 1 we can see the time overhead introduced by SAVED. The first row of the table refers to the first execution of the Mutual Authentication phase, while the second row refers to the subsequent executions. In the first case the more time required is justified by the need to update the keystore with the new digital certificates; this delay is paid once. The total framework overhead amounts to 1.5 second preserving the usability for real use cases.

We have chosen to test the framework on a HTC Magic that is one of the earliest models of Android, so the performance issues are more evident; clearly, new devices are more responsive and performance problem will always be less significant.

The testing phase has been really useful to verify the effectiveness of the framework and to solve some side issues due to the complexity of the interaction between processes in a real mobile device: for example, we noted that the device puts the application in stand-by mode when it notices a display rotation, this is because the OS calls the application that rotates the display, and later gives back the control of the system to the running application. This issue has been solved using onSaveInstanceState and onRestoreInstanceState when the application is put in stand-by and woken up from the OS. Implementing such methods prevents data loss due to changes in the state of the application.

7 Conclusion

In this paper we have presented a new framework called SAVED in which applications and services taking part can communicate and share safely functionalities and facilities. We have implemented a prototype of SAVED and we have tested it on a real device. The operational capacities of the framework have been verified.

Using the proposed framework enables complex use cases, with a range of value added services actually certified. Future extensions of the framework include: the enhancement of the discovery mechanism through which each Application receive information about SAVED VAS available on the device; a mechanism to audit the history of each VAS or Application, so as to keep track of all kinds of actions and information exchanged between the entities of the framework.

References

1. Android developers, "Security and Permission" (June 2010), http://developer.android.com/guide/topics/security/security.html
2. Gartner Inc., "Gartner Says Worldwide Mobile Phone Sales Grew 17 Per Cent in First Quarter 2010" (June 2010), http://www.gartner.com/it/page.jsp?id=1372013
3. Bachmann, F., et al.: Documenting Software Architecture: Documenting Interfaces. Sofware Enginerring Institute, Carniege Mellon (2002)
4. Lamb, D.A.: Sharing intermediate representations: the interface description language, Ph.D. Dissertation, Carnegie-Mellon University, Department of Computer Science (1983)
5. Gong, L., Mueller, M., Prafullchandra, H., Schemers, R.: Going Beyond the Sandbox: An Overview of the New Security Architecture in the Java Development Kit 1.2. In: Proceedings of the USENIX Symposium on Internet Technologies and Systems, Monterey, California (December 1997)
6. Burns, J.: Mobile application security on Android, Context on Android security, Black Hat (2009)

7. Burns, J.: Developing secure mobile applications for Android, iSEC Partners (October 2008)
8. PayPal, Adaptive Payments Guide (June 2009), PayPalIntegrationCenter `https://www.x.com/community/ppx/documentation`
9. PayPal Mobile Checkout, PayPal Mobile Checkout Developer Guide (October 2009), PayPalIntegrationCenter https://www.x.com/community/ppx/documentation
10. Google checkout, About Google Checkout, http://checkout.google.com

Author Index

Babu, Karishma 323
Bal, Henri 59, 302, 342
Battisti, Fabrizio 415
Beugnard, Antoine 100
Boneh, Dan 17
Burbey, Ingrid 156
Buthpitiya, Senaka 195, 211, 282

Campbell, Roy H. 1
Cao, Yiwei 361
Chabridon, Sophie 100
Chen, Peng-Wen 263
Cheng, Heng-Tze 195, 211, 282
Chennuru, Snehal 263
Collins, Patricia 211, 282
Cossalter, Michele 306

Dantu, Ram 137
DeBruhl, Bruce 306
Demeure, Isabelle 117
Dodson, Ben 17
Dorizzi, Bernadette 396

Elovici, Yuval 382, 401
El-Yacoubi, Mounim 396

Fledel, Yuval 382, 401

Garcia-Salicetti, Sonia 396
Gavrilovska, Ada 323
Gibbs, Simon 373
Gilbert, Benjamin 315
Goodney, Andrew 231
Grillo, Antonio 415
Griss, Martin 195, 211, 282

Ha Duong, Hoa Dung 117
Harkes, Jan 315
Holmstedt, Lasse 39
Houmani, Nesma 396
Huang, Dijiang 329

Iftode, Liviu 315
Italiano, Giuseppe F. 415

Jeon, Won 373
Jung, Jinho 231

Kallioinen, Olli 80
Kannan, Sudarsun 323
Kemp, Roelof 59, 302, 342
Khan, Abdul Malik 100
Kielmann, Thilo 59, 302, 342
Kjærgaard, Mikkel Baun 176
Klamma, Ralf 361
Kovachev, Dejan 361
Kunjithapatham, Anugeetha 373
Kuo, Cynthia 211, 282

Lam, Monica S. 17
Lee, Kwan Hong 243
Lentini, Alessandro 415
Liang, Hongbin 329
Lippman, Andrew 243

Manjunatha, Ashwin 349
Martin, Thomas L. 156
Mengshoel, Ole 306
Mikkonen, Tommi 39, 80

Nanavati, Vibhor 311
Needham, Scott 231

Ottaviani, Vittorio 415

Palmer, Nicholas 59, 302, 342
Peng, Daiyuan 329
Poduri, Sameera 231
Potashnik, Dennis 401

Ranabahu, Ajith 349
Rokach, Lior 382
Ross, Erik 243

Satyanarayanan, Mahadev 315
Scholten, Gert 302
Schwan, Karsten 323
Sengupta, Debangsu 17
Shabtai, Asaf 401
Shen, Dawei 243

Author Index

Sheth, Amit 349
Smaldone, Stephen 315
Stavrou, Angelos 137
Sun, Feng-Tso 195, 211, 282

Terho, Mikko 39
Thirunarayan, Krishnaprasad 349

Want, Roy 306
Weckemann, Kay 176

Weiss, Yael 382
Wijesekera, Duminda 137
Wu, Victor K.Y. 1

Yang, Xiaohui 137

Zhang, Joy Ying 263
Zhang, Pei 306
Zhang, Xinwen 373
Zhu, Jiang 263

GPSR Compliance

The European Union's (EU) General Product Safety Regulation (GPSR) is a set of rules that requires consumer products to be safe and our obligations to ensure this.

If you have any concerns about our products, you can contact us on ProductSafety@springernature.com

In case Publisher is established outside the EU, the EU authorized representative is:

Springer Nature Customer Service Center GmbH
Europaplatz 3
69115 Heidelberg, Germany

Batch number: 09478804

Printed by Printforce, the Netherlands